SCHAUM'S OUTLINE OF

THEORY AND PROBLEMS

OF

Calculus of

FINITE DIFFERENCES

and

DIFFERENCE EQUATIONS

•

BY

MURRAY R. SPIEGEL, Ph.D.

Professor of Mathematics
Rensselaer Polytechnic Institute

SCHAUM'S OUTLINE SERIES

McGRAW-HILL BOOK COMPANY

New York, St. Louis, San Francisco, Düsseldorf, Johannesburg, Kuala Lumpur, London, Mexico, Montreal, New Delhi, Panama, Rio de Janeiro, Singapore, Sydney, and Toronto

07-060218-2

3 4 5 6 7 8 9 1 0 SH SH 7 5 4 3 2 1 0 6 7

Preface

In recent years there has been an increasing interest in the calculus of finite differences and difference equations. There are several reasons for this. First, the advent of high speed computers has led to a need for fundamental knowledge of this subject. Second, there are numerous applications of the subject to fields ranging from engineering, physics and chemistry to actuarial science, economics, psychology, biology, probability and statistics. Third, the mathematical theory is of interest in itself especially in view of the remarkable analogy of the theory to that of differential and integral calculus and differential equations.

This book is designed to be used either as a textbook for a formal course in the calculus of finite differences and difference equations or as a comprehensive supplement to all current standard texts. It should also be of considerable value to those taking courses in which difference methods are employed or to those interested in the field for self-study.

Each chapter begins with a clear statement of pertinent definitions, principles and theorems, together with illustrative and other descriptive material. The solved problems serve to illustrate and amplify the theory, bring into sharp focus those fine points without which the student continually feels himself on unsafe ground, and provide the repetition of basic principles so vital to effective learning. Numerous proofs of theorems and derivations of formulas are included among the solved problems. The large number of supplementary problems with answers serves as a complete review of the material of each chapter.

Topics covered include the difference calculus, the sum calculus and difference equations [analogous to differential calculus, integral calculus and differential equations respectively] together with many applications.

Considerably more material has been included here than can be covered in most first courses. This has been done to make the book more flexible, to provide a more useful book of reference, and to stimulate further interest in the topics.

I wish to take this opportunity to thank Nicola Monti, David Beckwith and Henry Hayden for their splendid cooperation.

M. R. SPIEGEL

Rensselaer Polytechnic Institute
November 1970

CONTENTS

CONTENTS

Chapter 1

The Difference Calculus

OPERATORS

We are often concerned in mathematics with performing various operations such as squaring, cubing, adding, taking square roots, etc. Associated with these are *operators,* which can be denoted by letters of the alphabet, indicating the nature of the operation to be performed. The object on which the operation is to be performed, or on which the operator is to act, is called the *operand.*

Example 1.

If C or $[\quad]^3$ is the *cubing operator* and x is the operand then Cx or $[\quad]^3 x$ represents the cube of x, i.e. x^3.

Example 2.

If D or $\dfrac{d}{dx}$ is the *derivative operator* and the operand is the function of x given by $f(x) = 2x^4 - 3x^2 + 5$ then

$$Df(x) \quad = \quad D(2x^4 - 3x^2 + 5) \quad = \quad \frac{d}{dx}(2x^4 - 3x^2 + 5) \quad = \quad 8x^3 - 6x$$

Example 3.

If $\mathcal{J}$ or $\int (\quad)\,dx$ is the *integral operator* then

$$\mathcal{J}(2x^4 - 3x^2 + 5) \quad = \quad \left[\int (\quad)\,dx\right](2x^4 - 3x^2 + 5) \quad = \quad \int (2x^4 - 3x^2 + 5)\,dx \quad = \quad \frac{2x^5}{5} - x^3 + 5x + c$$

where c is an arbitrary constant.

Example 4.

The *doubling operator* can be represented by the ordinary symbol 2. Thus

$$2(2x^4 - 3x^2 + 5) \quad = \quad 4x^4 - 6x^2 + 10$$

It is assumed, unless otherwise stated, that the class of operands acted upon by a given operator is suitably restricted so that the results of the operation will have meaning. Thus for example with the operator D we would restrict ourselves to the set or class of *differentiable functions,* i.e. functions whose derivatives exist.

Note that if A is an operator and f is the operand then the result of the operation is indicated by Af. For the purposes of this book f will generally be some function belonging to a particular class of functions.

SOME DEFINITIONS INVOLVING OPERATORS

1. **Equality of operators.** Two operators are said to be *equal,* and we write $A = B$ or $B = A$, if and only if for an arbitrary function f we have $Af = Bf$.

2. **The identity or unit operator.** If for arbitrary f we have $If = f$ then I is called the *identity* or *unit operator*. For all practical purposes we can and will use 1 instead of I.

3. **The null or zero operator.** If for arbitrary f we have $Of = 0$ then O is called the *null* or *zero operator*. For all practical purposes we can and will use 0 instead of O.

4. Sum and difference of operators. We define

$$(A+B)f = Af + Bf, \qquad (A-B)f = Af - Bf \tag{1}$$

and refer to the operators $A+B$ and $A-B$ respectively as the *sum* and *difference* of operators A and B.

Example 5. $(C+D)x^2 = Cx^2 + Dx^2 = x^6 + 2x, \quad (C-D)x^2 = Cx^2 - Dx^2 = x^6 - 2x$

5. Product of operators. We define

$$(A \cdot B)f = (AB)f = A(Bf) \tag{2}$$

and refer to the operator AB or $A \cdot B$ as the *product* of operators A and B. If $A = B$ we denote AA or $A \cdot A$ as A^2.

Example 6. $(CD)x^2 = C(Dx^2) = C(2x) = 8x^3, \quad C^2x^2 = C(Cx^2) = C(x^6) = x^{18}$

6. Linear operators. If operator A has the property that for arbitrary functions f and g and an arbitrary constant α

$$A(f+g) = Af + Ag, \qquad A(\alpha f) = \alpha Af \tag{3}$$

then A is called a *linear operator*. If an operator is not a linear operator it is called a *non-linear operator*. See Problem 1.3.

7. Inverse operators. If A and B are operators such that $A(Bf) = f$ for an arbitrary function f, i.e. $(AB)f = f$ or $AB = I$ or $AB = 1$, then we say that B is an *inverse* of A and write $B = A^{-1} = 1/A$. Equivalently $A^{-1}f = g$ if and only if $Ag = f$.

ALGEBRA OF OPERATORS

We will be able to manipulate operators in the same manner as we manipulate algebraic quantities if the following laws of algebra hold for these operators. Here A, B, C denote any operators.

I-1.	$A + B = B + A$	Commutative law for sums
I-2.	$A + (B+C) = (A+B) + C$	Associative law for sums
I-3.	$AB = BA$	Commutative law for products
I-4.	$A(BC) = (AB)C$	Associative law for products
I-5.	$A(B+C) = AB + AC$	Distributive law

Special care must be taken in manipulating operators if these do not apply. If they do apply we can prove that other well-known rules of algebra also hold, for example the *index law* or *law of exponents* $A^mA^n = A^{m+n}$ where A^m denotes repeated application of operator A m times.

THE DIFFERENCE OPERATOR

Given a function $f(x)$ we define an operator Δ, called the *difference operator*, by

$$\Delta f(x) = f(x+h) - f(x) \tag{4}$$

where h is some given number usually positive and called the *difference interval* or *differencing interval*. If in particular $f(x) = x$ we have

$$\Delta x = (x+h) - x = h \quad \text{or} \quad h = \Delta x \tag{5}$$

Successive differences can also be taken. For example

$$\Delta^2 f(x) = \Delta[\Delta f(x)] = \Delta[f(x+h) - f(x)] = f(x+2h) - 2f(x+h) + f(x) \tag{6}$$

We call Δ^2 the *second order difference operator* or *difference operator of order 2*. In general we define the nth *order difference operator* by

$$\Delta^n f(x) = \Delta[\Delta^{n-1} f(x)] \tag{7}$$

THE TRANSLATION OR SHIFTING OPERATOR

We define the *translation* or *shifting operator* E by

$$Ef(x) = f(x+h) \tag{8}$$

By applying the operator twice we have

$$E^2 f(x) = E[Ef(x)] = E[f(x+h)] = f(x+2h)$$

In general if n is any integer [or in fact, any real number], we define

$$E^n f(x) = f(x+nh) \tag{9}$$

We can show [see Problem 1.10] that operators E and Δ are related by

$$\Delta = E - 1 \qquad \text{or} \qquad E = 1 + \Delta \tag{10}$$

using 1 instead of the unit operator I.

THE DERIVATIVE OPERATOR

From (*4*) and (*5*) we have

$$\frac{\Delta f(x)}{\Delta x} = \frac{f(x+h) - f(x)}{h} \tag{11}$$

where we can consider the operator acting on $f(x)$ to be $\Delta/\Delta x$ or Δ/h. The *first order derivative* or briefly *first derivative* or simply *derivative* of $f(x)$ is defined as the limit of the quotient in (*11*) as h or Δx approaches zero and is denoted by

$$Df(x) = f'(x) = \lim_{\Delta x \to 0} \frac{\Delta f(x)}{\Delta x} = \lim_{h \to 0} \frac{f(x+h) - f(x)}{h} \tag{12}$$

if the limit exists. The operation of taking derivatives is called *differentiation* and D is the *derivative* or *differentiation operator*.

The *second derivative* or *derivative of order two* is defined as the derivative of the first derivative, assuming it exists, and is denoted by

$$D^2 f(x) = D[Df(x)] = f''(x) \tag{13}$$

We can prove that the second derivative is given by

$$D^2 f(x) = \lim_{\Delta x \to 0} \frac{\Delta^2 f(x)}{(\Delta x)^2} = \lim_{h \to 0} \frac{f(x+2h) - 2f(x+h) + f(x)}{h^2} \tag{14}$$

and can in fact take this as a definition of the second derivative. Higher ordered derivatives can be obtained similarly.

THE DIFFERENTIAL OPERATOR

The *differential of first order* or briefly *first differential* or simply *differential* of a function $f(x)$ is defined by

$$df(x) = f'(x)\,\Delta x = f'(x)h \tag{15}$$

In particular if $f(x) = x$ we have $dx = \Delta x = h$ so that (*15*) becomes

$$df(x) = f'(x)\,dx = f'(x)h \qquad \text{or} \qquad df(x) = Df(x)\,dx = hDf(x) \tag{16}$$

We call d the *differential operator.* The second order differential of $f(x)$ can be defined as

$$d^2f(x) = f''(x)(\Delta x)^2 = f''(x)(dx)^2 \tag{17}$$

and higher ordered differentials are defined similarly.

Note that $df(x)$, $dx = \Delta x = h$, $d^2f(x)$, $(dx)^2 = (\Delta x)^2 = h^2$ are numbers which are not necessarily zero and not necessarily small.

It follows from (*16*) and (*17*) that

$$f'(x) = \frac{df(x)}{dx} = Df(x), \qquad f''(x) = \frac{d^2f(x)}{dx^2} = D^2f(x), \qquad \ldots \tag{18}$$

where in the denominator of (*18*) we have written $(dx)^2$ as dx^2 as indicated by custom or convention. It follows that we can consider the operator equivalence

$$D = \frac{d}{dx}, \qquad D^2 = \frac{d^2}{dx^2}, \qquad \ldots \tag{19}$$

Similarly we shall write $\quad \dfrac{\Delta^2}{(\Delta x)^2}f(x) \quad$ as $\quad \dfrac{\Delta^2}{\Delta x^2}f(x)$

RELATIONSHIP BETWEEN DIFFERENCE, DERIVATIVE AND DIFFERENTIAL OPERATORS

From (*16*) we see that the relationship between the derivative operator D and the differential operator d is

$$D = \frac{d}{dx} = \frac{d}{h} \qquad \text{or} \qquad d = hD \tag{20}$$

Similarly from (*12*) and (*16*) we see that the relationship between the difference operator Δ and the derivative operator D is

$$D = \lim_{\Delta x \to 0} \frac{\Delta}{\Delta x} = \lim_{h \to 0} \frac{\Delta}{h} = \frac{d}{dx} \tag{21}$$

with analogous relationships among higher ordered operators.

Because of the close relationship of the difference operator Δ with the operators D and d as evidenced by the above we would feel that it should be possible to develop a *difference calculus* or *calculus of differences* analogous to *differential calculus* which would give the results of the latter in the special case where h or Δx approaches zero. This is in fact the case as we shall see and since h is taken as some given constant, called *finite,* as opposed to a variable approaching zero, called *infinitesimal,* we refer to such a calculus as the *calculus of finite differences.*

To recognize the analogy more clearly we will first review briefly some of the results of differential calculus.

GENERAL RULES OF DIFFERENTIATION

It is assumed that the student is already familiar with the elementary rules for differentiation of functions. In the following we list some of the most important ones.

II-1. $\quad D[f(x) + g(x)] = Df(x) + Dg(x)$

II-2. $\quad D[\alpha f(x)] = \alpha Df(x) \qquad \alpha = \text{constant}$

II-3. $\quad D[f(x)\, g(x)] = f(x)\, Dg(x) + g(x)\, Df(x)$

II-4. $\quad D\left[\dfrac{f(x)}{g(x)}\right] = \dfrac{g(x)\, Df(x) - f(x)\, Dg(x)}{[g(x)]^2}$

II-5. $\quad D[f(x)]^m = m[f(x)]^{m-1}\, Df(x) \qquad m = \text{constant}$

II-6. $$D[f(g(x))] = D[f(u)] \quad \text{where} \quad u = g(x)$$
$$= \frac{df}{du} \cdot \frac{du}{dx} = f'(g(x))\, g'(x)$$
$$= D_u f \cdot Dg \quad \text{where} \quad D_u = d/du$$

The above results can equivalently be written in terms of d rather than D. Thus for example II-3 becomes

$$d[f(x)\, g(x)] = f(x)\, dg(x) + g(x)\, df(x) = [f(x)\, g'(x) + g(x)\, f'(x)]dx$$

DERIVATIVES OF SPECIAL FUNCTIONS

In the following we list derivatives of some of the more common functions. The constant $e = 2.71828\ldots$ is the *natural base of logarithms* and we write $\log_e x$ as $\ln x$, called the *natural logarithm*. All letters besides x denote given constants.

III-1. $D[c] = 0$

III-2. $D[x^m] = mx^{m-1}$

III-3. $D[(px+q)^m] = mp(px+q)^{m-1}$

III-4. $D[b^x] = b^x \ln b$

III-5. $D[e^{rx}] = re^{rx}$

III-6. $D[\sin rx] = r \cos rx$

III-7. $D[\cos rx] = -r \sin rx$

III-8. $D[\ln x] = 1/x$

III-9. $D[\log_b x] = \dfrac{\log_b e}{x} \qquad b > 0,\ b \neq 1$

These results can also be written in terms of d rather than D. Thus for example III-2 becomes

$$d(x^m) = mx^{m-1}\, dx$$

GENERAL RULES OF THE DIFFERENCE CALCULUS

The following results involving Δ bear close resemblances to the results II-1 to II-4 on page 4.

IV-1. $\Delta[f(x) + g(x)] = \Delta f(x) + \Delta g(x)$

IV-2. $\Delta[\alpha f(x)] = \alpha \Delta f(x) \qquad \alpha = \text{constant}$

IV-3. $$\Delta[f(x)\, g(x)] = f(x)\, \Delta g(x) + g(x+h)\, \Delta f(x)$$
$$= g(x)\, \Delta f(x) + f(x+h)\, \Delta g(x)$$
$$= f(x)\, \Delta g(x) + g(x)\, \Delta f(x) + \Delta f(x)\, \Delta g(x)$$

IV-4. $$\Delta\left[\frac{f(x)}{g(x)}\right] = \frac{g(x)\, \Delta f(x) - f(x)\, \Delta g(x)}{g(x)\, g(x+h)}$$

Note that if we divide by Δx and let $\Delta x \to 0$ the above results become those of II-1 to II-4.

FACTORIAL FUNCTIONS

Formula III-2 states that $Dx^m = mx^{m-1}$. In an effort to obtain an analogous formula involving Δ we write

$$\frac{\Delta x^m}{\Delta x} = \frac{(x+h)^m - x^m}{h}$$

which however does not resemble the formula for Dx^m. To achieve a resemblance we introduce the *factorial function* defined by

$$x^{(m)} = x(x-h)(x-2h)\cdots(x-[m-1]h) \qquad m = 1,2,3,\ldots \tag{22}$$

consisting of m factors. The name factorial arises because in the special case $x = m$, $h = 1$ we have $m^{(m)} = m(m-1)(m-2)\cdots 2\cdot 1 = m!$, i.e. *factorial m*.

If $m = 0$, we define $x^{(m)} = 1$, i.e.

$$x^{(0)} = 1 \tag{23}$$

For negative integers we define [see Problem 1.18(*a*)]

$$x^{(-m)} = \frac{1}{(x+h)(x+2h)\cdots(x+mh)} = \frac{1}{(x+mh)^{(m)}} \qquad m = 1,2,3,\ldots \tag{24}$$

Note that as $h \to 0$, $x^{(m)} \to x^m$, $x^{(-m)} \to x^{-m}$.

Using (*22*), (*23*), (*24*) it follows that [see Problems 1.17 and 1.18(*b*)] for all integers m

$$\frac{\Delta x^{(m)}}{\Delta x} = mx^{(m-1)} \qquad \text{or} \qquad \Delta x^{(m)} = mx^{(m-1)}h \tag{25}$$

in perfect analogy to $Dx^m = mx^{m-1}$ and $d(x^m) = mx^{m-1}\,dx$ respectively.

We can also define $x^{(m)}$ for nonintegral values of m in terms of the *gamma function* [see page 103].

FACTORIAL POLYNOMIALS

From (*22*) we find on putting $m = 1,2,3,\ldots$

$$\left.\begin{aligned} x^{(1)} &= x \\ x^{(2)} &= x^2 - xh \\ x^{(3)} &= x^3 - 3x^2h + 2xh^2 \\ x^{(4)} &= x^4 - 6x^3h + 11x^2h^2 - 6xh^3 \\ x^{(5)} &= x^5 - 10x^4h + 35x^3h^2 - 50x^2h^3 + 24xh^4 \end{aligned}\right\} \tag{26}$$

etc. If p is any positive integer, we define a *factorial polynomial* of degree p as

$$a_0x^{(p)} + a_1x^{(p-1)} + \cdots + a_p$$

where $a_0 \neq 0, a_1, \ldots, a_p$ are constants. Using (*26*) we see that a factorial polynomial of degree p can be expressed uniquely as an ordinary polynomial of degree p.

Conversely any ordinary polynomial of degree p can be expressed uniquely as a factorial polynomial of degree p. This can be accomplished by noting that

$$\left.\begin{aligned} x &= x^{(1)} \\ x^2 &= x^{(2)} + x^{(1)}h \\ x^3 &= x^{(3)} + 3x^{(2)}h + x^{(1)}h^2 \\ x^4 &= x^{(4)} + 7x^{(3)}h + 6x^{(2)}h^2 + x^{(1)}h^3 \\ x^5 &= x^{(5)} + 15x^{(4)}h + 25x^{(3)}h^2 + 10x^{(2)}h^3 + x^{(1)}h^4 \end{aligned}\right\} \tag{27}$$

etc. Another method for converting an ordinary polynomial into a factorial polynomial uses *synthetic division* [see Problems 1.23-1.25].

STIRLING NUMBERS

Any of the equations (*26*) can be written as

$$x^{(n)} = \sum_{k=1}^{n} s_k^n x^k h^{n-k} \tag{28}$$

where the coefficients s_k^n are called *Stirling numbers of the first kind.* A recursion formula for these numbers is

$$s_k^{n+1} = s_{k-1}^n - n s_k^n \tag{29}$$

where we define $s_n^n = 1, \ s_k^n = 0 \quad \text{for} \quad k \leqq 0, \ k \geqq n+1 \quad \text{where } n > 0$ (30)

Similarly any of the equations (27) can be written as

$$x^n = \sum_{k=1}^{n} S_k^n x^{(k)} h^{n-k} \tag{31}$$

where the coefficients S_k^n are called *Stirling numbers of the second kind.* A recursion formula for these numbers is

$$S_k^{n+1} = S_{k-1}^n + k S_k^n \tag{32}$$

where we define

$$S_n^n = 1, \ S_k^n = 0 \quad \text{for} \quad k \leqq 0, \ k \geqq n+1 \quad \text{where } n > 0 \tag{33}$$

In obtaining properties of the Stirling numbers results are greatly simplified by choosing $h = 1$ in (28) and (31).

For tables of Stirling numbers of the first and second kinds see pages 232 and 233.

GENERALIZED FACTORIAL FUNCTIONS

The *generalized factorial function* corresponding to any function $f(x)$ is defined by

$$[f(x)]^{(m)} = f(x)\,f(x-h)\,f(x-2h)\cdots f(x-[m-1]h) \qquad m = 1,2,3,\ldots \tag{34}$$

$$[f(x)]^{(-m)} = \frac{1}{f(x+h)\,f(x+2h)\cdots f(x+mh)} \qquad m = 1,2,3,\ldots \tag{35}$$

$$[f(x)]^{(0)} = 1 \tag{36}$$

The special case $f(x) = x$ yields (22), (24) and (23) respectively. It should be noted that as $h \to 0$, $[f(x)]^{(m)} \to [f(x)]^m$ and $[f(x)]^{(-m)} \to [f(x)]^{-m}$.

DIFFERENCES OF SPECIAL FUNCTIONS

In the following we list difference formulas for special functions analogous to the differentiation formulas on page 5.

V-1. $\Delta[c] = 0$

V-2. $\Delta[x^{(m)}] = mx^{(m-1)}h$

V-3. $\Delta[(px+q)^{(m)}] = mph(px+q)^{(m-1)}$

V-4. $\Delta[b^x] = b^x(b^h - 1)$

V-5. $\Delta[e^{rx}] = e^{rx}(e^{rh} - 1)$

V-6. $\Delta[\sin rx] = 2\sin(rh/2)\cos r(x + h/2)$

V-7. $\Delta[\cos rx] = -2\sin(rh/2)\sin r(x + h/2)$

V-8. $\Delta[\ln x] = \ln(1 + h/x)$

V-9. $\Delta[\log_b x] = \log_b(1 + h/x)$

Note that if we divide each of these results by $\Delta x = h$ and take the limit as Δx or h approaches zero we arrive at the corresponding formulas for derivatives given on page 5. For example V-5 gives

$$\lim_{\Delta x \to 0} \frac{\Delta[e^{rx}]}{\Delta x} = \lim_{h \to 0} e^{rx}\left(\frac{e^{rh}-1}{h}\right) = re^{rx} = \frac{d}{dx}(e^{rx}) = D(e^{rx})$$

TAYLOR SERIES

From the calculus we know that if all the derivatives of $f(x)$ up to order $n+1$ at least exist at a point a of an interval, then there is a number η between a and any point x of the interval such that

$$f(x) \quad = \quad f(a) + f'(a)(x-a) + \frac{f''(a)(x-a)^2}{2!} + \cdots + \frac{f^{(n)}(a)(x-a)^n}{n!} + R_n \qquad (37)$$

where the remainder R_n is given by

$$R_n \quad = \quad \frac{f^{(n+1)}(\eta)(x-a)^{n+1}}{(n+1)!} \qquad (38)$$

This is often called *Taylor's theorem* or *Taylor's formula with a remainder*. The case where $n=0$ is often called the *law of the mean* or *mean value theorem for derivatives.*

If $\lim_{n\to\infty} R_n = 0$ for all x in an interval, then

$$f(x) \quad = \quad f(a) + f'(a)(x-a) + \frac{f''(a)(x-a)^2}{2!} + \cdots \qquad (39)$$

is called the *Taylor series* of $f(x)$ about $x=a$ and converges in the interval. We can also write (*39*) with $x=a+h$ so that

$$f(a+h) \quad = \quad f(a) + f'(a)h + \frac{f''(a)h^2}{2!} + \cdots \qquad (40)$$

The special case where $a=0$ is often called a *Maclaurin series.*

Some important special series together with their intervals of convergence [i.e. the values of x for which the series converges] are as follows.

1. $$e^x \quad = \quad 1 + x + \frac{x^2}{2!} + \frac{x^3}{3!} + \cdots \qquad -\infty < x < \infty$$

2. $$\sin x \quad = \quad x - \frac{x^3}{3!} + \frac{x^5}{5!} - \frac{x^7}{7!} + \cdots \qquad -\infty < x < \infty$$

3. $$\cos x \quad = \quad 1 - \frac{x^2}{2!} + \frac{x^4}{4!} - \frac{x^6}{6!} + \cdots \qquad -\infty < x < \infty$$

4. $$\ln(1+x) \quad = \quad x - \frac{x^2}{2} + \frac{x^3}{3} - \frac{x^4}{4} + \cdots \qquad -1 < x \leq 1$$

TAYLOR SERIES IN OPERATOR FORM

If in (*40*) we replace a by x we have

$$f(x+h) \quad = \quad f(x) + f'(x)h + \frac{f''(x)h^2}{2!} + \cdots \qquad (41)$$

which can formally be written in terms of operators as

$$Ef(x) \quad = \quad \left[1 + hD + \frac{h^2D^2}{2!} + \frac{h^3D^3}{3!} + \cdots\right] f(x) \quad = \quad e^{hD}f(x)$$

using the series for e^{hD} obtained from the result 1 above on replacing x by hD. This leads to the operator equivalence

$$E \; = \; e^{hD} \qquad (42)$$

or using (*10*), page 3, $\qquad \Delta = e^{hD} - 1 \quad$ or $\quad e^{hD} = 1 + \Delta \qquad (43)$

Such formal series of operators often prove lucrative in obtaining many important results and the methods involved are often called *symbolic operator methods.*

THE GREGORY-NEWTON FORMULA

As may be expected there exists a formula in the difference calculus which is analogous to the Taylor series of differential calculus. The formula is called the *Gregory-Newton formula* and one of the ways in which it can be written is

$$f(x) \quad = \quad f(a) + \frac{\Delta f(a)}{\Delta x}\frac{(x-a)^{(1)}}{1!} + \frac{\Delta^2 f(a)}{\Delta x^2}\frac{(x-a)^{(2)}}{2!} + \cdots + \frac{\Delta^n f(a)}{\Delta x^n}\frac{(x-a)^{(n)}}{n!} + R_n \qquad (44)$$

which is analogous to (*37*). The remainder R_n is given by

$$R_n \quad = \quad \frac{f^{(n+1)}(\eta)\,(x-a)^{(n+1)}}{(n+1)!} \qquad (45)$$

where η lies between a and x. If $\lim_{n\to\infty} R_n = 0$ for all x in an interval, then (*44*) can be written as an infinite series which converges in the interval.

If $f(x)$ is any polynomial of degree n, the remainder $R_n = 0$ for all x.

LEIBNITZ'S RULE

In the calculus there exists a formula for the nth derivative of the product of two functions $f(x)$ and $g(x)$ known as *Leibnitz's rule*. This states that

$$D^n(fg) \quad = \quad (f)(D^n g) + \binom{n}{1}(Df)(D^{n-1}g) + \binom{n}{2}(D^2 f)(D^{n-2}g) + \cdots + \binom{n}{n}(D^n f)(g) \qquad (46)$$

where for brevity we have written f and g for $f(x)$ and $g(x)$ and where

$$\binom{n}{r} \quad = \quad \frac{n(n-1)\cdots(n-r+1)}{r!} \quad = \quad \frac{n!}{r!\,(n-r)!}, \qquad 0! = 1 \qquad (47)$$

are the *binomial coefficients*, i.e. the coefficients in the expansion of $(1+x)^n$.

An analogous formula exists for differences and is given by

$$\Delta^n(fg) \quad = \quad (f)(\Delta^n g) + \binom{n}{1}(\Delta f)(\Delta^{n-1}Eg) + \binom{n}{2}(\Delta^2 f)(\Delta^{n-2}E^2 g) + \cdots + \binom{n}{n}(\Delta^n f)(E^n g) \qquad (48)$$

which we shall refer to as *Leibnitz's rule for differences*. See Problem 1.36.

OTHER DIFFERENCE OPERATORS

Various other operators are sometimes used in the difference calculus although these can be expressed in terms of the fundamental operators Δ and E. Two such operators are ∇ and δ defined by

$$\nabla f(x) \quad = \quad f(x) - f(x-h) \qquad (49)$$

$$\delta f(x) \quad = \quad f\left(x + \frac{h}{2}\right) - f\left(x - \frac{h}{2}\right) \qquad (50)$$

We call ∇ the *backward difference operator* [in contrast with Δ which is then called the *forward difference operator*] and δ the *central difference operator*. These are related to Δ and E by

$$\nabla \;=\; \Delta E^{-1} \;=\; 1 - E^{-1} \qquad (51)$$

$$\delta \;=\; E^{1/2} - E^{-1/2} \;=\; \Delta E^{-1/2} \;=\; \nabla E^{1/2} \qquad (52)$$

See Problems 1.38 and 1.39.

Two other operators which are sometimes used are called *averaging operators* and are denoted by M and μ respectively. They are defined by the equations

$$Mf(x) \;=\; \tfrac{1}{2}[f(x+h)+f(x)] \tag{53}$$

$$\mu f(x) \;=\; \frac{1}{2}\left[f\left(x+\frac{h}{2}\right)+f\left(x-\frac{h}{2}\right)\right] \tag{54}$$

Some relationships of these operators to the other operators are given by

$$M \;=\; \tfrac{1}{2}(E+1) \;=\; 1+\tfrac{1}{2}\Delta \tag{55}$$

$$\mu \;=\; \tfrac{1}{2}(E^{1/2}+E^{-1/2}) \tag{56}$$

See Problem 1.40.

Solved Problems

OPERATORS

1.1. If C is the cubing operator and D is the derivative operator, determine each of the following.

(*a*) $C(2x+1)$ (*e*) $(C+D)(4x+2)$ (*h*) $(D^2-2D+1)(3x^2-5x+4)$

(*b*) $D(4x-5x^3)$ (*f*) $(D+C)(4x+2)$ (*i*) $(C-2)(D+3)x^2$

(*c*) $CD(2x-3)$ (*g*) $(C^2-4C+1)(\sqrt[3]{x-1})$ (*j*) $(xCxD)x^3$

(*d*) $DC(2x-3)$

(*a*) $C(2x+1) = (2x+1)^3$ (*b*) $D(4x-5x^3) = 4-15x^2$ (*c*) $CD(2x-3) = C\cdot 2 = 2^3 = 8$

(*d*) $DC(2x-3) = D(2x-3)^3 = 3(2x-3)^2\cdot 2 = 6(2x-3)^2$

Another method. $DC(2x-3) = D(2x-3)^3 = D[8x^3-36x^2+54x-27] = 24x^2-72x+54$

Note that from (*c*) and (*d*) it is seen that the result of operating with CD is not the same as operating with DC, i.e. $CD \neq DC$ so that C and D are *noncommutative* with respect to multiplication.

(*e*) $(C+D)(4x+2) = C(4x+2)+D(4x+2) = (4x+2)^3+4$

(*f*) $(D+C)(4x+2) = D(4x+2)+C(4x+2) = 4+(4x+2)^3$

The results (*e*) and (*f*) illustrate the *commutative law of addition* for operators C and D, i.e. $C+D=D+C$.

(*g*) $(C^2-4C+1)(\sqrt[3]{x-1}) = C^2(\sqrt[3]{x-1})-4C(\sqrt[3]{x-1})+1(\sqrt[3]{x-1}) = (x-1)^3-4(x-1)+\sqrt[3]{x-1}$

(*h*) $(D^2-2D+1)(3x^2-5x+4) = D^2(3x^2-5x+4)-2D(3x^2-5x+4)+1(3x^2-5x+4)$

$= 6-2(6x-5)+3x^2-5x+4 = 3x^2-17x+20$

(*i*) $(C-2)(D+3)x^2 = (C-2)[Dx^2+3x^2] = (C-2)(2x+3x^2)$

$= C(2x+3x^2)-2(2x+3x^2) = (2x+3x^2)^3-2(2x+3x^2)$

(*j*) $(xCxD)x^3 = xCx(Dx^3) = xCx(3x^2) = xC(3x^3) = x(27x^9) = 27x^{10}$

1.2. If D is the derivative operator and f is any *differentiable* function, prove the operator equivalence $Dx-xD=I=1$, i.e. the unit or identity operator.

We have

$$(Dx-xD)f \;=\; Dxf - xDf \;=\; \frac{d}{dx}(xf) - x\frac{df}{dx} \;=\; x\frac{df}{dx}+f-x\frac{df}{dx} \;=\; f \;=\; If$$

Then for any differentiable f, $(Dx-xD)f = If = 1$ or $Dx-xD=I=1$.

1.3. Prove that (*a*) the cubing operator C is a nonlinear operator while (*b*) the derivative operator D is a linear operator.

(*a*) If f and g are any functions we have

$$C(f+g) \;=\; (f+g)^3, \qquad Cf + Cg \;=\; f^3 + g^3$$

Then in general since $(f+g)^3 \neq f^3 + g^3$ it follows that $C(f+g) \neq Cf + Cg$ so that C cannot be a linear operator, i.e. C is a *nonlinear operator.*

(*b*) If f and g are any differentiable functions, we have

$$D(f+g) \;=\; \frac{df}{dx} + \frac{df}{dx} \;=\; Df + Dg$$

Also if α is any constant, then $\quad D(\alpha f) \;=\; \alpha\frac{df}{dx} \;=\; \alpha Df$

Then D is a linear operator.

1.4. (*a*) If C denotes the *cubing operator,* explain what is meant by the inverse operator C^{-1}. (*b*) Does C^{-1} exist always? (*c*) Is C^{-1} unique? Illustrate by considering $C^{-1}(8)$.

(*a*) By definition $B = C^{-1}$ is that operator such that $CBf(x) = f(x)$ or $CB = I = 1$, the unit or identity operator. Equivalently $C^{-1}f(x) = g(x)$ if and only if $Cg(x) = f(x)$ or $[g(x)]^3 = f(x)$, i.e. $g(x)$ is a *cube root* of $f(x)$. Thus C^{-1} is the *operation of taking cube roots.* As a particular case $C^{-1}(8)$ is the result of taking cube roots of 8.

(*b*) If $f(x)$ is real, then $C^{-1}[f(x)]$ will always exist but may or may not be real. For example if $C^{-1}(8)$ is denoted by x then $x^3 = 8$, i.e. $(x-2)(x^2+2x+4) = 0$ and $x = 2, -1 \pm \sqrt{3}\,i$ so that there are both real and complex cube roots. The only case where $C^{-1}f(x)$ is always real when $f(x)$ is real is if $f(x) = 0$.

If $f(x)$ is complex, $C^{-1}f(x)$ always exists.

(*c*) It is clear from (*b*) that $C^{-1}f(x)$ is not always unique since for example $C^{-1}(8)$ has three different values. However if $f(x)$ is real and if we restrict ourselves to real values only, then $C^{-1}f(x)$ will be unique. Thus if we restrict ourselves to real values we have for example $C^{-1}(8) = 2$.

THE DIFFERENCE AND TRANSLATION OPERATORS

1.5. Find each of the following.

(*a*) $\Delta(2x^2+3x)$ (*b*) $E(4x-x^2)$ (*c*) $\Delta^2(x^3-x^2)$ (*d*) $E^3(3x-2)$ (*e*) $(2\Delta^2+\Delta-1)(x^2+2x+1)$ (*f*) $(E^2-3E+2)(2^{x/h}+x)$ (*g*) $(\Delta+1)(2\Delta-1)(x^2+2x+1)$ (*h*) $(E-2)(E-1)(2^{x/h}+x)$

(*a*)
$$\begin{aligned}\Delta(2x^2+3x) &= [2(x+h)^2+3(x+h)] - [2x^2+3x]\\ &= 2x^2 + 4hx + 2h^2 + 3x + 3h - 2x^2 - 3x \;=\; 4hx + 2h^2 + 3h\end{aligned}$$

(*b*)
$$E(4x-x^2) = 4(x+h) - (x+h)^2 \;=\; 4x + 4h - x^2 - 2hx - h^2$$

(*c*)
$$\begin{aligned}\Delta^2(x^3-x^2) &= \Delta[\Delta(x^3-x^2)] \;=\; \Delta[\{(x+h)^3-(x+h)^2\} - (x^3-x^2)]\\ &= \Delta[3x^2h + 3xh^2 + h^3 - 2hx - h^2]\\ &= [3(x+h)^2h + 3(x+h)h^2 + h^3 - 2h(x+h) - h^2] - [3x^2h + 3xh^2 + h^3 - 2hx - h^2]\\ &= 6h^2x + 6h^3 - 2h^2\end{aligned}$$

(*d*)
$$\begin{aligned}E^3(3x-2) &= E^2E(3x-2) \;=\; E^2[3(x+h)-2] \;=\; E\cdot E[3(x+h)-2]\\ &= E[3(x+2h)-2] \;=\; 3(x+3h) - 2\end{aligned}$$

(*e*)
$$\begin{aligned}(2\Delta^2+\Delta-1)(x^2+2x+1) &= 2\Delta^2(x^2+2x+1) + \Delta(x^2+2x+1) - 1(x^2+2x+1)\\ &= 2\Delta[\{(x+h)^2+2(x+h)+1\} - \{x^2+2x+1\}]\\ &\qquad + [\{(x+h)^2+2(x+h)+1\} - \{x^2+2x+1\}] - x^2 - 2x - 1\\ &= 2\Delta[2hx+h^2+2h] + [2hx+h^2+2h] - x^2 - 2x - 1\\ &= 2[\{2h(x+h)+h^2+2h\} - \{2hx+h^2+2h\}]\\ &\qquad + 2hx + h^2 + 2h - x^2 - 2x - 1\\ &= 5h^2 + 2hx + 2h - x^2 - 2x - 1\end{aligned}$$

Note that we use 1 in $2\Delta^2+\Delta-1$ instead of the unit or identity operator I.

$$(f)\quad (E^2-3E+2)(2^{x/h}+x) = E^2(2^{x/h}+x) - 3E(2^{x/h}+x) + 2(2^{x/h}+x)$$
$$= (2^{(x+2h)/h}+x+2h) - 3(2^{(x+h)/h}+x+h) + 2^{(x+h)/h}+2x$$
$$= (4\cdot 2^{x/h}+x+2h) - 3(2\cdot 2^{x/h}+x+h) + 2\cdot 2^{x/h}+2x$$
$$= -h$$

$$(g)\quad (\Delta+1)(2\Delta-1)(x^2+2x+1) = (\Delta+1)[2\Delta(x^2+2x+1) - 1(x^2+2x+1)]$$
$$= (\Delta+1)[2\{(x+h)^2+2(x+h)+1\} - 3(x^2+2x+1)]$$
$$= (\Delta+1)[4hx+2h^2+4h-x^2-2x-1]$$
$$= \Delta[4hx+2h^2+4h-x^2-2x-1] + [4hx+2h^2+4h-x^2-2x-1]$$
$$= 4h(x+h) + 2h^2 + 4h - (x+h)^2 - 2(x+h) - 1$$
$$= 5h^2 + 2hx + 2h - x^2 - 2x - 1$$

Note that this result is the same as that of (*e*) and illustrates the operator equivalence $(\Delta+1)(2\Delta-1) = 2\Delta^2+\Delta-1$.

$$(h)\quad (E-2)(E-1)(2^{x/h}+x) = (E-2)[E(2^{x/h}+x) - (2^{x/h}+x)]$$
$$= (E-2)[(2^{(x+h)/h}+x+h) - (2^{x/h}+x)]$$
$$= (E-2)[2^{x/h}+h]$$
$$= E[2^{x/h}+h] - 2[2^{x/h}+h]$$
$$= 2^{(x+h)/h} + h - 2^{(x+h)/h} - 2h$$
$$= -h$$

Note that this result is the same as that of (*f*) and illustrates the operator equivalence $(E-2)(E-1) = E^2-3E+2$.

1.6. Prove that (*a*) $\Delta[f(x)+g(x)] = \Delta f(x) + \Delta g(x)$ and (*b*) $\Delta[\alpha f(x)] = \alpha f(x)$, α = constant, and thus show that Δ is a linear operator.

$$(a)\quad \Delta[f(x)+g(x)] = [f(x+h)+g(x+h)] - [f(x)+g(x)]$$
$$= [f(x+h)-f(x)] + [g(x+h)-g(x)] = \Delta f(x) + \Delta g(x)$$

$$(b)\quad \Delta[\alpha f(x)] = [\alpha f(x+h)] - [\alpha f(x)] = \alpha[f(x+h)-f(x)] = \alpha\Delta f(x)$$

Then from the definition 6 on page 2 it follows that Δ is a linear operator.

1.7. Prove that (*a*) $\Delta[f(x)\,g(x)] = f(x)\,\Delta g(x) + g(x+h)\,\Delta f(x)$

$$(b)\quad \Delta\left[\frac{f(x)}{g(x)}\right] = \frac{g(x)\,\Delta f(x) - f(x)\,\Delta g(x)}{g(x)\,g(x+h)}$$

$$(a)\quad \Delta[f(x)\,g(x)] = f(x+h)\,g(x+h) - f(x)\,g(x)$$
$$= f(x+h)\,g(x+h) - f(x)\,g(x+h) + f(x)\,g(x+h) - f(x)\,g(x)$$
$$= g(x+h)[f(x+h)-f(x)] + f(x)[g(x+h)-g(x)]$$
$$= g(x+h)\,\Delta f(x) + f(x)\,\Delta g(x)$$
$$= f(x)\,\Delta g(x) + g(x+h)\,\Delta f(x)$$

$$(b)\quad \Delta\left[\frac{f(x)}{g(x)}\right] = \frac{f(x+h)}{g(x+h)} - \frac{f(x)}{g(x)}$$
$$= \frac{g(x)\,f(x+h) - f(x)\,g(x+h)}{g(x)\,g(x+h)}$$
$$= \frac{g(x)[f(x+h)-f(x)] - f(x)[g(x+h)-g(x)]}{g(x)\,g(x+h)}$$
$$= \frac{g(x)\,\Delta f(x) - f(x)\,\Delta g(x)}{g(x)\,g(x+h)}$$

1.8. Prove formulas (*a*) V-4 (*b*) V-5 (*c*) V-6, page 7.

(*a*) $$\Delta[b^x] = b^{x+h} - b^x = b^x(b^h - 1)$$

(*b*) $$\Delta[e^{rx}] = e^{r(x+h)} - e^{rx} = e^{rx}(e^{rh} - 1)$$

(*c*) $$\Delta[\sin rx] = \sin r(x+h) - \sin rx = 2\sin\frac{rh}{2}\cos r\left(x + \frac{h}{2}\right)$$

on using the trigonometric formula

$$\sin\theta_1 - \sin\theta_2 = 2\sin\left(\frac{\theta_1 - \theta_2}{2}\right)\cos\left(\frac{\theta_1 + \theta_2}{2}\right)$$

with $\theta_1 = r(x+h)$ and $\theta_2 = rx$.

1.9. Prove that $\Delta E = E\Delta$, i.e. the operators Δ and E are commutative with respect to multiplication.

We have for arbitrary $f(x)$

$$\Delta E f(x) = \Delta[Ef(x)] = \Delta[f(x+h)] = f(x+2h) - f(x+h)$$
$$E\Delta f(x) = E[f(x+h) - f(x)] = f(x+2h) - f(x+h)$$

Then $\Delta E f(x) = E\Delta f(x)$, i.e. $\Delta E = E\Delta$.

1.10. Prove that (*a*) $\Delta = E - 1$, (*b*) $E = 1 + \Delta$, (*c*) $\Delta^2 = (E-1)^2 = E^2 - 2E + 1$.

(*a*) We have for arbitrary $f(x)$

$$\Delta f(x) = f(x+h) - f(x) = Ef(x) - f(x) = (E-1)f(x)$$

Thus $\Delta = E - 1$ where we use 1 instead of the unit or identity operator I.

(*b*) We have for arbitrary $f(x)$

$$Ef(x) = f(x+h) = f(x) + [f(x+h) - f(x)] = f(x) + \Delta f(x) = (1+\Delta)f(x)$$

Thus $E = 1 + \Delta$. Note that this illustrates the fact that we can treat operators Δ and E as ordinary algebraic quantities. In particular we can transpose the 1 in $\Delta = E - 1$ to obtain $E = 1 + \Delta$.

(*c*) We have for arbitrary $f(x)$

$$\begin{aligned}\Delta^2 f(x) = \Delta[\Delta f(x)] &= \Delta[f(x+h) - f(x)]\\ &= [f(x+2h) - f(x+h)] - [f(x+h) - f(x)]\\ &= f(x+2h) - 2f(x+h) + f(x)\\ &= E^2 f(x) - 2Ef(x) + f(x)\\ &= (E^2 - 2E + 1)f(x)\\ &= (E-1)^2 f(x)\end{aligned}$$

Thus $\Delta^2 = (E-1)^2 = E^2 - 2E + 1$.

1.11. Obtain a generalization of Problem 1.10 for Δ^n where n is any positive integer.

Method 1.

Using the fact that Δ and E can be manipulated as ordinary algebraic quantities we have by the binomial theorem

$$\Delta^n = (E-1)^n = E^n - \binom{n}{1}E^{n-1} + \binom{n}{2}E^{n-2} - \binom{n}{3}E^{n-3} + \cdots + (-1)^n \qquad (1)$$

where $\binom{n}{r} = \dfrac{n!}{r!(n-r)!}$ are the binomial coefficients.

By operating on $f(x)$ the result (*1*) is equivalent to

$$\Delta^n f(x) = f(x+nh) - \binom{n}{1}f(x+(n-1)h) + \binom{n}{2}f(x+(n-2)h) - \cdots + (-1)^n f(x)$$

Method 2.

Assume that

$$\Delta^n f(x) = f(x+nh) - \binom{n}{1} f(x+(n-1)h) + \cdots + (-1)^n f(x) \qquad (2)$$

is true for a positive integer n, i.e. assume the operator equivalence

$$\Delta^n = (E-1)^n = E^n - \binom{n}{1} E^{n-1} + \binom{n}{2} E^{n-2} - \cdots + (-1)^n$$

Then operating on both sides of (2) with Δ we have

$$\Delta^{n+1} f(x) = f(x+(n+1)h) - f(x+nh) - \binom{n}{1}[f(x+nh) - f(x+(n-1)h)]$$
$$+ \cdots + (-1)^n[f(x+h) - f(x)]$$
$$= f(x+(n+1)h) - \left[\binom{n}{1} + 1\right] f(x+nh) + \left[\binom{n}{1} + \binom{n}{2}\right] f(x+(n-1)h)$$
$$- \cdots + (-1)^{n+1} f(x)$$

or since [see Problem 1.105] $\displaystyle\binom{n}{r} + \binom{n}{r+1} = \binom{n+1}{r+1}$

we have $$\Delta^{n+1} f(x) = f(x+(n+1)h) - \binom{n+1}{1} f(x+nh) + \binom{n+1}{2} f(x+(n-1)h)$$
$$- \cdots + (-1)^{n+1} f(x) \qquad (3)$$

It follows that if (2) is true then (3) is true, i.e. if the result is true for n it is also true for $n+1$. Now since (2) is true for $n=1$ [because then (2) reduces to $\Delta f(x) = f(x+h) - f(x)$], it follows that it is also true for $n=2$ and thus for $n=3$ and so on. It is thus true for all positive integer values of n. The method of proof given here is called *mathematical induction.*

1.12. Prove that (*a*) $E^{-1}f(x) = f(x-h)$, (*b*) $E^{-n}f(x) = f(x-nh)$ for any integer n.

(*a*) By definition if $E^{-1}f(x) = g(x)$, then $Eg(x) = f(x)$, i.e. $g(x+h) = f(x)$ or $g(x) = f(x-h)$ on replacing x by $x-h$. Thus $E^{-1}f(x) = f(x-h)$.

(*b*) Case 1. If n is a negative integer or zero, let $n = -m$ where m is a positive integer or zero. Then

$$E^m f(x) = f(x+mh), \quad \text{i.e.} \quad E^{-n}f(x) = f(x-nh)$$

Case 2. If n is a positive integer, then by definition if

$$E^{-n}f(x) = g(x) \quad \text{then} \quad E^n g(x) = f(x) \quad \text{or} \quad g(x+nh) = f(x)$$

Then replacing x by $x-nh$ we have

$$E^{-n}f(x) = g(x) = f(x-nh)$$

In general we shall define $$E^n f(x) = f(x+nh)$$
for all real numbers n.

THE DERIVATIVE AND DIFFERENTIAL OPERATORS

1.13. Show that (*a*) $\displaystyle\lim_{\Delta x \to 0} \frac{\Delta}{\Delta x}(2x^2+3x) = \frac{d}{dx}(2x^2+3x) = 4x+3$

(*b*) $\displaystyle\lim_{\Delta x \to 0} \frac{\Delta^2}{\Delta x^2}(x^3-x^2) = \frac{d^2}{dx^2}(x^3-x^2) = 6x-2$

directly from the definition.

(*a*) From Problem 1.5(*a*) we have since $h = \Delta x$

$$\frac{\Delta}{\Delta x}(2x^2+3x) = \frac{4hx + 2h^2 + 3h}{h} = 4x + 2h + 3$$

Then $$\lim_{\Delta x \to 0} \frac{\Delta}{\Delta x}(2x^2+3x) = \lim_{h\to 0}(4x+2h+3) = 4x+3 = \frac{d}{dx}(2x^2+3x)$$

(*b*) From Problem 1.5(*c*) we have since $h = \Delta x$ and $h^2 = (\Delta x)^2$

$$\frac{\Delta^2}{\Delta x^2}(x^3 - x^2) = \frac{6h^2x + 6h^3 - 2h^2}{h^2} = 6x + 6h - 2$$

Then $$\lim_{\Delta x \to 0} \frac{\Delta^2}{\Delta x^2}(x^3 - x^2) = \lim_{h \to 0} (6x + 6h - 2) = 6x - 2 = \frac{d^2}{dx^2}(x^3 - x^2)$$

The results illustrate the fact that $\lim_{\Delta x \to 0} \frac{\Delta^n}{\Delta x^n} f(x) = \frac{d^n}{dx^n} f(x)$. Note the operator equivalence $\lim_{\Delta x \to 0} \frac{\Delta^n}{\Delta x^n} = \frac{d^n}{dx^n}$. The notation $D = \frac{d}{dx}$, $D^2 = \frac{d^2}{dx^2}$, ..., $D^n = \frac{d^n}{dx^n}$ can also be used. Strictly speaking we should write $\frac{\Delta^n}{\Delta x^n}$ as $\frac{\Delta^n}{(\Delta x)^n}$ and $\frac{d^n}{dx^n}$ as $\frac{d^n}{(dx)^n}$. However by custom we leave off the parentheses in the denominator.

1.14. Find (*a*) $d(2x^2 + 3x)$, (*b*) $d^2(x^3 - x^2)$.

(*a*) By definition, using $dx = \Delta x = h$ we have

$$d(2x^2 + 3x) = \frac{d}{dx}(2x^2 + 3x)\,dx = (4x + 3)\,dx = (4x + 3)h$$

(*b*) By definition,

$$d^2(x^3 - x^2) = \frac{d^2}{dx^2}(x^3 - x^2)(dx)^2 = (6x - 2)(dx)^2 = (6x - 2)h^2$$

1.15. Prove that $D[f(x)\,g(x)] = f(x)\,Dg(x) + g(x)\,Df(x)$.

From Problem 1.7(*a*) we have on dividing by $h = \Delta x$

$$\frac{\Delta}{\Delta x}[f(x)\,g(x)] = f(x)\frac{\Delta}{\Delta x}g(x) + g(x+h)\frac{\Delta}{\Delta x}f(x)$$

Then taking the limit as $h = \Delta x \to 0$ we have

$$\frac{d}{dx}[f(x)\,g(x)] = f(x)\frac{d}{dx}g(x) + g(x)\frac{d}{dx}f(x)$$

which gives the required result on writing $d/dx = D$.

1.16. Use Problem 1.8(*c*) and the fact that $\lim_{\theta \to 0} \frac{\sin\theta}{\theta} = 1$ to prove that $\frac{d}{dx}\sin rx = r\cos x$.

From Problem 1.8(*c*) we have since $\Delta x = h$

$$\frac{\Delta}{\Delta x}[\sin rx] = \frac{2\sin(rh/2)\cos r(x + h/2)}{h} = r\frac{\sin(rh/2)}{(rh/2)}\cos r(x + h/2)$$

Then
$$\begin{aligned}\frac{d}{dx}[\sin rx] &= \lim_{\Delta x \to 0}\frac{\Delta}{\Delta x}[\sin rx] = \lim_{h \to 0}\left\{r\frac{\sin(rh/2)}{(rh/2)}\cos r(x + h/2)\right\} \\ &= r\lim_{h \to 0}\frac{\sin(rh/2)}{(rh/2)}\lim_{h \to 0}\cos r(x + h/2) \\ &= r \cdot 1 \cdot \cos rx \\ &= r\cos rx\end{aligned}$$

We have used here the theorem familiar from the calculus that the limit of a product of functions is equal to the product of the limits of the functions whenever these limits exist.

FACTORIAL FUNCTIONS

1.17. If $x^{(m)} = x(x-h)(x-2h)\cdots(x-[m-1]h)$, $m = 1, 2, 3, \ldots$, prove that $\dfrac{\Delta}{\Delta x}x^{(m)} = mx^{(m-1)}$ or equivalently $\Delta x^{(m)} = mx^{(m-1)}h$ where $h = \Delta x$.

We have

$$\begin{aligned}
\Delta x^{(m)} &= (x+h)^{(m)} - x^{(m)} \\
&= (x+h)(x)(x-h)\cdots(x-[m-2]h) - x(x-h)(x-2h)\cdots(x-[m-1]h) \\
&= x(x-h)\cdots(x-[m-2]h)\,\{(x+h)-(x-[m-1]h)\} \\
&= mx^{(m-1)}h
\end{aligned}$$

Note that for $m = 1$ this formally reduces to $\Delta x^{(1)} = x^{(0)}h$

However since $\Delta x^{(1)} = \Delta x = h$ we are led to define $x^{(0)} = 1$.

1.18. (*a*) Motivate the definition of $x^{(-m)}, m = 1, 2, 3, \ldots$, given on page 6. (*b*). Using the definition on page 6 prove that

$$\frac{\Delta}{\Delta x}x^{(-m)} = -mx^{(-m-1)} \qquad \text{or} \qquad \Delta x^{(-m)} = -mx^{(-m-1)}h$$

(*a*) From the definition of $x^{(m)}$ for $m = 1, 2, 3, \ldots$ we have

$$x^{(m)} = x(x-h)\cdots(x-[m-1]h)$$

$$x^{(m+1)} = x(x-h)\cdots(x-mh)$$

so that

$$x^{(m+1)} = x^{(m)}(x-mh) \tag{1}$$

If now we *formally* put $m = -1$ in this last result, we are led to

$$x^{(0)} = x^{(-1)}(x+h)$$

Using $x^{(0)} = 1$, from Problem 1.17 we are thus led to define

$$x^{(-1)} = \frac{1}{x+h} \tag{2}$$

Similarly, putting $m = -2$ in (*1*) we find

$$x^{(-1)} = x^{(-2)}(x+2h)$$

so that we are led to define

$$x^{(-2)} = \frac{x^{(-1)}}{x+2h} = \frac{1}{(x+h)(x+2h)}$$

Proceeding in this way we are thus led to define

$$x^{(-m)} = \frac{1}{(x+h)(x+2h)\cdots(x+mh)} \qquad m = 1, 2, 3, \ldots$$

as on page 6.

(*b*) We have

$$\begin{aligned}
\Delta x^{(-m)} &= (x+h)^{(-m)} - x^{(-m)} \\
&= \frac{1}{(x+2h)(x+3h)\cdots(x+[m+1]h)} - \frac{1}{(x+h)(x+2h)\cdots(x+mh)} \\
&= \frac{1}{(x+2h)\cdots(x+mh)}\left[\frac{1}{x+(m+1)h} - \frac{1}{x+h}\right] \\
&= \frac{-mh}{(x+h)(x+2h)\cdots(x+mh)(x+[m+1]h)} \\
&= -mx^{(-m-1)}h
\end{aligned}$$

or equivalently

$$\frac{\Delta x^{(-m)}}{\Delta x} = -mx^{(-m-1)}$$

Note that the results of this problem and Problem 1.17 enable us to write for all integers m

$$\frac{\Delta x^{(m)}}{\Delta x} = mx^{(m-1)} \qquad \text{or} \qquad \Delta x^{(m)} = mx^{(m-1)}h$$

1.19. If m is any integer and c is any constant prove that

(a) $\dfrac{\Delta}{\Delta x}[cx^{(m)}] = c\dfrac{\Delta}{\Delta x}x^{(m)} = mcx^{(m-1)}$

(b) $\Delta[cx^{(m)}] = c\Delta x^{(m)} = mchx^{(m-1)}$

Method 1.

(a) $\Delta[cx^{(m)}] = c(x+h)^{(m)} - cx^{(m)} = c[(x+h)^{(m)} - x^{(m)}] = c\Delta x^{(m)}$

Thus by Problems 1.17 and 1.18,

$$\frac{\Delta}{\Delta x}[cx^{(m)}] = c\frac{\Delta x^{(m)}}{\Delta x} = cmx^{(m-1)} = mcx^{(m-1)}$$

(b) Since $\Delta x = h$ we have from part (a)

$$\Delta[cx^{(m)}] = c\Delta x^{(m)} = mchx^{(m-1)}$$

Method 2.

Since Δ and thus $\Delta/\Delta x$ are linear operators we have

$$\Delta[cx^{(m)}] = c\Delta x^{(m)} \quad \text{and} \quad \frac{\Delta}{\Delta x}[cx^{(m)}] = c\frac{\Delta x^{(m)}}{\Delta x}$$

and the required result follows at once from Problems 1.17 and 1.18.

1.20. Find (a) $\dfrac{\Delta}{\Delta x}[3x^{(4)}]$, (b) $\Delta[10x^{(3)}]$, (c) $\dfrac{\Delta}{\Delta x}[5x^{(-2)}]$, (d) $\Delta[-6x^{(-3)}]$ expressing all results in terms of h.

From the results of Problem 1.19 we have

(a) $\dfrac{\Delta}{\Delta x}[3x^{(4)}] = 4\cdot 3x^{(3)} = 12x^{(3)} = 12x(x-h)(x-2h)$

(b) $\Delta[10x^{(3)}] = 3\cdot 10x^{(2)}h = 30hx^{(2)} = 30hx(x-h)$

(c) $\dfrac{\Delta}{\Delta x}[5x^{(-2)}] = (-2)(5)x^{(-3)} = -10x^{(-3)} = \dfrac{-10}{(x+h)(x+2h)(x+3h)}$

(d) $\Delta[-6x^{(-3)}] = (-3)(-6)x^{(-4)} = 18x^{(-4)} = \dfrac{18}{(x+h)(x+2h)(x+3h)(x+4h)}$

1.21. Find (a) $\dfrac{\Delta}{\Delta x}(2x^{(4)} - 3x^{(2)} + x - 4)$, (b) $\Delta(3x^{(-2)} - 2x^{(2)} - 5x^{(-1)})$.

By Problem 1.19 and the fact that Δ and thus $\Delta/\Delta x$ are linear operators we have

(a) $\dfrac{\Delta}{\Delta x}(2x^{(4)} - 3x^{(2)} + x - 4) = 8x^{(3)} - 6x^{(1)} + 1$

(b) $\dfrac{\Delta}{\Delta x}(3x^{-2} - 2x^{(2)} - 5x^{(-1)}) = (-6x^{(-3)} - 4x^{(1)} + 5x^{(-2)})h$

1.22. Find $\dfrac{\Delta^2}{\Delta x^2}[2x^{(3)} - 8x^{(-2)}]$.

$$\frac{\Delta^2}{\Delta x^2}[2x^{(3)} - 8x^{(-2)}] = \frac{\Delta}{\Delta x}\frac{\Delta}{\Delta x}[2x^{(3)} - 8x^{(-2)}] = \frac{\Delta}{\Delta x}[6x^{(2)} + 16x^{(-3)}] = 12x^{(1)} - 48x^{(-4)}$$

1.23. Express $2x^3 - 3x^2 + 5x - 4$ as a factorial polynomial in which the difference interval is h.

Method 1.

From page 6 we have

$$\begin{aligned} 2x^3 - 3x^2 + 5x - 4 &= 2(x^{(3)} + 3x^{(2)}h + x^{(1)}h^2) - 3(x^{(2)} + x^{(1)}h) + 5x^{(1)} - 4 \\ &= 2x^{(3)} + (6h-3)x^{(2)} + (2h^2 - 3h + 5)x^{(1)} - 4 \end{aligned}$$

Method 2.

Write
$$\begin{aligned} 2x^3 - 3x^2 + 5x - 4 &= A_0x^{(3)} + A_1x^{(2)} + A_2x^{(1)} + A_3 \\ &= A_0x(x-h)(x-2h) + A_1x(x-h) + A_2x + A_3 \\ &= A_0x^3 + (A_1 - 3A_0h)x^2 + (2A_0h^2 - A_1h + A_2)x + A_3 \end{aligned}$$

where A_0, A_1, A_2, A_3 are constants to be determined. Equating coefficients of like powers of x we have

$$A_0 = 2, \quad A_1 - 3A_0h = -3, \quad 2A_0h^2 - A_1h + A_2 = 5, \quad A_3 = -4$$

from which
$$A_0 = 2, \quad A_1 = 6h - 3, \quad A_2 = 2h^2 - 3h + 5, \quad A_3 = -4$$

Thus
$$2x^3 - 3x^2 + 5x - 4 = 2x^{(3)} + (6h-3)x^{(2)} + (2h^2 - 3h + 5)x^{(1)} - 4$$

Method 3.

As in Method 2 we have

$$2x^3 - 3x^2 + 5x - 4 = A_0x(x-h)(x-2h) + A_1x(x-h) + A_2x + A_3$$

Let $x = 0$. Then $A_3 = -4$ so that

$$2x^3 - 3x^2 + 5x = A_0x(x-h)(x-2h) + A_1x(x-h) + A_2x$$

Then dividing by x we find

$$2x^2 - 3x + 5 = A_0(x-h)(x-2h) + A_1(x-h) + A_2$$

Now let $x = h$. Then $A_2 = 2h^2 - 3h + 5$ so that

$$2(x^2 - h^2) - 3(x-h) = A_0(x-h)(x-2h) + A_1(x-h)$$

Then dividing by $x - h$ we find
$$2(x+h) - 3 = A_0(x-2h) + A_1$$

Letting $x = 2h$ we then find $A_1 = 6h - 3$ so that $2x - 4h = A_0(x-2h)$ and $A_0 = 2$. Thus we obtain the same result as in Method 1.

Method 4.

As suggested by Method 3 we see that A_3 is the remainder on dividing $2x^3 - 3x^2 + 5x - 4$ by x yielding a first quotient, A_2 is the remainder on dividing the first quotient by $x - h$ yielding a second quotient, A_3 is the remainder on dividing the second quotient by $x - 2h$ yielding a third quotient and finally A_4 is the remainder on dividing the third quotient by $x - 3h$ yielding a quotient which should be zero. The results can be illustrated as follows

Divisor	Dividend / quotient	Remainder	
x	$)\,2x^3 - 3x^2 + 5x - 4$		
$x - h$	$)\,2x^2 - 3x + 5$	-4	← First remainder
$x - 2h$	$)\,2x + (2h - 3)$	$+\ 2h^2 - 3h + 5$	← Second remainder
$x - 3h$	$)\,2$	$6h - 3$	← Third remainder
		2	← Fourth remainder

The required coefficients are given by these remainders reading upwards. Thus we have

$$2x^3 - 3x^2 + 5x - 4 = 2x^{(3)} + (6h-3)x^{(2)} + (2h^2 - 3h + 5)x^{(1)} - 4$$

Method 5.

Since the quotients and remainders can be found by use of *synthetic division* in which only coefficients of the various powers of x are used we can arrange the computation of Method 4 in the following form

h	2	-3	5	(-4)
		$2h$	$2h^2 - 3h$	
$2h$	2	$2h - 3$	($2h^2 - 3h + 5$)	
		$4h$		
	(2)	($6h - 3$)		

where the required coefficients are shown encircled. The numbers at the extreme left indicate multipliers to be used at each stage.

1.24. Express $2x^3 - 3x^2 + 5x - 4$ as a factorial polynomial in which the differencing interval $h = 1$.

In this case any of the methods of Problem 1.23 can of course be used and the results are greatly simplified because h is replaced by 1. For example Method 5 which is the simplest method reduces to the following

1	2	−3	5	−4
		2	−1	
2	2	−1	4	
		4		
	2	3		

Thus
$$2x^3 - 3x^2 + 5x - 4 = 2x^{(3)} + 3x^{(2)} + 4x^{(1)} - 4$$

Check.
$$\begin{aligned} 2x^{(3)} + 3x^{(2)} + 4x^{(1)} - 4 &= 2x(x-1)(x-2) + 3x(x-1) + 4x - 4 \\ &= 2x^3 - 6x^2 + 4x + 3x^2 - 3x + 4x - 4 \\ &= 2x^3 - 3x^2 + 5x - 4 \end{aligned}$$

1.25. Express $x^4 + x - 2$ as a factorial polynomial in which $h = 1$.

We must be careful in using Method 5 of Problem 1.23 to introduce zero coefficients where needed. Thus we have the following

1	1	0	0	1	−2
		1	1	1	
2	1	1	1	2	
		2	6		
3	1	3	7		
		3			
	1	6			

Thus
$$x^4 + x - 2 = x^{(4)} + 6x^{(3)} + 7x^{(2)} + 2x^{(1)} - 2$$

Check.
$$\begin{aligned} x^{(4)} + 6x^{(3)} + 7x^{(2)} + 2x^{(1)} - 2 &= x(x-1)(x-2)(x-3) + 6x(x-1)(x-2) + 7x(x-1) + 2x - 2 \\ &= x^4 - 6x^3 + 11x^2 - 6x + 6x^3 - 18x^2 + 12x + 7x^2 - 7x + 2x - 2 \\ &= x^4 + x - 2 \end{aligned}$$

1.26. Show that if $m = 1, 2, 3, \ldots$
$$(ax+b)^{(m)} = (ax+b)(ax+b-ah)(ax+b-2ah)\cdots(ax+b-mah+ah)$$

The result follows from definition *(34)*, page 7, on letting $f(x) = ax + b$ so that
$$f(x-h) = a(x-h) + b = ax + b - ah, \quad f(x-2h) = a(x-2h) + b = ax + b - 2ah, \quad \ldots,$$
$$f(x-[m-1]h) = a(x-[m-1]h) + b = ax + b - a[m-1]h = ax + b - mah + ah$$

1.27. Show that if $m = 1, 2, 3, \ldots$
$$(ax+b)^{(-m)} = \frac{1}{(ax+b+ah)(ax+b+2ah)\cdots(ax+b+mah)}$$

The result follows from definition *(35)*, page 7, on letting $f(x) = ax + b$ so that
$$f(x+h) = a(x+h) + b = ax + b + ah, \quad f(x+2h) = a(x+2h) + b = ax + b + 2ah, \quad \ldots,$$
$$f(x+mh) = a(x+mh) + b = ax + b + mah$$

1.28. Write (a) $(2x+5)(2x+7)(2x+9)$, (b) $\dfrac{1}{(3x-2)(3x+1)(3x+4)(3x+7)}$ as factorials.

(a) Letting $h = 1$, $a = 2$, $b = 9$ and $m = 3$ in Problem 1.26 we have

$$(2x+9)^{(3)} = (2x+9)(2x+7)(2x+5)$$

(b) Letting $h = 1$, $a = 3$, $b = -5$ *and* $m = 4$ in Problem 1.27 we have

$$(3x-5)^{(-4)} = \frac{1}{(3x-2)(3x+1)(3x+4)(3x+7)}$$

STIRLING NUMBERS

1.29. Derive the recursion formula (*29*), page 7, for Stirling numbers of the first kind.

Using $h = 1$ in (*28*), page 6, we can write the result as

$$x^{(n)} = \sum_{k=-\infty}^{\infty} s_k^n x^k \tag{1}$$

where we take $\quad s_k^n = 0 \quad$ for $k \leqq 0$, $k \geqq n+1$ where $n > 0$ (*2*)

since the series (*1*) is actually finite and

$$s_n^n = 1$$

since both sides of (*1*) are polynomials of degree n.

Replacing n by $n+1$ in (*1*) we have

$$x^{(n+1)} = \sum_{k=-\infty}^{\infty} s_k^{n+1} x^k \tag{3}$$

Now if $h = 1$ we have $\quad x^{(n+1)} = x^{(n)}(x-n)$ (*4*)

Then substituting (*1*) and (*3*) into (*4*) we have

$$\begin{aligned}\sum_{k=-\infty}^{\infty} s_k^{n+1} x^k &= (x-n)\sum_{k=-\infty}^{\infty} s_k^n x^k \\ &= \sum_{k=-\infty}^{\infty} s_k^n x^{k+1} - \sum_{k=-\infty}^{\infty} n s_k^n x^k \\ &= \sum_{k=-\infty}^{\infty} s_{k-1}^n x^k - \sum_{k=-\infty}^{\infty} n s_k^n x^k\end{aligned}$$

Equating coefficients of x^k we find as required

$$s_k^{n+1} = s_{k-1}^n - n s_k^n$$

1.30. Derive the recursion formula (*32*), page 7, for Stirling numbers of the second kind.

Let $$x^n = \sum_{k=-\infty}^{\infty} S_k^n x^{(k)} \tag{1}$$

where we use (*31*), page 7, with $h = 1$. Since both sides are polynomials of degree n we have

$$S_k^n = 0 \quad \text{for } k \leqq 0,\ k \geqq n+1 \quad \text{where } n > 0 \tag{2}$$

so that the series (*1*) is actually finite. From (*1*) we have on replacing n by $n+1$

$$x^{n+1} = \sum_{k=-\infty}^{\infty} S_k^{n+1} x^{(k)} \tag{3}$$

Since $x^{n+1} = x^n \cdot x$, we have from (*1*) and (*4*)

$$\sum_{k=-\infty}^{\infty} S_k^{n+1} x^{(k)} \;=\; x \sum_{k=-\infty}^{\infty} S_k^n x^{(k)} \;=\; \sum_{k=-\infty}^{\infty} S_k^n x \cdot x^{(k)}$$

But

$$x \cdot x^{(k)} \;=\; x \cdot x(x-1)\cdots(x-k+1) \;=\; (x-k+k)x(x-1)\cdots(x-k+1) \;=\; x^{(k+1)} + kx^{(k)}$$

Thus

$$\begin{aligned}\sum_{k=-\infty}^{\infty} S_k^{n+1} x^{(k)} &= \sum_{k=-\infty}^{\infty} S_k^n (x^{(k+1)} + kx^{(k)}) \\ &= \sum_{k=-\infty}^{\infty} S_k^n x^{(k+1)} + \sum_{k=-\infty}^{\infty} kS_k^n x^{(k)} \\ &= \sum_{k=-\infty}^{\infty} S_{k-1}^n x^{(k)} + \sum_{k=-\infty}^{\infty} kS_k^n x^{(k)}\end{aligned}$$

Equating coefficients of $x^{(k)}$ we find as required

$$S_k^{n+1} \;=\; S_{k-1}^n + kS_k^n$$

1.31. Show how the method of Problem 1.25 can be used to obtain Stirling numbers of the second kind by using the polynomial $f(x) = x^4$.

We use the same method as that of Problem 1.25 as indicated in the following

1	1	0	0	0	0
		1	1	1	
2	1	1	1	1	
		2	6		
3	1	3	7		
		3			
	1	6			

It follows that

$$x^4 \;=\; x^{(4)} + 6x^{(3)} + 7x^{(2)} + x^{(1)}$$

and the Stirling numbers are 1, 6, 7, 1.

THE GREGORY-NEWTON FORMULA AND TAYLOR SERIES

1.32. Prove the Gregory-Newton formula (*44*), page 9, for the case where $f(x)$ is a polynomial and $a = 0$.

If $f(x)$ is a polynomial of degree n we can write it as a factorial polynomial, i.e.

$$f(x) \;=\; A_0 + A_1 x^{(1)} + A_2 x^{(2)} + A_3 x^{(3)} + \cdots + A_n x^{(n)} \tag{1}$$

Then

$$\frac{\Delta f(x)}{\Delta x} \;=\; A_1 + 2A_2 x^{(1)} + 3A_3 x^{(2)} + \cdots + nA_n x^{(n-1)}$$

$$\frac{\Delta^2 f(x)}{\Delta x^2} \;=\; 2!\,A_2 + 3\cdot 2A_3 x^{(1)} + \cdots + n(n-1)A_n x^{(n-2)}$$

$$\frac{\Delta^n f(x)}{\Delta x^n} \;=\; n!\,A_n$$

Putting $x = 0$ in the above equations we find

$$A_0 = f(0), \quad A_1 = \left.\frac{\Delta f}{\Delta x}\right|_{x=0}, \quad A_2 = \frac{1}{2!}\left.\frac{\Delta^2 f}{\Delta x^2}\right|_{x=0}, \quad \ldots, \quad A_n = \frac{1}{n!}\left.\frac{\Delta^n f}{\Delta x^n}\right|_{x=0}$$

which we can agree to write as

$$A_0 = f(0), \quad A_1 = \frac{\Delta f(0)}{\Delta x}, \quad A_2 = \frac{1}{2!}\frac{\Delta^2 f(0)}{\Delta x^2}, \quad \ldots, \quad A_n = \frac{1}{n!}\frac{\Delta^n f(0)}{\Delta x^n}$$

Using these in *(1)* we obtain the required formula for $a = 0$.

If we use the same method above with x replaced by $x - a$ and then put $x = a$ the more general formula *(44)* on page 9 is obtained.

1.33. Prove the Gregory-Newton formula with a remainder given by *(44)* and *(45)*, page 9.

Let $p_n(x)$ denote the polynomial of degree n as determined in Problem 1.32. Writing

$$p_n(x) = a_0 x^n + a_1 x^{n-1} + \cdots + a_n$$

we see that it has $n+1$ coefficients $a_0, a_1, \ldots, a_n$. It follows that if $f(x)$ is some given function we can determine these $n+1$ coefficients uniquely in terms of the values of $f(x)$ at $n+1$ different values of x, say $x_0, x_1, \ldots, x_n$. We shall suppose that this is done. In such case

$$f(x_0) = p_n(x_0), \quad f(x_1) = p_n(x_1), \quad \ldots, \quad f(x_n) = p_n(x_n) \tag{1}$$

Now let

$$f(x) = p_n(x) + g(x) \tag{2}$$

Putting $x = x_0, x_1, \ldots, x_n$ we have

$$f(x_0) = p_n(x_0) + g(x_0)$$
$$f(x_1) = p_n(x_1) + g(x_1)$$
$$\cdots\cdots\cdots\cdots\cdots\cdots$$
$$f(x_n) = p_n(x_n) + g(x_n)$$

Then using *(1)* we see that

$$g(x_0) = 0, \quad g(x_1) = 0, \quad \ldots, \quad g(x_n) = 0$$

It thus follows that unless $g(x)$ is identically zero [in which case $f(x)$ is a polynomial and we will have $f(x) = p_n(x)$] we must have

$$g(x) = K(x)\,(x - x_0)(x - x_1)\cdots(x - x_n) \tag{3}$$

Thus from *(2)*,

$$f(x) = p_n(x) + K(x)\,(x - x_0)(x - x_0)\cdots(x - x_n) \tag{4}$$

To obtain $K(x)$ in terms of $f(x)$ let us consider the function

$$U(t) = f(t) - p_n(t) - K(x)\,(t - x_0)(t - x_1)\cdots(t - x_n) \tag{5}$$

It follows from *(4)* that this equation has the $n+2$ roots $t = x, x_0, x_1, \ldots, x_n$. Then by Problem 1.117 the $(n+1)$st derivative of $U(t)$, i.e. $U^{(n+1)}(t)$, is zero for at least one value, say $t = \eta$, between the smallest and largest of $x, x_0, x_1, \ldots, x_n$. But from *(5)*

$$U^{(n+1)}(t) = f^{(n+1)}(t) - (n+1)!\,K(x) \tag{6}$$

since $p_n^{(n+1)}(x) = 0$. Putting $t = \eta$ in *(6)* and setting it equal to zero it thus follows that

$$K(x) = \frac{f^{(n+1)}(\eta)}{(n+1)!}$$

Thus we have from *(4)*

$$f(x) = p_n(x) + \frac{f^{(n+1)}(\eta)}{(n+1)!}(x - x_0)\cdots(x - x_n)$$

Since we can choose any values for $x_0, x_1, \ldots, x_n$, let us choose

$$x_0 = a, \quad x_1 = a+h, \quad x_2 = a+2h, \quad \ldots, \quad x_n = a+nh$$

Then

$$f(x) = p_n(x) + \frac{f^{(n+1)}(\eta)\,(x-a)^{(n+1)}}{(n+1)!}$$

which is Gregory-Newton's formula with a remainder.

1.34. Work Problem 1.24 by using the Gregory-Newton formula.

For any polynomial $f(x)$ of degree n we have

$$f(x) = f(0) + \frac{\Delta f(0)}{\Delta x}x^{(1)} + \frac{\Delta^2 f(0)}{2!\,\Delta x^2}x^{(2)} + \cdots + \frac{\Delta^n f(0)}{n!\,\Delta x^n}x^{(n)}$$

or using $\Delta x = h = 1$,

$$f(x) = f(0) + \Delta f(0)x^{(1)} + \frac{\Delta^2 f(0)}{2!}x^{(2)} + \cdots + \frac{\Delta^n f(0)}{n!}x^{(n)} \tag{1}$$

Now $f(x) = 2x^3 - 3x^2 + 5x - 4$ so that

$$f(0) = -4, \quad f(1) = 0, \quad f(2) = 10, \quad f(3) = 38, \quad f(4) = 96$$

Then

$$\Delta f(0) = f(1) - f(0) = 4, \quad \Delta f(1) = f(2) - f(1) = 10, \quad \Delta f(2) = f(3) - f(2) = 28$$

$$\Delta f(3) = f(4) - f(3) = 58$$

From these we find

$$\Delta^2 f(0) = \Delta f(1) - \Delta f(0) = 6, \quad \Delta^2 f(1) = \Delta f(2) - \Delta f(1) = 18,$$

$$\Delta^2 f(2) = \Delta f(3) - \Delta f(2) = 30$$

Similarly,

$$\Delta^3 f(0) = \Delta^2 f(1) - \Delta^2 f(0) = 12, \quad \Delta^3 f(1) = \Delta^2 f(2) - \Delta^2 f(1) = 12$$

and finally

$$\Delta^4 f(0) = \Delta^3 f(1) - \Delta^3 f(0) = 0$$

From these we see that

$$f(0) = -4, \quad \Delta f(0) = 4, \quad \Delta^2 f(0) = 6, \quad \Delta^3 f(0) = 12, \quad \Delta^4 f(0) = 0$$

and so (*1*) becomes

$$2x^3 - 3x^2 + 5x - 4 = -4 + 4x^{(1)} + 3x^{(2)} + 2x^{(3)}$$

in agreement with the result of Problem 1.24.

The computation of the differences given above can be arranged quite simply in a table as shown in Fig. 1-1, where each entry in a column after the second is obtained by subtracting the entries immediately below and above it in the preceding column. For example $28 = 38 - 10$, $30 = 58 - 28$, etc. This table is called a *difference table*. Further uses of such tables are given in Chapter 2.

x	$f(x)$	$\Delta f(x)$	$\Delta^2 f(x)$	$\Delta^3 f(x)$	$\Delta^4 f(x)$
0	−4				
		4			
1	0		6		
		10		12	
2	10		18		0
		28		12	
3	38		30		
		58			
4	96				

Fig. 1-1

1.35. Show how to arrive at the Gregory-Newton formula for the case where $h = 1$, $a = 0$ by using symbolic operator methods.

We have for any value of n [see Problem 1.12]

$$E^n f(u) = f(u + nh) \tag{1}$$

Then in particular if we choose $u = 0$, $n = x$, $h = 1$, this becomes

$$E^x f(0) = f(x) \tag{2}$$

Using $E = 1 + \Delta$ in (2) and expanding formally by the binomial theorem, (2) becomes

$$\begin{aligned} f(x) &= E^x f(0) = (1+\Delta)^x f(0) \\ &= \left[1 + x\Delta + \frac{x(x-1)}{2!}\Delta^2 + \frac{x(x-1)(x-2)}{3!}\Delta^3 + \cdots\right] f(0) \\ &= f(0) + x\Delta f(0) + \frac{x(x-1)}{2!}\Delta^2 f(0) + \frac{x(x-1)(x-2)}{3!}\Delta^3 f(0) + \cdots \end{aligned}$$

The result is the same as the infinite series obtained from (44), page 9, with $h = \Delta x = 1$ and $a = 0$.

Extensions to the case where $h \neq 1$, $a \neq 0$ can also be obtained symbolically [see Problem 1.83].

LEIBNITZ'S RULE

1.36. Prove Leibnitz's rule for the nth difference of a product of functions.

Let us define operators E_1 and E_2 which operate only on $f(x)$ and $g(x)$ respectively, i.e.

$$E_1[f(x)\,g(x)] = f(x+h)\,g(x), \qquad E_2[f(x)\,g(x)] = f(x)\,g(x+h)$$

Then $E_1E_2[f(x)\,g(x)] = f(x+h)\,g(x+h) = E[f(x)\,g(x)]$ so that $E = E_1E_2$. Associate the difference operators Δ_1, Δ_2 with E_1, E_2 respectively, i.e. $E_1 = 1 + \Delta_1$, $E_2 = 1 + \Delta_2$. Then

$$\Delta = E - 1 = E_1E_2 - 1 = (1+\Delta_1)E_2 - 1 = E_2 + \Delta_1E_2 - 1 = \Delta_2 + \Delta_1E_2$$

and so

$$\begin{aligned} \Delta^n[fg] &= (\Delta_2 + \Delta_1E_2)^n[fg] \\ &= \left(\Delta_2^n + \binom{n}{1}\Delta_2^{n-1}\Delta_1E_2 + \binom{n}{2}\Delta_2^{n-2}\Delta_1^2E_2^2 + \cdots + \binom{n}{n}\Delta_1^nE_2^n\right)[fg] \\ &= f\Delta^n g + \binom{n}{1}(\Delta f)(\Delta^{n-1}Eg) + \binom{n}{2}(\Delta^2 f)(\Delta^{n-2}E^2g) + \cdots + \binom{n}{n}(\Delta^n f)(E^n g) \end{aligned}$$

1.37. Find $\Delta^n[x^2a^x]$.

Let $f = x^2$, $g = a^x$. Then by Leibnitz's rule we have

$$\begin{aligned} \Delta^n[x^2a^x] &= x^2\Delta^n(a^x) + \binom{n}{1}(\Delta x^2)(\Delta^{n-1}a^{x+h}) + \binom{n}{2}(\Delta^2x^2)(\Delta^{n-2}a^{x+2h}) + \cdots \\ &= x^2a^x(a^h-1)^n + \binom{n}{1}(2hx+h^2)a^{x+h}(a^h-1)^{n-1} + \binom{n}{2}(2h^2)a^{x+2h}(a^h-1)^{n-2} \\ &= a^x(a^h-1)^{n-2}[x^2(a^h-1)^2 + n(2hx+h^2)a^h(a^h-1) + n(n-1)h^2a^{2h}] \end{aligned}$$

OTHER DIFFERENCE OPERATORS

1.38. Find (a) $\nabla(x^2+2x)$, (b) $\delta(x^2+2x)$.

(a)
$$\begin{aligned} \nabla(x^2+2x) &= [x^2+2x] - [(x-h)^2 + 2(x-h)] \\ &= x^2 + 2x - [x^2 - 2hx + h^2 + 2x - 2h] \\ &= 2hx + 2h - h^2 \end{aligned}$$

(b)
$$\delta(x^2+2x) = \left[\left(x+\frac{h}{2}\right)^2 + 2\left(x+\frac{h}{2}\right)\right] - \left[\left(x-\frac{h}{2}\right)^2 + 2\left(x-\frac{h}{2}\right)\right]$$
$$= \left(x^2 + hx + \frac{h^2}{4} + 2x + h\right) - \left(x^2 - hx + \frac{h^2}{4} + 2x - h\right)$$
$$= 2hx + 2h$$

1.39. Prove that (a) $\nabla = \Delta E^{-1} = E^{-1}\Delta = 1 - E^{-1}$
(b) $\delta = E^{1/2} - E^{-1/2} = \Delta E^{-1/2} = \nabla E^{1/2}$

(a) Given any function $f(x)$ we have
$$\nabla f(x) = f(x) - f(x-h) = \Delta f(x-h) = \Delta E^{-1} f(x)$$
so that $\nabla = \Delta E^{-1}$. Similarly
$$\nabla f(x) = f(x) - f(x-h) = E^{-1}[f(x+h) - f(x)] = E^{-1}\Delta f(x)$$
so that $\nabla = E^{-1}\Delta$. Finally
$$\nabla f(x) = f(x) - f(x-h) = f(x) - E^{-1}f(x) = (1 - E^{-1})f(x)$$
so that $\nabla = 1 - E^{-1}$.

(b)
$$\delta f(x) = f\left(x+\frac{h}{2}\right) - f\left(x-\frac{h}{2}\right) = E^{1/2}f(x) - E^{-1/2}f(x) = (E^{1/2} - E^{-1/2})f(x)$$
so that $\delta = E^{1/2} - E^{-1/2}$. Similarly
$$\delta f(x) = f\left(x+\frac{h}{2}\right) - f\left(x-\frac{h}{2}\right) = \Delta f\left(x-\frac{h}{2}\right) = \Delta E^{-1/2} f(x)$$
so that $\delta = \Delta E^{-1/2}$. Finally
$$\delta f(x) = f\left(x+\frac{h}{2}\right) - f\left(x-\frac{h}{2}\right) = \nabla f\left(x+\frac{h}{2}\right) = \nabla E^{1/2} f(x)$$
so that $\delta = \nabla E^{1/2}$.

1.40. Find (a) $M(4x^2 - 8x)$, (b) $\mu(4x^2 - 8x)$.

(a) $$M(4x^2 - 8x) = \tfrac{1}{2}[4(x+h)^2 - 8(x+h) + 4x^2 - 8x] = 4x^2 + 4hx + 2h^2 - 8x - 4h$$

(b) $$\mu(4x^2 - 8x) = \frac{1}{2}\left[4\left(x+\frac{h}{2}\right)^2 - 8\left(x+\frac{h}{2}\right) + 4\left(x-\frac{h}{2}\right)^2 - 8\left(x-\frac{h}{2}\right)\right] = 4x^2 - 8x + h^2$$

MISCELLANEOUS PROBLEMS

1.41. If $f(x) = a_0x^n + a_1x^{n-1} + \cdots + a_n$, i.e. a polynomial of degree n, prove that
(a) $\Delta^n f(x) = n!\,a_0h^n$ (b) $\Delta^{n+1}f(x) = 0,\ \Delta^{n+2}f(x) = 0,\ \ldots$

Method 1.

We have
$$\Delta f(x) = [a_0(x+h)^n + a_1(x+h)^{n-1} + \cdots + a_n] - [a_0x^n + a_1x^{n-1} + \cdots + a_n]$$
$$= [a_0nx^{n-1}h + \text{terms involving } x^{n-2}, x^{n-3}, \ldots]$$

It follows that if Δ operates on a polynomial of degree n the result is a polynomial of degree $n-1$. From this we see that $\Delta^n f(x)$ must be independent of x, i.e. a constant, and so $\Delta^{n+1}f(x) = 0$, $\Delta^{n+2}f(x) = 0, \ldots$, which proves (b). To find the constant value of $\Delta^n f(x)$ note that we need only consider the term of highest degree. Thus we have
$$\Delta^2 f(x) = [a_0n(x+h)^{n-1}h + \cdots] - [a_0nx^{n-1}h + \cdots]$$
$$= a_0n[(x+h)^{n-1} - x^{n-1}]h + \cdots$$
$$= a_0n[(n-1)x^{n-2}h + \cdots]h + \cdots$$
$$= a_0n(n-1)x^{n-2}h^2 + \cdots$$

Proceeding in this manner we see finally that

$$\Delta^n f(x) = a_0 n(n-1)(n-2)\cdots(1)h^n = n!\, a_0 h^n$$

For another method see Problem 1.84.

Method 2.

Since every polynomial of degree n can be written as a factorial polynomial of degree n, we have

$$a_0 x^n + a_1 x^{n-1} + \cdots + a_n = b_0 x^{(n)} + b_1 x^{(n-1)} + \cdots + b_n$$

Equating coefficients of x^n on both sides we find $b_0 = a_0$. The required result then follows since

$$\Delta^n[a_0 x^{(n)} + b_1 x^{(n-1)} + \cdots + b_n] = n!\, a_0 h^n, \qquad \Delta^{n+1}[a_0 x^{(n)} + b_1 x^{(n-1)} + \cdots + b_n] = 0$$

1.42. Prove that

$$(x+r)^n - \binom{r}{1}(x+r-1)^n + \binom{r}{2}(x+r-2)^n - \cdots (-1)^r\binom{r}{r}x^n = \begin{cases} 0 & r > n \\ n! & r = n \end{cases}$$

We have

$$\Delta^r x^n = (E-1)^r x^n = \left[E^r - \binom{r}{1}E^{r-1} + \binom{r}{2}E^{r-2} - \cdots + (-1)^r\binom{r}{r}\right]x^n \tag{1}$$

Now from Problem 1.41 with $h = 1$ we have for $r > n$

$$0 = (x+r)^n - \binom{r}{1}(x+r-1)^n + \binom{r}{2}(x+r-2)^n - \cdots + (-1)^n\binom{r}{r}x^n \tag{2}$$

while if $r = n$,

$$n! = (x+n)^n - \binom{n}{1}(x+n-1)^n + \binom{n}{2}(x+n-2)^n - \cdots + (-1)^n\binom{n}{n}x^n \tag{3}$$

Thus the required result follows.

1.43. Show that (*a*) $\Delta^r 0^n = 0$ if $r > n$, (*b*) $\Delta^n 0^n = n!$

(*a*) Putting $x = 0$ in (*2*) of Problem 1.42 we have for $r > n$, $\Delta^r 0^n = 0$.

(*b*) Putting $x = 0$ in (*3*) of Problem 1.42 we have for $r = n$, $\Delta^n 0^n = n!$

We call $\Delta^r 0^n$ the *differences of zero.*

1.44. Prove *Rolle's theorem*: If $f(x)$ is continuous in $a \leqq x \leqq b$, has a derivative in $a < x < b$, and if $f(a) = 0$, $f(b) = 0$, then there is at least one value η between a and b such that $f'(\eta) = 0$.

We assume that $f(x)$ is not identically zero since in such case the result is immediate. Suppose then that $f(x) > 0$ for some value between a and b. Then it follows since $f(x)$ is continuous that it attains its maximum value somewhere between a and b, say at η. Consider now

$$\frac{\Delta f(x)}{\Delta x} = \frac{f(\eta+h) - f(\eta)}{h}$$

where we choose $h = \Delta x$ so small that $\eta + h$ is between a and b. Since $f(\eta)$ is a maximum value, it follows that $\Delta f(x)/\Delta x \geqq 0$ for $h < 0$ and $\Delta f(x)/\Delta x \leqq 0$ for $h > 0$. Then taking the limit as $h \to 0$ through positive values of h, we have $f'(\eta) \leqq 0$, while if the limit is taken through negative values of h, we have $f'(\eta) \geqq 0$. Thus $f'(\eta) = 0$. A similar proof holds if $f(x) < 0$ for some value between a and b.

1.45. (*a*) Prove the *mean value theorem for derivatives*: If $f(x)$ is continuous in $a \leqq x \leqq b$ and has a derivative in $a < x < b$, then there is at least one value η between a and b such that

$$\frac{f(b) - f(a)}{b - a} = f'(\eta)$$

(*b*) Use (*a*) to show that for any value of x such that $a \leqq x \leqq b$,

$$f(x) = f(a) + (x - a)f'(\eta)$$

where η is between a and x.

(*c*) Give a geometric interpretation of the result.

(*a*) Consider the function

$$F(x) = f(x) - f(a) - (x - a)\frac{f(b) - f(a)}{b - a} \tag{1}$$

From this we see that $F(a) = 0$, $F(b) = 0$ and that $F(x)$ satisfies the conditions of Rolle's theorem [Problem 1.44]. Then there is at least one point η between a and b such that $F'(\eta) = 0$. But from (*1*)

$$F'(x) = f'(x) - \frac{f(b) - f(a)}{b - a} \tag{2}$$

so that

$$F'(\eta) = f'(\eta) - \frac{f(b) - f(a)}{b - a} = 0$$

or

$$f'(\eta) = \frac{f(b) - f(a)}{b - a} \tag{3}$$

(*b*) Replacing b by x in (*3*) we find as required

$$f(x) = f(a) + (x - a)f'(\eta) \tag{4}$$

where η is between x and a.

Note that (*4*) is a special case of Taylor's series with a remainder for $n = 1$. The general Taylor series can be proved by extensions of this method.

(*c*) The theorem can be illustrated geometrically with reference to Fig. 1-2 where it is geometrically evident that there is at least one point R on the curve $y = f(x)$ where the tangent line PRQ is parallel to the secant line ACB. Since the slope of the tangent line at R is $f'(\eta)$ and the slope of the secant line is $[f(b) - f(a)]/(b - a)$, the result follows. It is of interest to note that $F(x)$ in equation (*1*) represents geometrically RC of Fig. 1-2. In the case where $f(a) = 0$, $f(b) = 0$, RC represents the maximum value of $f(x)$ in the interval $a \leqq x \leqq b$.

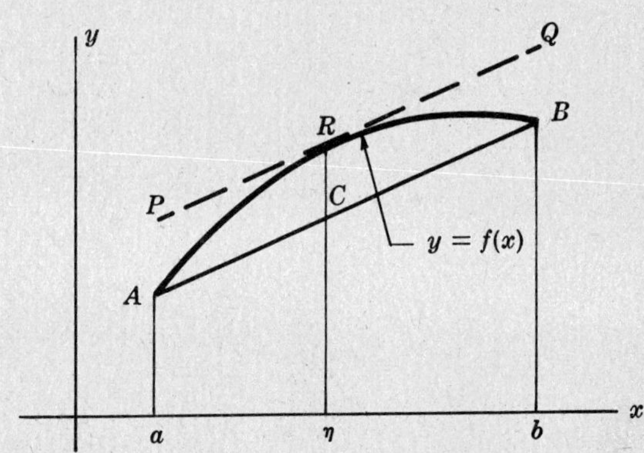

Fig. 1-2

Supplementary Problems

OPERATORS

1.46. Let $\mathcal{S} = [\quad]^2$ be the squaring operator and D the derivative operator. Determine each of the following.

(a) $\mathcal{S}(1+\sqrt{x})$ (e) $(\mathcal{S}^2 + 2\mathcal{S} - 3)(2x-1)$ (i) $x^3D^3\mathcal{S}(x+1)$

(b) $(2\mathcal{S} + 3D)(x^2 - x)$ (f) $(D+2)(\mathcal{S}-3)x^2$ (j) $(x\mathcal{S} - \mathcal{S}x)D\mathcal{S}x^2$

(c) $\mathcal{S}D(3x+2)$ (g) $(\mathcal{S}-3)(D+2)x^2$

(d) $D\mathcal{S}(3x+2)$ (h) $(xD)^3\mathcal{S}(x+1)$

1.47. (a) Prove that the operator $\mathcal{S}$ of Problem 1.46 is a nonlinear operator. (b) Explain the significance of $\mathcal{S}^{-1}$ and determine whether it always exists and is unique.

1.48. Prove that $\mathcal{S}^3 = C^2$ where $\mathcal{S}$ and C are the squaring and cubing operators respectively.

1.49. Let $\mathcal{S}$ be the squaring operator and α be any real number. (a) Explain the meaning of the operators $\alpha\mathcal{S}$ and $\mathcal{S}\alpha$. (b) Do the operators α and $\mathcal{S}$ obey the commutative law? Illustrate by an example.

1.50. Is the operator $(xD)^4$ the same as the operator x^4D^4? Justify your answer.

1.51. Prove that (a) $D^2x - xD^2 = 2D$, (b) $D^3x - xD^3 = 3D^2$. Obtain a generalization of these results and prove it.

THE DIFFERENCE AND TRANSLATION OPERATORS

1.52. Find each of the following.

(a) $\Delta(2x-1)^2$ (e) $(\Delta+1)^2(x+1)^2$ (i) $(2E-1)(3\Delta+2)x^2$

(b) $E(\sqrt[3]{5x-4})$ (f) $(xE^2 + 2xE + 1)x^2$ (j) $(x\Delta E)^2x^2$

(c) $\Delta^2(2x^2 - 5x)$ (g) Δ^2E^3x

(d) $3E^2(x^2+1)$ (h) $(3\Delta+2)(2E-1)x^2$

1.53. Determine whether (a) $(E-2)(\Delta+3) = (\Delta+3)(E-2)$, (b) $(E-2x)(\Delta+3x) = (\Delta+3x)(E-2)$ and discuss the significance of the results.

1.54. Prove that E is a linear operator.

1.55. Determine whether (a) Δ^2 and (b) E^2 are linear operators. Do your conclusions apply to Δ^n and E^n? Explain.

1.56. Verify directly that $\Delta^3 = (E-1)^3 = E^3 - 3E^2 + 3E - 1$.

1.57. Prove formulas (a) V-7, (b) V-8, (c) V-9 on page 7.

1.58. Prove that the commutative law for the operators D and Δ holds (a) with respect to addition, (b) with respect to multiplication.

1.59. (a) Does the associative law with respect to multiplication hold D, Δ and E?

(b) Does the commutative law with respect to multiplication hold for D and E?

1.60. Show that (a) $\lim_{\Delta x \to 0} \dfrac{\Delta}{\Delta x}[x(2-x)] = D[x(2-x)] = 2(1-x)$

(b) $\lim_{\Delta x \to 0} \dfrac{\Delta^2}{\Delta x^2}[x(2-x)] = \dfrac{d^2}{dx^2}[x(2-x)] = -2$

directly from the definition.

1.61. Find (a) $d(x^3 - 3x^2 + 2x - 1)$, (b) $d^2(3x^2 + 2x - 5)$.

1.62. Prove that $D[f(x) + g(x)] = Df(x) + Dg(x)$ giving restrictions if any.

1.63. Prove that $D\left[\dfrac{f(x)}{g(x)}\right] = \dfrac{g(x)\,Df(x) - f(x)\,Dg(x)}{[g(x)]^2}$ giving restrictions if any.

1.64. Prove that $\lim_{\Delta x \to 0} \frac{\Delta}{\Delta x}[\cos rx] = -r \sin rx$ by using the fact that $\lim_{\theta \to 0} \frac{\sin \theta}{\theta} = 1$.

1.65. Prove that $\lim_{\Delta x \to 0} \frac{\Delta}{\Delta x}[b^x] = b^x \ln b$ stating assumptions made.

1.66. Prove equation *(14)*, page 3, giving suitable restrictions.

1.67. Obtain a relationship similar to that of equation *(14)*, page 3, between $D^3 f(x)$ and $\Delta^3 f(x)/\Delta x^3$.

FACTORIAL FUNCTIONS

1.68. Find (*a*) $\Delta(3x^{(5)} + 5x^{(4)} - 7x^{(2)} + 3x^{(1)} + 6)$, (*b*) $\frac{\Delta}{\Delta x}(x^{(-3)} - 3x^{(-2)})$, (*c*) $\Delta\left(\frac{x^{(2)} + x^{(-2)}}{2}\right)$.

1.69. Find (*a*) $\Delta^2(2x^{(-3)} - 3x^{(-2)} + 4x^{(2)})$, (*b*) $\frac{\Delta^3}{\Delta x^3}(x^{(4)} + x^{(-4)})$.

1.70. Express each of the following as factorial polynomials for $h = 1$ and for $h \neq 1$.
(*a*) $3x^2 - 5x + 2$, (*b*) $2x^4 + 5x^2 - 4x + 7$.

1.71. Find (*a*) $\frac{\Delta}{\Delta x}(x^4 - 2x^2 + 5x - 3)$, (*b*) $\frac{\Delta^2}{\Delta x^2}(x^4 - 2x^2 + 5x - 3)$.

1.72. Express each of the following as a product of suitable factors using the indicated values of h.
(*a*) $(2x-1)^{(4)}$ if $h = 2$, (*b*) $(3x+5)^{(3)}$ if $h = 1$, (*c*) $(4x-5)^{(-2)}$ if $h = 1$, (*d*) $(5x+2)^{-(4)}$ if $h = 2$.

1.73. Write each of the following as a factorial function.

(*a*) $(3x-2)(3x+5)(3x+12)$, (*b*) $(2+2x)(5+2x)(8+2x)(11+2x)$

(*c*) $\dfrac{1}{x(x+2)(x+4)}$, (*d*) $\dfrac{1}{(2x-1)(2x+3)(2x+7)(2x+11)}$

1.74. Prove that $\frac{\Delta}{\Delta x}(px+q)^{(m)} = mp(px+q)^{(m-1)}$ for (*a*) $m = 0, 1, 2, \ldots$, (*b*) $m = -1, -2, -3, \ldots$.

1.75. Express in terms of factorial functions (*a*) $\dfrac{x^2 - 1}{(x+2)(x+4)(x+6)}$, (*b*) $\dfrac{2x+1}{(2x+3)(2x+5)(2x+9)}$

STIRLING NUMBERS

1.76. Obtain Stirling numbers of the first kind s_k^n for $n, k = 1, 2, 3$ by using the recursion formula *(29)*, page 7.

1.77. Obtain Stirling numbers of the second kind S_k^n for $n, k = 1, 2, 3$ using the recursion formula *(32)*, page 7.

1.78. Find S_k^6 for $k = 1, 2, \ldots, 6$ by using the method of Problem 1.25.

1.79. Prove that
$$s_1^n + s_2^n + \cdots + s_n^n = 0$$
and illustrate by referring to the table of Stirling numbers of the first kind in Appendix A, page 232.

1.80. Prove that

(*a*) $s_1^n - s_2^n + s_3^n - \cdots + (-1)^{n-1} s_n^n = (-1)^{n-1} n!$

(*b*) $|s_1^n| + |s_2^n| + \cdots + |s_n^n| = n!$

THE GREGORY-NEWTON FORMULA AND TAYLOR SERIES

1.81. Express each of the following as factorial polynomials for the cases $h = 1$ and $h \neq 1$ by using the Gregory-Newton formula. (*a*) $3x^2 - 5x + 2$, (*b*) $2x^4 + 5x^2 - 4x + 7$.

1.82. Under what conditions is the remainder R_n in (45), page 9, equal to zero?

1.83. Show how to generalize Problem 1.35 by obtaining the Gregory-Newton formula for the case where $h \neq 1,\ a \neq 0$.

1.84. Prove that
$$\Delta^n[a_0x^n + a_1x^{n-1} + \cdots + a_n] \;=\; n!\, a_0h^n$$
by using the Gregory-Newton formula.

1.85. Obtain the Taylor series expansion for $f(x)$ from the Gregory-Newton formula by using a limiting procedure.

1.86. Obtain the Taylor series expansions (a) 1, (b) 2, (c) 3 and (d) 4 on page 8 and verify the intervals of convergence in each case.

LEIBNITZ'S RULE

1.87. Use Leibnitz's rule to find $\Delta^3(x^2 \cdot 2^x)$ if $h = 1$.

1.88. Find $\Delta^n(xa^x)$.

1.89. Find $\Delta^n(x^2a^x)$.

1.90. Obtain Leibnitz's rule for derivatives from Leibnitz's rule for differences by using an appropriate limiting procedure.

OTHER DIFFERENCE OPERATORS

1.91. If $f(x) = 2x^2 + 3x - 5$ find (a) $\nabla f(x)$, (b) $\delta f(x)$, (c) $\nabla^2 f(x)$, (d) $\delta^2 f(x)$.

1.92. Evaluate $(\nabla^2 - 3\nabla\delta + 2\delta^2)(x^2 + 2x)$.

1.93. Prove that (a) $\nabla^2 = (\Delta E^{-1})^2 = \Delta^2E^{-2}$, (b) $\nabla^n = \Delta^nE^{-n}$.

1.94. Determine whether the operators ∇ and δ are commutative.

1.95. Demonstrate the operator equivalence $E \;=\; \left(\dfrac{\delta}{2} + \sqrt{1 + \dfrac{\delta^2}{4}}\right)^2$.

1.96. Prove that $\nabla\Delta = \Delta\nabla = \delta^2$.

1.97. Is it true that (a) $\displaystyle\lim_{\delta x \to 0} \frac{\delta y}{\delta x} = \frac{dy}{dx}$, (b) $\displaystyle\lim_{\delta x \to 0} \frac{\delta^n y}{\delta x^n} = \frac{d^n y}{dx^n}$? Explain.

1.98. Show that (a) $M = \frac{1}{2}(1 + E) = E - \frac{1}{2}\Delta$, (b) $\mu = M/E^{1/2}$.

1.99. Determine whether (a) M and (b) μ commutes with Δ, D and E.

1.100. Show that (a) $\Delta = \mu\delta + \frac{1}{2}\delta^2$, (b) $\Delta^{2m+1} = E^m[\mu\delta^{2m+1} + \frac{1}{2}\delta^{m+2}]$.

MISCELLANEOUS PROBLEMS

1.101. (a) If A and B are any operators show that $(A - B)(A + B) = A^2 - B^2 + AB - BA$. (b) Under what conditions will it be true that $(A - B)(A + B) = A^2 - B^2$? (c) Illustrate the results of parts (a) and (b) by considering $(\Delta^2 - D^2)x^2$ and $(\Delta - D)(\Delta + D)x^2$.

1.102. Prove that
$$(a)\quad \Delta \sin(px + q) \;=\; 2 \sin\frac{ph}{2} \sin\left[px + q + \tfrac{1}{2}(ph + \pi)\right]$$
$$(b)\quad \Delta \cos(px + q) \;=\; 2 \sin\frac{ph}{2} \cos\left[px + q + \tfrac{1}{2}(ph + \pi)\right]$$

1.103. Prove that (a) $\Delta^m \sin(px+q) = \left[2\sin\frac{ph}{2}\right]^m \sin\left[px+q+\frac{m}{2}(ph+\pi)\right]$

(b) $\Delta^m \cos(px+q) = \left[2\sin\frac{ph}{2}\right]^m \cos\left[px+q+\frac{m}{2}(ph+\pi)\right]$

1.104. Use Problem 1.103 to show that

(a) $\frac{d^m}{dx^m}\sin x = \sin\left(x+\frac{m\pi}{2}\right)$ (b) $\frac{d^m}{dx^m}\cos x = \cos\left(x+\frac{m\pi}{2}\right)$

1.105. Verify that $\binom{n}{r}+\binom{n}{r+1} = \binom{n+1}{r+1}$ and thus complete the proof of Problem 1.11, Method 2.

1.106. (a) Show that $\Delta \tan x = \frac{\sec^2 x \tan h}{1-\tan x \tan h}$ and (b) deduce that $\frac{d}{dx}\tan x = \sec^2 x$.

1.107. (a) Show that $\Delta \tan^{-1} x = \tan^{-1}\frac{h}{x^2+hx+1}$ and (b) deduce that $\frac{d}{dx}\tan^{-1} x = \frac{1}{x^2+1}$ and $\frac{d}{dx}\tan^{-1}\frac{x}{a} = \frac{a}{x^2+a^2}$.

1.108. (a) Show that $\Delta \sin^{-1} x = (x+h)\sqrt{1-x^2} - x\sqrt{1-(x+h)^2}$.

(b) Deduce from (a) that $\frac{d}{dx}\sin^{-1} x = \frac{1}{\sqrt{1-x^2}}$ and $\frac{d}{dx}\sin^{-1}\frac{x}{a} = \frac{1}{\sqrt{a^2-x^2}}$.

1.109. Prove that for $h=1$ (a) $s_k^n = \frac{1}{n!}D^n x^{(k)}\Big|_{x=0}$, (b) $S_k^n = \frac{1}{k!}\Delta^k x^n\Big|_{x=0}$

1.110. Prove that $S_k^n = \frac{(-1)^k}{k!}\sum_{p=0}^{k}(-1)^p\binom{k}{p}p^n$ and illustrate by using the table on page 233.

1.111. Prove that (a) $E[f(x)\,g(x)] = [Ef(x)][Eg(x)]$

(b) $E[f(x)]^n = [Ef(x)]^n$

(c) $E^m[f_1(x)\cdots f_n(x)] = [E^m f_1(x)]\cdots[E^m f_n(x)]$

1.112. Show that the index law for factorial functions, i.e. $x^{(m)}x^{(n)} = x^{(m+n)}$, does not hold.

1.113. Prove that

$$\Delta^n[f(x)\,g(x)] = \sum_{k=0}^{n}(-1)^{k+n}\binom{n}{k}f(x+kh)\,g(x+kh)$$

and discuss the relationship with Leibnitz's rule.

1.114. Prove that

$$n! = n^n - n(n-1)^n + \frac{n(n-1)}{2!}(n-2)^n - \frac{n(n-1)(n-2)}{3!}(n-3)^n + \cdots$$

1.115. Show that (a) $\Delta = hD + \frac{h^2D^2}{2!} + \frac{h^3D^3}{3!} + \frac{h^4D^4}{4!} + \cdots$

(b) $\Delta^2 = h^2D^2 + h^3D^3 + \frac{7}{12}h^4D^4 + \frac{h^5D^5}{4} + \frac{31h^6D^6}{360} + \cdots$

(c) $\Delta^3 = h^3D^3 + \frac{3h^4D^4}{2} + \frac{5h^5D^5}{4} + \frac{3h^6D^6}{4} + \frac{903h^7D^7}{2520} + \cdots$

1.116. Find (a) $\Delta^2(3x^3-2x^2+4x-6)$, (b) $\Delta^3(x^2+x)^2$ by using Problem 1.115 and compare results by direct evaluation.

1.117. If $U(t)$ is the function defined by (5) of Problem 1.33, prove that $U^{(n+1)}(t) = 0$ for at least one value $t=\eta$ between the smallest and largest of $x, x_0, x_1, \ldots, x_n$. [*Hint.* Apply Rolle's theorem successively.]

Chapter 2

Applications of the Difference Calculus

SUBSCRIPT NOTATION

Suppose that in $f(x)$ we make the transformation $x = a + kh$ from the variable x to the variable k. Let us use the notation $y = f(x)$ and

$$y_k = f(a + kh) \tag{1}$$

It follows that [see Problem 2.1]

$$\Delta y_k = y_{k+1} - y_k, \qquad Ey_k = y_{k+1} \tag{2}$$

and so as on page 3

$$\Delta = E - 1 \quad \text{or} \quad E = 1 + \Delta \tag{3}$$

Using this subscript notation it is clear that a unit change in the subscript k actually corresponds to a change of h in the argument x of $f(x)$ and conversely. In addition all of the basic rules of the difference calculus obtained in Chapter 1 can be written with subscript notation. Thus for example formula IV-4, page 5, becomes

$$\Delta\left(\frac{y_k}{z_k}\right) = \frac{z_k\,\Delta y_k - y_k\,\Delta z_k}{z_k\,z_{k+1}} \tag{4}$$

where $y_k = f(a + kh)$, $z_k = g(a + kh)$. Also since $x = a + kh$ becomes $x = k$ if $a = 0$ and $h = 1$, results involving x remain valid if we replace x by k and put $h = 1$. Thus for example equations (22) and (25), page 6, become respectively

$$k^{(m)} = k(k-1)(k-2)\cdots(k-m+1), \qquad \Delta k^{(m)} = mk^{(m-1)} \tag{5}$$

It should be noted that k *need not be an integer*. In fact k is a variable which is discrete or continuous according as x is. An important special case with which we shall be mainly concerned arises however if $k = 0, 1, 2, \ldots$ so that the variable x is equally spaced, i.e. $x = a, a+h, a+2h, \ldots$. In such case we have $y_0 = f(a)$, $y_1 = f(a+h)$, $y_2 = f(a+2h)$, $\ldots$.

DIFFERENCE TABLES

A table such as that shown in Fig. 2-1 below which gives successive differences of $y = f(x)$ for $x = a, a+h, a+2h, \ldots$, i.e. $\Delta y, \Delta^2 y, \Delta^3 y, \ldots$, is called a *difference table.*

Note that the entries in each column after the second are located between two successive entries of the preceding column and are equal to the differences between these entries. Thus for example $\Delta^2 y_1$ in the fourth column is between Δy_1 and Δy_2 of the third column and $\Delta^2 y_1 = \Delta y_2 - \Delta y_1$. Similarly $\Delta^3 y_2 = \Delta^2 y_3 - \Delta^2 y_2$

x	y	Δy	$\Delta^2 y$	$\Delta^3 y$	$\Delta^4 y$	$\Delta^5 y$
a	$y_0 = f(a)$					
		Δy_0				
$a+h$	$y_1 = f(a+h)$		$\Delta^2 y_0$			
		Δy_1		$\Delta^3 y_0$		
$a+2h$	$y_2 = f(a+2h)$		$\Delta^2 y_1$		$\Delta^4 y_0$	
		Δy_2		$\Delta^3 y_1$		$\Delta^5 y_0$
$a+3h$	$y_3 = f(a+3h)$		$\Delta^2 y_2$		$\Delta^4 y_1$	
		Δy_3		$\Delta^3 y_2$		
$a+4h$	$y_4 = f(a+4h)$		$\Delta^2 y_3$			
		Δy_4				
$a+5h$	$y_5 = f(a+5h)$					

Fig. 2-1

Example 1.

The difference table corresponding to $y = f(x) = x^3$ for $x = 1, 2, \ldots, 6$ is as follows.

x	$y = f(x) = x^3$	Δy	$\Delta^2 y$	$\Delta^3 y$	$\Delta^4 y$
1	1				
		7			
2	8		12		
		19		6	
3	27		18		0
		37		6	
4	64		24		0
		61		6	
5	125		30		
		91			
6	216				

Fig. 2-2

The first entry in each column beyond the second is called the *leading difference* for that column. In the table of Fig. 2-1 the leading differences for the successive columns are $\Delta y_0, \Delta^2 y_0, \Delta^3 y_0, \ldots$. The leading differences in the table of Fig. 2-2 are $7, 12, 6, 0$. It is often desirable also to include the first entry of the second column called a *leading difference of order zero.*

It is of interest that a difference table is completely determined when only one entry in each column beyond the first is known [see Problem 2.4].

DIFFERENCES OF POLYNOMIALS

It will be noticed that for $f(x) = x^3$ the difference table of Example 1 indicates that the third differences are all constant, i.e. 6, and the fourth differences [and thus all higher differences] are all zero. The result is a special case of the following theorem already proved in Problem 1.41, page 25.

Theorem 2-1. If $f(x)$ is a polynomial of degree n, then $\Delta^n f(x)$ is a constant and $\Delta^{n+1} f(x)$, $\Delta^{n+2} f(x), \ldots$ are all zero.

GREGORY-NEWTON FORMULA IN SUBSCRIPT NOTATION

If we put $x = a + kh$ in the Gregory-Newton formula (*44*) on page 9, it becomes

$$f(a+kh) = f(a) + \frac{\Delta f(a)\, k^{(1)}}{1!} + \frac{\Delta^2 f(a)\, k^{(2)}}{2!} + \cdots + \frac{\Delta^n f(a)\, k^{(n)}}{n!} + R_n \tag{6}$$

where the remainder is given by

$$R_n = \frac{h^{n+1} f^{(n+1)}(\eta)\, k^{(n+1)}}{(n+1)!} \tag{7}$$

the quantity η being between a and $a+kh$ and where

$$k^{(1)} = k, \quad k^{(2)} = k(k-1), \quad k^{(3)} = k(k-1)(k-2), \quad \ldots$$

In subscript notation (*6*) can be written as

$$y_k = y_0 + \frac{\Delta y_0 k^{(1)}}{1!} + \frac{\Delta^2 y_0 k^{(2)}}{2!} + \cdots + \frac{\Delta^n y_0 k^{(n)}}{n!} + R_n \tag{8}$$

If $\Delta^{n+1}y_0, \Delta^{n+2}y_0, \ldots$ are all zero, then $R_n = 0$ and y_k is a polynomial of degree n in k.

GENERAL TERM OF A SEQUENCE OR SERIES

The Gregory-Newton formula is often useful in finding the general law of formation of terms in a sequence or series [see Problems 2.8 and 2.9].

INTERPOLATION AND EXTRAPOLATION

Often in practice we are given a table showing values of y or $f(x)$ corresponding to various values of x as indicated in the table of Fig. 2-3.

x	x_0	x_1	$\ldots$	x_p
y	y_0	y_1	$\ldots$	y_p

Fig. 2-3

We assume that the values of x are increasing, i.e. $x_0 < x_1 < \cdots < x_p$.

An important practical problem involves obtaining values of y [usually approximate] corresponding to certain values of x which are not in the table. It is assumed of course that we can justify seeking such values, i.e. we suspect some underlying law of formation which may be mathematical or physical in nature.

Finding the [approximate] value of y corresponding to an untabulated value of x *between* x_0 and x_p is often called *interpolation* and can be thought of as a "reading between the lines of the table". The process of obtaining the [approximate] value of y corresponding to a value of x which is either less than x_0 or greater than x_p, i.e. lies outside the table, is often called *extrapolation*. If x represents the time, this can involve a problem in *prediction* or *forecasting*.

Suppose that the x values are equally spaced and the nth differences of y or $f(x)$ as obtained from the table can be considered as small or zero for some value of n. Then we can obtain a suitable interpolation or extrapolation formula in the form of an approximating polynomial by using the Gregory-Newton formula. See Problem 2.11.

If the x values are not equally spaced we can use the *Lagrange interpolation formula* [see page 38].

Because formulas for interpolation can also in general be used for extrapolation we shall refer to such formulas collectively as *interpolation formulas*.

CENTRAL DIFFERENCE TABLES

In the table of Fig. 2-1 it was assumed that the first value of x was $x = a$, the second $x = a + h$, and so on. We could however have extended the table backwards by considering $x = a - h, a - 2h, \ldots$. By doing this we obtain the table of Fig. 2-4 which we call a *central difference table.*

x	y	Δy	$\Delta^2 y$	$\Delta^3 y$	$\Delta^4 y$	$\Delta^5 y$	$\Delta^6 y$
$a - 3h$	y_{-3}						
		Δy_{-3}					
$a - 2h$	y_{-2}		$\Delta^2 y_{-3}$				
		Δy_{-2}		$\Delta^3 y_{-3}$			
$a - h$	y_{-1}		$\Delta^2 y_{-2}$		$\Delta^4 y_{-3}$		
		Δy_{-1}		$\Delta^3 y_{-2}$		$\Delta^5 y_{-3}$	
a	y_0		$\Delta^2 y_{-1}$		$\Delta^4 y_{-2}$		$\Delta^6 y_{-3}$
		Δy_0		$\Delta^3 y_{-1}$		$\Delta^5 y_{-2}$	
$a + h$	y_1		$\Delta^2 y_0$		$\Delta^4 y_{-1}$		
		Δy_1		$\Delta^3 y_0$			
$a + 2h$	y_2		$\Delta^2 y_1$				
		Δy_2					
$a + 3h$	y_3						

Fig. 2-4

The table of Fig. 2-4 can also be written equivalently in terms of central differences as shown in Fig. 2-5.

x	y	δy	$\delta^2 y$	$\delta^3 y$	$\delta^4 y$	$\delta^5 y$	$\delta^6 y$
$a - 3h$	y_{-3}						
		$\delta y_{-5/2}$					
$a - 2h$	y_{-2}		$\delta^2 y_{-2}$				
		$\delta y_{-3/2}$		$\delta^3 y_{-3/2}$			
$a - h$	y_{-1}		$\delta^2 y_{-1}$		$\delta^4 y_{-1}$		
		$\delta y_{-1/2}$		$\delta^3 y_{-1/2}$		$\delta^5 y_{-1/2}$	
a	y_0		$\delta^2 y_0$		$\delta^4 y_0$		$\delta^6 y_0$
		$\delta y_{1/2}$		$\delta^3 y_{1/2}$		$\delta^5 y_{1/2}$	
$a + h$	y_1		$\delta^2 y_1$		$\delta^4 y_1$		
		$\delta y_{3/2}$		$\delta^3 y_{3/2}$			
$a + 2h$	y_2		$\delta^2 y_2$				
		$\delta y_{5/2}$					
$a + 3h$	y_3						

Fig. 2-5

Note that the entries of this table can be related to those in the table of Fig. 2-4 by simply using the operator equivalence $\delta = \Delta E^{-1/2}$ on page 9. Thus for example,

$$\delta^3 y_{-3/2} = (\Delta E^{-1/2})^3 y_{-3/2} = \Delta^3 E^{-3/2} y_{-3/2} = \Delta^3 y_{-3}$$

Other tables can be made using the backward difference operator ∇.

GENERALIZED INTERPOLATION FORMULAS

In using the Gregory-Newton formula *(6)* or *(8)* for interpolation, greater accuracy is obtained when x or $a+kh$ is near the beginning of the table rather than near the middle or end of the table. This is to be expected since in this formula use is made of the leading differences $\Delta f(a), \Delta^2 f(a), \ldots$.

To obtain more accuracy near the middle or end of the table we need to find interpolation formulas which use differences near the middle or end of the table. The following are formulas which do this. All of these formulas are exact and give the same result when the function is a polynomial. If it is not a polynomial, a remainder or error term can be added. The results can also be expressed in terms of the operators ∇ or δ of page 9.

For the purposes of completeness and reference we include in the list the result *(8)*.

1. Gregory-Newton forward difference interpolation formula.

$$y_k = y_0 + \Delta y_0 \frac{k^{(1)}}{1!} + \Delta^2 y_0 \frac{k^{(2)}}{2!} + \Delta^3 y_0 \frac{k^{(3)}}{3!} + \cdots$$

2. Gregory-Newton backward difference interpolation formula.

$$y_k = y_0 + \Delta y_{-1} \frac{k^{(1)}}{1!} + \Delta^2 y_{-2} \frac{(k+1)^{(2)}}{2!} + \Delta^3 y_{-3} \frac{(k+2)^{(3)}}{3!} + \cdots$$

3. Gauss' interpolation formulas.

$$y_k = y_0 + \Delta y_0 \frac{k^{(1)}}{1!} + \Delta^2 y_{-1} \frac{k^{(2)}}{2!} + \Delta^3 y_{-1} \frac{(k+1)^{(3)}}{3!} + \Delta^4 y_{-2} \frac{(k+1)^{(4)}}{4!} + \cdots$$

$$y_k = y_0 + \Delta y_{-1} \frac{k^{(1)}}{1!} + \Delta^2 y_{-1} \frac{(k+1)^{(2)}}{2!} + \Delta^3 y_{-2} \frac{(k+1)^{(3)}}{3!} + \Delta^4 y_{-2} \frac{(k+2)^{(4)}}{4!} + \cdots$$

4. Stirling's interpolation formula.

$$y_k = y_0 + \frac{1}{2}(\Delta y_{-1} + \Delta y_0) \frac{k^{(1)}}{2!} + \Delta^2 y_{-1} \left[\frac{1}{2}\left(\frac{k^{(2)}}{2!} + \frac{(k+1)^{(2)}}{2!}\right)\right]$$
$$+ \frac{1}{2}(\Delta^3 y_{-2} + \Delta^3 y_{-1}) \frac{(k+1)^{(3)}}{3!} + \cdots$$

5. Bessel's interpolation formula.

$$y_k = \frac{1}{2}(y_0 + y_1) + \Delta y_0 \left[\frac{1}{2}\left(\frac{k^{(1)}}{1!} + \frac{(k-1)^{(1)}}{1!}\right)\right]$$
$$+ \frac{1}{2}(\Delta^2 y_{-1} + \Delta^2 y_0) \frac{k^{(2)}}{2!} + \Delta^3 y_{-1} \left[\frac{1}{2}\left(\frac{(k+1)^{(3)}}{3!} + \frac{k^{(3)}}{3!}\right)\right] + \cdots$$

ZIG-ZAG PATHS AND LOZENGE DIAGRAMS

There exists a simple technique for not only writing down all of the above interpolation formulas but developing others as well. To accomplish this we express the central difference table of Fig. 2-4 in the form of a diagram called a *lozenge diagram* as shown in Fig. 2-6 below. A path from left to right such as indicated by the heavy solid line in Fig. 2-6 or the heavy dashed line is called a *zig-zag path*.

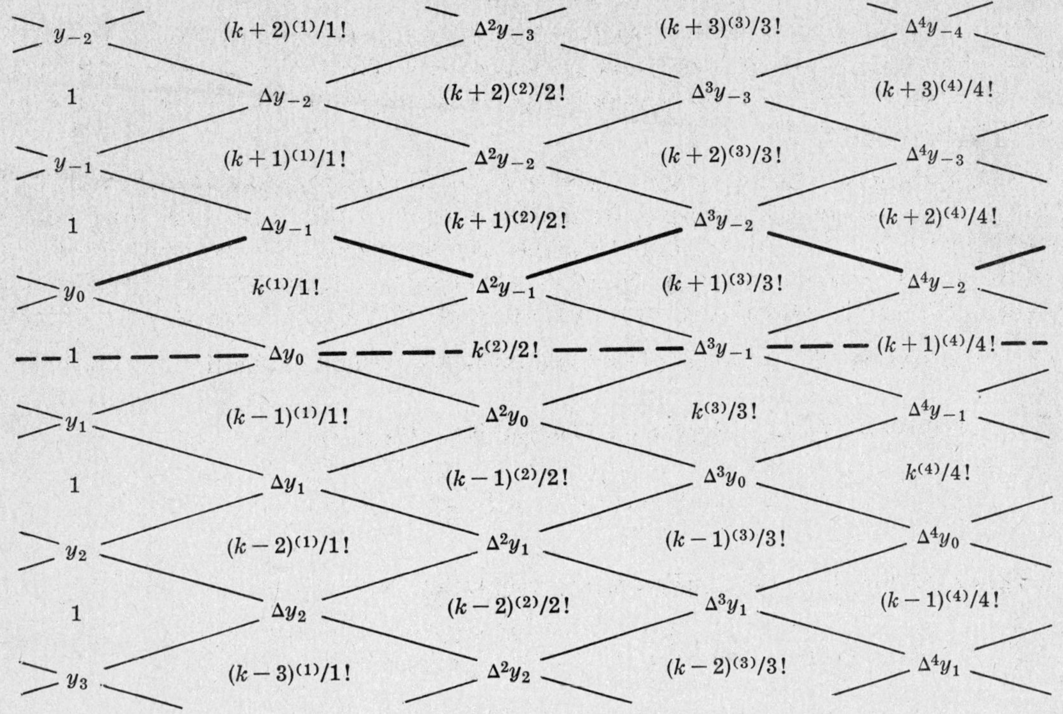

Fig. 2-6

The following rules are observed in obtaining an interpolation formula.

1. A term is to be added whenever any column containing differences is crossed from left to right. The first column in Fig. 2-6 is to be considered as containing differences of order zero.

2. If the path enters a difference column from the left with *positive slope* [as for example from y_0 to Δy_{-1} in Fig. 2-6], the term which is to be added is equal to the product of the difference and the coefficient which is indicated directly *below* the difference.

3. If the path enters a difference column from the left with *negative* slope [as for example from Δy_{-1} to $\Delta^2 y_{-1}$ in Fig. 2-6], the term which is to be added is equal to the product of the difference and the coefficient which is indicated directly *above* the difference.

4. If the path enters a difference column with *zero slope* [as for example from 1 to Δy_0 in Fig. 2-6], the term which is to be added is equal to the product of the difference and the arithmetic mean of the coefficients directly *above* and *below* the difference.

5. If the path crosses with zero slope a column between two differences [as for example in the path from Δy_0 to $\Delta^3 y_{-1}$ in Fig. 2-6], the term which is to be added is equal to the product of the coefficient between the two differences and the arithmetic mean of the two differences.

6. A reversal of path changes the sign of the corresponding term to be added.

LAGRANGE'S INTERPOLATION FORMULA

In case the table of Fig. 2-1 either has nonequally spaced values of x or if the nth differences of y are not small or zero the above interpolation formulas cannot be used.

In such case we can use the formula

$$y = y_0 \frac{(x-x_1)(x-x_2)\cdots(x-x_p)}{(x_0-x_1)(x_0-x_2)\cdots(x_0-x_p)} + y_1 \frac{(x-x_0)(x-x_2)\cdots(x-x_p)}{(x_1-x_0)(x_1-x_2)\cdots(x_1-x_p)}$$

$$+ \cdots + y_p \frac{(x-x_0)(x-x_1)\cdots(x-x_{p-1})}{(x_p-x_0)(x_p-x_1)\cdots(x_p-x_{p-1})}$$

which is called *Lagrange's interpolation formula* [although it was known to Euler].

The result holds whether the values $x_0, x_1, \ldots, x_p$ are equally spaced or not and whether nth differences are small or not. Note that y is a polynomial in x of degree p.

TABLES WITH MISSING ENTRIES

Suppose that in a table the x values are equally spaced, i.e. $x = a, a+h, a+2h, \ldots$ and that all the corresponding entries $f(x)$, except for a few missing ones, are given. Then there are various methods by which these missing entries can be found as shown in Problems 2.28-2.30.

DIVIDED DIFFERENCES

If a table does not have equally spaced values of x, it is convenient to introduce the idea of *divided differences*. Assuming that the values of x are $x_0, x_1, x_2, \ldots$ and that the function is $f(x)$ we define successive divided differences by

$$f(x_0, x_1) = \frac{f(x_1) - f(x_0)}{x_1 - x_0} \tag{9}$$

$$f(x_0, x_1, x_2) = \frac{f(x_1, x_2) - f(x_0, x_1)}{x_2 - x_0} \tag{10}$$

$$f(x_0, x_1, x_2, x_3) = \frac{f(x_1, x_2, x_3) - f(x_0, x_1, x_2)}{x_3 - x_0} \tag{11}$$

etc. They are called divided differences of *orders* 1, 2, 3, etc. We can represent these differences in a *divided difference table* as in Fig. 2-7.

x_0	$f(x_0)$				
		$f(x_0, x_1)$			
x_1	$f(x_1)$		$f(x_0, x_1, x_2)$		
		$f(x_1, x_2)$		$f(x_0, x_1, x_2, x_3)$	
x_2	$f(x_2)$		$f(x_1, x_2, x_3)$		$f(x_0, x_1, x_2, x_3, x_4)$
		$f(x_2, x_3)$		$f(x_1, x_2, x_3, x_4)$	
x_3	$f(x_3)$		$f(x_2, x_3, x_4)$		
		$f(x_3, x_4)$			
x_4	$f(x_4)$				

Fig. 2-7

Various other notations are used for divided differences. For example $f(x_0, x_1, x_2)$ is sometimes denoted by $[x_0, x_1, x_2]$ or $\Delta[x_0, x_1, x_2]$.

The following are some important results regarding divided differences.

1. The divided differences are symmetric. For example

$$f(x_0, x_1) = f(x_1, x_0), \qquad f(x_0, x_1, x_2) = f(x_1, x_0, x_2) = f(x_2, x_0, x_1), \qquad \text{etc.}$$

2. If $f(x)$ is a polynomial of degree n then $f(x, x_0)$ is a polynomial of degree $n-1$, $f(x, x_0, x_1)$ a polynomial of degree $n-2, \ldots$. Thus $f(x, x_0, x_1, \ldots, x_{n-1})$ is a constant and $f(x, x_0, x_1, \ldots, x_n) = 0$.

3. Divided differences can be expressed in terms of first differences. For example,

$$f(x_0, x_1, x_2) = \frac{f(x_0)}{(x_0 - x_1)(x_0 - x_2)} + \frac{f(x_1)}{(x_1 - x_0)(x_1 - x_2)} + \frac{f(x_2)}{(x_2 - x_0)(x_2 - x_1)}$$

NEWTON'S DIVIDED DIFFERENCE INTERPOLATION FORMULA

From the above results we obtain

$$f(x) = f(x_0) + (x - x_0)f(x_0, x_1) + (x - x_0)(x - x_1)f(x_0, x_1, x_2) + \cdots + (x - x_0)(x - x_1)\cdots(x - x_{n-1})f(x_0, x_1, \ldots, x_n) + R \tag{12}$$

where

$$R = \frac{f^{(n)}(\eta)}{n!}\prod_{k=0}^{n}(x - x_k) \tag{13}$$

$$\prod_{k=0}^{n}(x - x_k) = (x - x_0)(x - x_1)\cdots(x - x_n) \tag{14}$$

and η is some value between the smallest and largest of $x, x_0, x_1, \ldots, x_n$. The result *(12)* can be written as

$$f(x) = p_n(x) + R \tag{15}$$

where $p_n(x)$ is a polynomial in x of degree n often called an *interpolating polynomial.*

INVERSE INTERPOLATION

We have assumed up to now that the values of x in a table were specified and that we wanted to interpolate [or extrapolate] for a value of $y = f(x)$ corresponding to a value of x not present in the table. It may happen however that we want a value of x corresponding to a value $y = f(x)$ which is not in the table. The process of finding such a value of x is often called *inverse interpolation* and a solution for x in terms of y is often designated by $x = f^{-1}(y)$ where f^{-1} is called the *inverse function.* In this case the Lagrange interpolation formula or Newton's divided difference formula are especially useful. See Problem 2.38 and 2.39.

APPROXIMATE DIFFERENTIATION

From the formula $e^{hD} = 1 + \Delta$ [equation *(43)*, page 8] we obtain the operator equivalence

$$D = \frac{1}{h}\ln(1 + \Delta) = \frac{1}{h}\left(\Delta - \frac{\Delta^2}{2} + \frac{\Delta^3}{3} - \frac{\Delta^4}{4} + \cdots\right) \tag{16}$$

by formally expanding $\ln(1 + \Delta)$ using the series on page 8.

By using *(16)* we can obtain approximate derivatives of a function from a difference table. Higher derivatives can be obtained by finding series expansions for $D^2, D^3, \ldots$ from *(16)*. We find for example

$$D^2 = \frac{1}{h^2}\left(\Delta^2 - \Delta^3 + \frac{11}{12}\Delta^4 - \frac{5}{6}\Delta^5 + \frac{137}{180}\Delta^6 - \cdots\right) \tag{17}$$

$$D^3 = \frac{1}{h^3}\left(\Delta^3 - \frac{3}{2}\Delta^4 + \frac{7}{4}\Delta^5 - \cdots\right) \tag{18}$$

Use can also be made of different interpolation formulas to find approximate derivatives. In these cases a remainder term can be included enabling us to estimate the accuracy.

Solved Problems

SUBSCRIPT NOTATION

2.1. If $y = f(x)$ and $y_k = f(a+kh)$, show that

$$(a)\quad \Delta y_k = y_{k+1} - y_k \qquad (b)\quad Ey_k = y_{k+1} \qquad (c)\quad E = 1 + \Delta$$

Since $y_k = f(a+kh)$ we have

(a) $\Delta y_k = \Delta f(a+kh) = f(a+kh+h) - f(a+kh) = y_{k+1} - y_k$

(b) $Ey_k = f(a+kh) = f(a+kh+h) = y_{k+1}$

(c) From parts (a) and (b), $y_k + \Delta y_k = Ey_k$ or $(1+\Delta)y_k = Ey_k$

Then since y_k is arbitrary we have $E = 1 + \Delta$.

2.2. Prove that if $y_k = f(a+kh)$, $z_k = g(a+kh)$ then

$$(a)\quad \Delta(y_k + z_k) = \Delta y_k + \Delta z_k \qquad (b)\quad \Delta y_k z_k = y_k \Delta z_k + z_{k+1}\Delta y_k$$

(a)
$$\begin{aligned}\Delta(y_k + z_k) &= \Delta[f(a+kh) + g(a+kh)] \\ &= [f(a+kh+h) + g(a+kh+h)] - [f(a+kh) + g(a+kh)] \\ &= [y_{k+1} + z_{k+1}] - [y_k + z_k] \\ &= \Delta y_k + \Delta z_k\end{aligned}$$

(b)
$$\begin{aligned}\Delta(y_k z_k) &= \Delta[f(a+kh)\,g(a+kh)] \\ &= f(a+kh+h)\,g(a+kh+h) - f(a+kh)\,g(a+kh) \\ &= y_{k+1}z_{k+1} - y_k z_k \\ &= y_k(z_{k+1} - z_k) + z_{k+1}(y_{k+1} - y_k) \\ &= y_k \Delta z_k + z_{k+1}\Delta y_k\end{aligned}$$

DIFFERENCE TABLES

2.3. Let $f(x) = 2x^2 - 7x + 9$ where $x = 3, 5, 7, 9, 11$. Set up a table for the differences of $f(x)$.

The required table is shown in Fig. 2-8 below. Note that in column 2 the various values of $f(x)$ are computed from $f(x) = 2x^2 - 7x + 9$. Thus for example $f(5) = 2(5)^2 - 7(5) + 9 = 24$, etc.

x	$y = f(x)$	Δy	$\Delta^2 y$	$\Delta^3 y$	$\Delta^4 y$
3	6				
		18			
5	24		16		
		34		0	
7	58		16		0
		50		0	
9	108		16		
		66			
11	174				

Fig. 2-8

From this table we see that the second differences are constant and equal to 16 while the higher differences are zero illustrating Theorem 2-1, page 33.

2.4. Find the numerical values corresponding to the letters in the following difference table where it is assumed that there is equal spacing of the independent variable x.

y	Δy	$\Delta^2 y$	$\Delta^3 y$	$\Delta^4 y$	$\Delta^5 y$
A					
	F				
B		-1			
	G		M		
3		J		-3	
	H		N		-5
C		K		P	
	-2		4		
D		L			
	I				
E					

Fig. 2-9

Starting at the right end it is clear that $P-(-3) = -5$ or $P = -8$. Similarly we find in succession the equations $4 - N = P$ or $N = 12$, $N - M = -3$ or $M = 15$, $J-(-1) = M$ or $J = 14$, $K - J = N$ or $K = 26$, $L - K = 4$ or $L = 30$, $I-(-2) = L$ or $I = 28$, $-2 - H = K$ or $H = -28$, $H - G = J$ or $G = -42$, $G - F = -1$ or $F = -41$, $3 - B = G$ or $B = 45$, $B - A = F$ or $A = 86$, $C - 3 = H$ or $C = -25$, $D - C = -2$ or $D = -27$, $E - D = I$ or $E = 1$.

The final difference table with the required numbers replacing letters is as follows:

y	Δy	$\Delta^2 y$	$\Delta^3 y$	$\Delta^4 y$	$\Delta^5 y$
86					
	-41				
45		-1			
	-42		15		
3		14		-3	
	-28		12		-5
-25		26		-8	
	-2		4		
-27		30			
	28				
1					

Fig. 2-10

The problem illustrates the fact that a difference table is completely determined when one entry in each column is known. From the manner in which the letters were obtained it is clear that they can each have only one possible value, i.e. the difference table is *unique*.

2.5. By using only the difference table of Problem 2.3 [and not $f(x) = 2x^2 - 7x + 9$] find the values of $y = f(x)$ for $x = 13, 15, 17, 19, 21$.

We must extend the difference table of Problem 2.3 as shown in Fig. 2-11. To do this we first note that the second differences are all constant and equal to 16. Thus we add the entries 16 in this column as shown in Fig. 2-11. We then fill in the remaining entries of the table.

From the table it is seen that the values of y corresponding to $x = 13, 15, 17, 19, 21$ are 256, 354, 468, 598, 744 respectively.

x	y	Δy	$\Delta^2 y$
3	6		
		18	
5	24		16
		34	
7	58		16
		50	
9	108		16
		66	
11	174		16
		82	
13	256		16
		98	
15	354		16
		114	
17	468		16
		130	
19	598		16
		146	
21	744		16

Fig. 2-11

THE GREGORY-NEWTON FORMULA AND APPLICATIONS

2.6. Prove that $(kh)^{(n)} = h^n k^{(n)}$ for $n = 1, 2, 3, \ldots$.

Since $(kh)^{(1)} = kh = hk^{(1)}$, the result is true for $n = 1$.

Since $(kh)^{2} = (kh)(kh - h) = h^2 k(k-1) = h^2 k^{(2)}$, the result is true for $n = 2$.

Using mathematical induction [Problem 2.67] we can prove for all n

$$(kh)^{(n)} = (kh)(kh-h)\cdots(kh - nh + h) = h^n k(k-1)\cdots(k-n+1) = h^n k^{(n)}$$

2.7. Show that if we put $x = a + kh$ in the Gregory-Newton formula given by equations (*44*) and (*45*) on page 9 we obtain (*6*) and (*7*) or (*8*) on page 34.

From equations (*44*) and (*45*) on page 9 we have

$$f(x) = f(a) + \frac{\Delta f(a)}{\Delta x}\frac{(x-a)^{(1)}}{1!} + \frac{\Delta^2 f(a)}{\Delta x^2}\frac{(x-a)^{(2)}}{2!} + \cdots + \frac{\Delta^n f(a)}{\Delta x^n}\frac{(x-a)^{(n)}}{n!} + R_n \tag{1}$$

where

$$R_n = \frac{f^{(n+1)}(\eta)\,(x-a)^{(n+1)}}{(n+1)!}$$

Then putting $x = a + kh$ in (*1*) we find using Problem 2.6 and $\Delta x = h$

$$f(a+kh) = f(a) + \frac{\Delta f(a)}{h}\frac{(kh)^{(1)}}{1!} + \frac{\Delta^2 f(a)}{h^2}\frac{(kh)^{(2)}}{2!} + \cdots + \frac{\Delta^n f(a)}{h^n}\frac{(kh)^{(n)}}{n!} + R_n$$

$$= f(a) + \frac{\Delta f(a) k^{(1)}}{1!} + \frac{\Delta^2 f(a) k^{(2)}}{2!} + \cdots + \frac{\Delta^n f(a) k^{(n)}}{n!} + R_n \tag{2}$$

where $$R_n = \frac{f^{(n+1)}(\eta)(kh)^{(n+1)}}{(n+1)!} = \frac{h^{(n+1)}f^{(n+1)}(\eta)k^{(n+1)}}{(n+1)!}$$

Since $f(a+kh) = y_k$, $f(a) = y_0$, $\Delta f(a) = \Delta y_0, \ldots,$ (2) yields the required result.

2.8. The first 5 terms of a sequence are given by 2, 7, 16, 35, 70. (a) Find the general term of the sequence. (b) What is the tenth term of the sequence? (c) What assumptions are you making in these cases?

(a) Let us represent the sequence by $u_0, u_1, u_2, \ldots$ where the general term is u_k. The given terms can be represented in the table of Fig. 2-12.

k	0	1	2	3	4
u_k	2	7	16	35	70

Fig. 2-12

The difference table corresponding to this is shown in Fig. 2-13.

u_k	Δu_k	$\Delta^2 u_k$	$\Delta^3 u_k$	$\Delta^4 u_k$
2				
	5			
7		4		
	9		6	
16		10		0
	19		6	
35		16		
	35			
70				

Fig. 2-13

Then by the Gregory-Newton formula with u_k replacing y_k, we have since $u_0 = 2$, $\Delta u_0 = 5$, $\Delta^2 u_0 = 4$, $\Delta^3 u_0 = 6$, $\Delta^4 u_0 = 0$

$$u_k = u_0 + \frac{\Delta u_0 k^{(1)}}{1!} + \frac{\Delta^2 u_0 k^{(2)}}{2!} + \frac{\Delta^3 u_0 k^{(3)}}{3!} + \frac{\Delta^4 u_0 k^{(4)}}{4!}$$

$$= 2 + 5k + \frac{4k(k-1)}{2} + \frac{6k(k-1)(k-2)}{6}$$

$$= k^3 - k^2 + 5k + 2$$

It should be noted that if the first 5 terms of the sequence are represented by $u_1, u_2, \ldots, u_5$ rather than $u_0, u_1, \ldots, u_4$, we can accomplish this by replacing k by $k-1$ in the above polynomial to obtain

$$u_k = (k-1)^3 - (k-1)^2 + 5(k-1) + 2 = k^3 - 4k^2 + 10k - 5$$

(b) The 10th term is obtained by putting $k = 9$ and we find

$$u_9 = 9^3 - 9^2 + 5(9) + 2 = 695$$

(c) In obtaining the above results we are assuming that there exists some law of formation of the terms in the sequence and that we can find this law on the basis of limited information provided by the first 5 terms, namely that the fourth and higher order differences are zero. Theoretically any law of formation is possible [see Problem 2.9].

2.9. Are there other formulas for the general term of the sequence in Problem 2.8? Explain.

In obtaining the general term for the sequence in Problem 2.8 we assumed first the *existence* of some underlying law or formula giving these terms. Second we assumed that the function of k describing this law was such that the fourth and higher differences were all zero. A natural consequence of this was that the function was a third degree polynomial in k which we found by use of the Gregory-Newton formula. In such case the general term not only exists but is *unique*.

By using other assumptions we can obtain many other formulas. For example if we write as general term

$$u_k = k^3 - k^2 + 5k + 2 + k(k-1)(k-2)(k-3)(k-4) \tag{1}$$

we obtain all the data in the table of Problem 2.8. On putting $k = 9$ however we would obtain for the 10th term the value $u_9 = 15{,}815$ differing from that of Problem 2.8. The formula (*1*) does not however have the fourth and higher differences equal to zero.

From these remarks it is clear that one must give careful consideration as to the meaning of the three dots in writing a sequence such as $2, 7, 16, 35, 70, \ldots$.

2.10. Find a polynomial which fits the data in the following table.

x	3	5	7	9	11
y	6	24	58	108	174

Method 1.

The difference table corresponding to this is given in Fig. 2-8, page 41. In this case $a = 3$, $h = 2$ and we can use the Gregory-Newton formula (*8*) on page 34. We have for the leading differences

$$y_0 = 6, \quad \Delta y_0 = 18, \quad \Delta^2 y_0 = 16, \quad \Delta^3 y_0 = 0, \quad \Delta^4 y_0 = 0, \quad \ldots$$

Then $$y_k = 6 + 18k^{(1)} + \frac{16k^{(2)}}{2!} + 0 = 6 + 18k + 8k(k-1) = 8k^2 + 10k + 6$$

To obtain this in the form $y = f(x)$ we can write it equivalently as

$$f(x) = f(a+kh) = f(3+2k) = 8k^2 + 10k + 6$$

Thus using $x = 3 + 2k$, i.e. $k = (x-3)/2$, we find

$$f(x) = 8[(x-3)/2]^2 + 10[(x-3)/2] + 6 = 2(x-3)^2 + 5(x-3) + 6 = 2x^2 - 7x + 9$$

That this is correct is seen from the fact that the entries in the table of Fig. 2-8 were actually obtained by using $f(x) = 2x^2 - 7x + 9$ [see Problem 2.3].

Method 2.

Letting $a = 3$, $h = \Delta x = 2$ in the Gregory-Newton formula (*6*), page 34, we have

$$\begin{aligned} f(x) &= f(a) + \frac{\Delta f(a)}{\Delta x}\frac{(x-a)^{(1)}}{1!} + \frac{\Delta^2 f(a)}{\Delta x^2}\frac{(x-a)^{(2)}}{2!} + \cdots \\ &= 6 + \frac{18}{2}\frac{(x-3)^{(1)}}{1!} + \frac{16}{2^2}\frac{(x-3)^{(2)}}{2!} \\ &= 6 + 9(x-3) + 2(x-3)(x-3-2) \\ &= 2x^2 - 7x + 9 \end{aligned}$$

using the fact that $(x-a)^{(2)} = (x-a)(x-a-h)$.

It should be noted that there are other polynomials of higher degree which also fit the data [see Problem 2.12]. Consequently it would have been more precise to ask for the polynomial of least degree which fits the data.

INTERPOLATION AND EXTRAPOLATION

2.11. Use the table of Problem 2.10 to find the value of y corresponding to (*a*) $x=8$, (*b*) $x=5.5$, (*c*) $x=15$, (*d*) $x=0$. (*e*) What assumptions are you making in these cases?

Method 1.

From Method 1 of Problem 2.10 we have $y_k = 8k^2+10k+6$. To find the value of k corresponding to $x=8$ we use $x=a+kh$ with $a=3$, $h=2$ to obtain $k=(x-a)/h=(8-3)/2=2.5$. Thus the required value of y corresponding to $x=8$ is

$$y_{2.5} = 8(2.5)^2+10(2.5)+6 = 81$$

In a similar manner the values of k corresponding to $x=5.5$, $x=15$, $x=0$ are $(5.5-3)/2$, $(15-3)/2$, $(0-3)/2$ or $1.25, 6, -1.5$ respectively. Then the required values of y corresponding to $x = 5.5, 15$ and 0 are respectively

$$y_{1.25} = 8(1.25)^2+10(1.25)+6 = 31, \qquad y_6 = 8(6)^2+10(6)+6 = 354$$

$$y_{-1.5} = 8(-1.5)^2+10(-1.5)+6 = 9$$

Method 2.

From either Method 1 or Method 2 of Problem 2.10 we have

$$y = f(x) = 2x^2-7x+9$$

Then the required values are

$$f(8) = 2(8)^2-7(8)+9 = 81$$

$$f(5.5) = 2(5.5)^2-7(5.5)+9 = 31$$

$$f(15) = 2(15)^2-7(15)+9 = 354$$

$$f(0) = 2(0)^2-7(0)+9 = 9$$

Note that in parts (*a*) and (*b*) we are *interpolating*, i.e. finding values *within* the table, while in parts (*c*) and (*d*) we are *extrapolating*, i.e. finding values *outside* the table.

(*e*) In using the method we are assuming that there *exists* some underlying law which the data follows and that we can find this law on the basis of the limited information supplied in the table which suggests that all third and higher differences are actually zero so that the data is fitted by a polynomial of second degree. It is possible however that even if an underlying law exists, it is not unique. The analogy with finding general terms of sequences or series as discussed in Problem 2.9 is apparent.

2.12. (*a*) Give an example of a polynomial of degree higher than two which fits the data of Problem 2.3. (*b*) Discuss the relationship of this to problems of interpolation and extrapolation.

(*a*) One example of a polynomial fitting the data of Problem 2.3 is obtained by using

$$F(x) = 2x^2-7x+9+(x-3)(x-5)(x-7)(x-9)(x-11) \tag{1}$$

which is a polynomial of degree 5. For $x = 3,5,7,9,11$ the values agree with those of the table or equivalently those obtained from $f(x)=2x^2-7x+9$.

Other examples can easily be made up, for example the polynomial of degree 11 given by

$$F_1(x) = 2x^2-7x+9+3(x-3)^2(x-5)(x-7)^4(x-9)(x-11)^3 \tag{2}$$

For an example of a function which fits the data but is not a polynomial we can consider

$$F_2(x) = 2x^2-7x+9+(x-3)(x-5)(x-7)(x-9)(x-11)e^{-x} \tag{3}$$

(*b*) Since there are many examples of functions which fit the data, it is clear that *any* interpolation formula based on data from a table will not be unique. Thus for example if we put $x=8$ in (*1*) of part (*a*), we find the value corresponding to it is 126 rather than 81 as in Problem 2.11.

Uniqueness is obtained only when we restrict ourselves in some way as for example requiring a polynomial of smallest degree which fits the data.

2.13. The table of Fig. 2-14 gives the values of $\sin x$ from $x = 25°$ to $x = 30°$ in steps of $1°$. (*a*) Using interpolation find $\sin 28°24'$ and (*b*) compare with the exact value.

x	25°	26°	27°	28°	29°	30°
$\sin x$	0.42262	0.43837	0.45399	0.46947	0.48481	0.50000

Fig. 2-14

To avoid decimals consider $f(x) = 10^5 \sin x$ and let $x = a + kh$ where $a = 25°$ and $h = 1°$. Then we can write $y_k = f(a+kh) = 10^5 \sin(25° + k \cdot 1°)$. The difference table is given in Fig. 2-15. In this table $k = 0$ corresponds to $x = 25°$, $k = 1$ to $26°$, etc.

k	y_k	Δy_k	$\Delta^2 y_k$	$\Delta^3 y_k$
0	42,262			
		1,575		
1	43,837		−13	
		1,562		−1
2	45,399		−14	
		1,548		0
3	46,947		−14	
		1,534		−1
4	48,481		−15	
		1,519		
5	50,000			

Fig. 2-15

It should be noted that the third differences are very small in comparison with the y_k and for all practical purposes can be taken as zero.

When $x = 28°24' = 28.4°$, we have $k = (28.4° - 25°)/1° = 3.4$. Thus by the Gregory-Newton formula

$$\begin{aligned} y_{3.4} &= 10^5 \sin(28°24') \\ &= y_0 + \frac{\Delta y_0 k^{(1)}}{1!} + \frac{\Delta^2 y_0 k^{(2)}}{2!} + \cdots \\ &= 42{,}262 + (1{,}575)(3.4) + \frac{(-13)(3.4)(3.4-1)}{2!} + \cdots \\ &= 47{,}564 \end{aligned}$$

This yields a value of $\sin 28°24' = 0.47564$. For an estimate of the error see Problem 2.14.

(*b*) The exact value to 5 decimal places is 0.47562. Thus the absolute error made is 0.00002 and the percent error is 0.004%.

2.14. Estimate the error term in Problem 2.13.

The error is given by the remainder (*7*), page 34, with $n = 2$

$$R_2 = \frac{h^3 f'''(\eta)\, k^{(3)}}{3!} \tag{1}$$

To evaluate this note that in formulas of calculus involving trigonometric functions, angles must be expressed in radians rather than degrees. This is accomplished by using $1° = \pi/180$ radians. If x is in radians we have

$$f(x) = 10^5 \sin x, \quad f'(x) = 10^5 \cos x, \quad f''(x) = -10^5 \sin x, \quad f'''(x) = -10^5 \cos x$$

so that (*1*) becomes

$$R_2 = -10^5 \left(\frac{\pi}{180}\right)^3 \frac{(\cos\eta)(3.4)(3.4-1)(3.4-2)}{3!} \tag{2}$$

Since the largest value of $\cos \eta$ is $\sqrt{1-\sin^2 25^\circ} = 0.90631$ we see from (2) that $R_2 = -1$ approximately, i.e. the best value we can get for $\sin 28^\circ 24'$ is 0.47563. The fact that this is still not correct is due to *rounding errors* in the given table.

CENTRAL DIFFERENCE TABLES AND INTERPOLATION FORMULAS

2.15. (*a*) Construct a central difference table corresponding to the data of Problem 2.13 by choosing $a = 28^\circ$. (*b*) Use Stirling's interpolation formula to work Problem 2.13 and (*c*) compare with the exact value and the value obtained from the Gregory-Newton formula.

(*a*) The central difference table is shown in Fig. 2-16.

k	y_k	Δy_k	$\Delta^2 y_k$	$\Delta^3 y_k$
−3	42,262			
		1,575		
−2	43,837		−13	
		1,562		−1
−1	45,399		−14	
		1,548		0
0	46,947		−14	
		1,534		−1
1	48,481		−15	
		1,519		
2	50,000			

Fig. 2-16

(*b*) Here $k = 0$ corresponds to 28°, $k = 1$ corresponds to 29°, $k = -1$ corresponds to 26°, etc. The value of k corresponding to 28.4° is $k = (28.4^\circ - 28^\circ)/1^\circ = 0.4$.

From the table we see that

$$y_0 = 46{,}947, \quad \Delta y_{-1} = 1{,}548, \quad \Delta y_0 = 1{,}534, \quad \Delta^2 y_{-1} = -14, \quad \Delta^3 y_{-2} = -1, \quad \Delta^3 y_{-1} = 0$$

Then by Stirling's formula

$$\begin{aligned} y_{0.4} &= 10^5 \sin(28^\circ 24') \\ &= y_0 + \frac{1}{2}(\Delta y_{-1} + \Delta y_0)\frac{k^{(1)}}{1!} + \Delta^2 y_{-1}\left[\frac{1}{2}\left(\frac{k^{(2)}}{2!} + \frac{(k+1)^{(2)}}{2!}\right)\right] + \cdots \\ &= 46{,}947 + \frac{1}{2}(1{,}548 + 1{,}534)\frac{(0.4)}{1!} + (-14)\left[\frac{1}{2}\left(\frac{(0.4)(-0.6)}{2!} + \frac{(1.4)(0.4)}{2!}\right)\right] + \cdots \\ &= 47{,}562 \end{aligned}$$

Thus $\sin 28^\circ 24' = 0.47562$.

(*c*) The result obtained in (*b*) is correct to 5 decimal places. The additional accuracy of Stirling's formula over the Gregory-Newton formula is explained by the fact that $28^\circ 24' = 28.4^\circ$ occurs in the central part of the table. Thus the Stirling formula, which uses the differences near the center, is expected to be more accurate than the Gregory-Newton formula, which uses differences near the beginning of the table.

2.16. (*a*) Explain why you might expect Bessel's interpolation formula to give a better approximation to $\sin 28°24'$ than Stirling's formula. (*b*) Can you confirm your expectation given in part (*a*)?

(*a*) Since the differences used in Bessel's interpolation formula lie to one side of the center it might be expected that since 28.4° lies to the same side, a better approximation would be obtained.

(*b*) From the table of Fig. 2-16,

$$y_0 = 46{,}947, \quad y_1 = 48{,}481, \quad \Delta y_0 = 1{,}534, \quad \Delta^2 y_{-1} = -14, \quad \Delta^2 y_0 = -15$$

Thus Bessel's interpolation formula gives

$$y_k = \tfrac{1}{2}(y_0+y_1) + \Delta y_0\left[\frac{1}{2}\left(\frac{k^{(1)}}{1!}+\frac{(k-1)^{(1)}}{1!}\right)\right] + \tfrac{1}{2}(\Delta^2 y_{-1}+\Delta^2 y_0)\frac{k^{(2)}}{2!} + \cdots$$

or $$y_{0.4} = \tfrac{1}{2}(46{,}947+48{,}481) + 1{,}534[\tfrac{1}{2}(0.4-0.6)] + \tfrac{1}{2}(-14-15)\left[\frac{(0.4)(-0.6)}{2}\right] + \cdots$$

$$= 47{,}562$$

Thus we obtain $\sin 28°24' = 0.47562$. Although Bessel's formula might have been expected to give a better approximation than Stirling's formula we cannot confirm the expectation in this case since the result obtained from Stirling's formula is already accurate to 5 decimal places.

2.17. Derive the Gregory-Newton backward difference formula by (*a*) using symbolic operator methods, (*b*) using any other method.

(*a*) We have since $E - \Delta = 1$

$$y_k = E^k y_0 = \left(\frac{E}{E-\Delta}\right)^k y_0 = (1-\Delta E^{-1})^{-k} y_0$$

$$= \left[1 + (-k)(-\Delta E^{-1}) + \frac{(-k)(-k-1)(-\Delta E^{-1})^2}{2!} + \frac{(-k)(-k-1)(-k-2)(-\Delta E^{-1})^3}{3!} + \cdots\right] y_0$$

$$= \left[1 + k\Delta E^{-1} + \frac{k(k+1)}{2!}\Delta^2 E^{-2} + \frac{k(k+1)(k+2)}{3!}\Delta^3 E^{-3} + \cdots\right] y_0$$

$$= y_0 + k\Delta y_{-1} + \frac{k(k+1)}{2!}\Delta^2 y_{-2} + \frac{k(k+1)(k+2)}{3!}\Delta^3 y_{-3} + \cdots$$

$$= y_0 + \Delta y_{-1}\frac{k^{(1)}}{1!} + \Delta^2 y_{-2}\frac{k^{(2)}}{2!} + \Delta^3 y_{-3}\frac{k^{(3)}}{3!} + \cdots$$

where we have used the binomial theorem.

(*b*) Assume $$y_k = A_0 + A_1 k^{(1)} + A_2(k+1)^{(2)} + A_3(k+2)^{(3)} + \cdots + A_n(k+n-1)^{(n)}$$

Then
$$\Delta y_k = A_1 + 2A_2(k+1)^{(1)} + 3A_3(k+2)^{(2)} + \cdots + nA_n(k+n-1)^{(n-1)}$$
$$\Delta^2 y_k = 2A_2 + 6A_3(k+2)^{(1)} + \cdots + n(n-1)A_n(k+n-1)^{(n-2)}$$
$$\cdots\cdots\cdots\cdots\cdots\cdots\cdots\cdots\cdots\cdots$$
$$\Delta^n y_k = n!\,A_n$$

Putting $k=0$ in the expression for y_k, $k=-1$ in Δy_k, $k=-2$ in $\Delta^2 y_k$, ... we find

$$A_0 = y_0, \quad A_1 = \Delta y_{-1}, \quad A_2 = \frac{\Delta^2 y_{-2}}{2!}, \quad \ldots, \quad A_n = \frac{\Delta^n y_{-n}}{n!}$$

and so $$y_k = y_0 + \Delta y_{-1}\frac{k^{(1)}}{1!} + \Delta^2 y_{-2}\frac{(k+1)^{(2)}}{2!} + \cdots + \Delta^n y_{-n}\frac{(k+n-1)^{(n)}}{n!}$$

2.18. **Derive the Gauss interpolation formulas from the Gregory-Newton formula.**

The Gregory-Newton formula is

$$y_k = y_0 + \Delta y_0 \frac{k^{(1)}}{1!} + \Delta^2 y_0 \frac{k^{(2)}}{2!} + \Delta^3 y_0 \frac{k^{(3)}}{3!} + \cdots$$

Now from the generalized difference table of Fig. 2-4, page 35, we have

$$\Delta^2 y_0 - \Delta^2 y_{-1} = \Delta^3 y_{-1} \qquad \text{or} \qquad \Delta^2 y_0 = \Delta^2 y_{-1} + \Delta^3 y_{-1}$$

Similarly,

$$\Delta^3 y_0 - \Delta^3 y_{-1} = \Delta^4 y_{-1}, \qquad \Delta^4 y_{-1} - \Delta^4 y_{-2} = \Delta^5 y_{-2}$$

from which

$$\Delta^3 y_0 = \Delta^3 y_{-1} + \Delta^4 y_{-1} = \Delta^3 y_{-1} + \Delta^4 y_{-2} + \Delta^5 y_{-2}$$

Then by substituting in the Gregory-Newton formula we find

$$\begin{aligned} y_k &= y_0 + \Delta y_0 \frac{k^{(1)}}{1!} + (\Delta^2 y_{-1} + \Delta^3 y_{-1}) \frac{k^{(2)}}{2!} + (\Delta^3 y_{-1} + \Delta^4 y_{-2} + \Delta^5 y_{-2}) \frac{k^{(3)}}{3!} + \cdots \\ &= y_0 + \Delta y_0 \frac{k^{(1)}}{1!} + \Delta^2 y_{-1} \frac{k^{(2)}}{2!} + \Delta^3 y_{-1} \left[\frac{k^{(2)}}{2!} + \frac{k^{(3)}}{3!}\right] + \Delta^4 y_{-2} \left[\frac{k^{(3)}}{3!} + \frac{k^{(4)}}{4!}\right] + \cdots \\ &= y_0 + \Delta y_0 \frac{k^{(1)}}{1!} + \Delta^2 y_{-1} \frac{k^{(2)}}{2!} + \Delta^3 y_{-1} \frac{(k+1)^{(3)}}{3!} + \Delta^4 y_{-2} \frac{(k+1)^{(4)}}{4!} + \cdots \end{aligned}$$

The second Gauss formula can be derived in a similar manner [see Problem 2.94].

2.19. **Derive Stirling's interpolation formula.**

From the Gauss formulas we have

$$y_k = y_0 + \Delta y_0 \frac{k^{(1)}}{1!} + \Delta^2 y_{-1} \frac{k^{(2)}}{2!} + \Delta^3 y_{-1} \frac{(k+1)^{(3)}}{3!} + \cdots$$

$$y_k = y_0 + \Delta y_{-1} \frac{k^{(1)}}{1!} + \Delta^2 y_{-1} \frac{(k+1)^{(2)}}{2!} + \Delta^3 y_{-2} \frac{(k+1)^{(3)}}{3!} + \cdots$$

Then taking the arithmetic mean of these two results we have

$$y_k = y_0 + \tfrac{1}{2}(\Delta y_{-1} + \Delta y_0) \frac{k^{(1)}}{1!} + \Delta^2 y_{-1} \left[\frac{1}{2}\left(\frac{k^{(2)}}{2!} + \frac{(k+1)^{(2)}}{2!}\right)\right] + \cdots$$

2.20. **Show how to express the Gregory-Newton backward difference interpolation formula in terms of the backward difference operator ∇.**

By Problems 1.39(*a*), page 25, and 1.93, page 30, we have

$$\nabla = \Delta E^{-1}, \qquad \nabla^2 = \Delta^2 E^{-2}, \qquad \ldots, \qquad \nabla^n = \Delta^n E^{-n}$$

Thus

$$\Delta^n E^{-n} f(a) = \nabla^n f(a) \qquad n = 1, 2, 3, \ldots$$

or equivalently

$$\Delta^n E^{-n} y_0 = \nabla^n y_0$$

Using this in (*8*), page 34, it becomes

$$y_k = y_0 + \nabla y_0 \frac{k^{(1)}}{1!} + \nabla^2 y_0 \frac{(k+1)^{(2)}}{2!} + \nabla^3 y_0 \frac{(k+2)^{(3)}}{3!} + \cdots$$

ZIG-ZAG PATHS AND LOZENGE DIAGRAMS

2.21. Demonstrate the rules for writing interpolation formulas by obtaining (*a*) Bessel's formula, (*b*) Gauss' second formula on page 36.

(*a*) We follow the path shown dashed in Fig. 2-5, page 37. Since the path starts with 1 in the first column of the difference table, the first term of the interpolation formula by rule 5, page 37, is 1 multiplied by the mean of the numbers y_0 and y_1 above and below 1, i.e.

$$\tfrac{1}{2}(y_0 + y_1)$$

Similarly by rule 4, page 37, since the path goes through Δy_0 the next term to be added is Δy_0 multiplied by the mean of the coefficients directly above and below, i.e.

$$\Delta y_0 \frac{1}{2}\left(\frac{k^{(1)}}{1!} + \frac{(k-1)^{(1)}}{1!}\right)$$

Again using rule 5 since the path passes through the coefficient $k^{(2)}/2!$, the next term to be added is $k^{(2)}/2!$ multiplied by the mean of the differences above and below the coefficient, i.e.

$$\frac{k^{(2)}}{2!}\frac{1}{2}(\Delta^2 y_{-1} + \Delta^2 y_0)$$

Thus continuing in this manner we find the required formula

$$y_k = \tfrac{1}{2}(y_0 + y_1) + \Delta y_0 \frac{1}{2}\left(\frac{k^{(1)}}{1!} + \frac{(k-1)^{(1)}}{1!}\right) + \tfrac{1}{2}(\Delta^2 y_{-1} + \Delta^2 y_0)\frac{k^{(2)}}{2!} + \cdots$$

(*b*) In this case we follow the zig-zag path indicated by the solid lines of Fig. 2-5, page 37.

By rule 2 if we assume that the path enters y_0 with negative slope, the first term is y_0 multiplied by the coefficient 1 directly above, i.e. the first term is

$$y_0 \cdot 1 = y_0$$

The same term is obtained if we assume that the path enters y_0 with positive or zero slope.

By rule 3 since the path enters Δy_{-1} with positive slope, the second term is Δy_{-1} multiplied by the coefficient directly below, i.e.

$$\Delta y_{-1}\frac{k^{(1)}}{1!}$$

By rule 2 since the path enters $\Delta^2 y_{-1}$ with negative slope, the third term is $\Delta^2 y_{-1}$ multiplied by the coefficient directly above, i.e.

$$\Delta^2 y_{-1}\frac{(k+1)^{(2)}}{2!}$$

Continuing in this manner we obtain the required formula

$$y_k = y_0 + \Delta y_{-1}\frac{k^{(1)}}{1!} + \Delta^2 y_{-1}\frac{(k+1)^{(2)}}{2!} + \Delta^3 y_{-2}\frac{(k+1)^{(3)}}{3!} + \cdots$$

2.22. Let P denote a zig-zag path around any closed lozenge or cell in Fig. 2-5, page 37, i.e. the path begins and ends on the *same* difference. Prove that according to the rules on page 37 the contribution is equal to zero.

A general cell or lozenge is shown in Fig. 2-17.

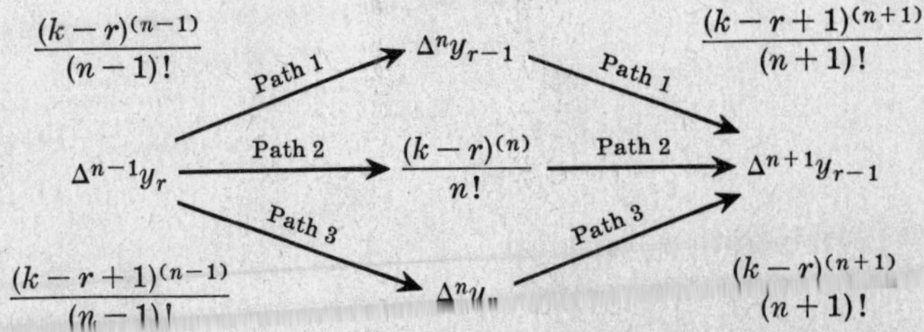

Fig. 2-17

By rule 6 a reversal of path changes the sign of the corresponding term added. Thus we need only show that the contributions corresponding to the three paths from $\Delta^{n-1}y_r$ to $\Delta^{n+1}y_{r-1}$ in Fig. 2-17 are all equal.

Now if path 1 is chosen the contribution is

$$C_1 = \Delta^n y_{r-1}\frac{(k-r)^{(n)}}{n!} + \Delta^{n+1}y_{r-1}\frac{(k-r+1)^{(n+1)}}{(n+1)!} \tag{1}$$

Similarly if paths 2 and 3 are chosen the contributions are respectively

$$C_2 = \tfrac{1}{2}(\Delta^n y_{r-1} + \Delta^n y_r)\frac{(k-r)^{(n)}}{n!} + \Delta^{n+1}y_{r-1}\frac{1}{2}\left(\frac{(k-r+1)^{(n+1)}}{(n+1)!} + \frac{(k-r)^{(n+1)}}{(n+1)!}\right) \tag{2}$$

$$C_3 = \Delta^n y_r\frac{(k-r)^{(n)}}{n!} + \Delta^{n+1}y_{r-1}\frac{(k-r)^{(n+1)}}{(n+1)!} \tag{3}$$

From (1) and (3) we see that

$$\begin{aligned} C_1 - C_3 &= \frac{(k-r)^{(n)}}{n!}(\Delta^n y_{r-1} - \Delta^n y_r) + \Delta^{n+1}y_{r-1}\left(\frac{(k-r+1)^{(n+1)}}{(n+1)!} - \frac{(k-r)^{(n+1)}}{(n+1)!}\right) \\ &= \frac{(k-r)^{(n)}}{n!}(-\Delta^{n+1}y_{r-1}) + \Delta^{n+1}y_{r-1}\frac{(k-r)^{(n)}}{n!}\left(\frac{(k-r-1)-(k-r-n)}{n+1}\right) \\ &= -\frac{(k-r)^{(n)}}{n!}\Delta^{n+1}y_{r-1} + \Delta^{n+1}y_{r-1}\frac{(k-r)^{(n)}}{n!} \\ &= 0 \end{aligned}$$

so that $C_1 = C_3$. Also we note that the mean of C_1 and C_3 is

$$\tfrac{1}{2}(C_1 + C_3) = \frac{(k-r)^{(n)}}{n!}\frac{1}{2}(\Delta^n y_{r-1} + \Delta^n y_r) + \Delta^{n+1}y_{r-1}\frac{1}{2}\left(\frac{(k-r+1)^{(n+1)}}{(n+1)!} + \frac{(k-r)^{(n+1)}}{(n+1)!}\right) = C_2$$

Thus $C_1 = C_2 = C_3$ and the required result is proved.

2.23. Prove that the sum of the contributions around any closed zig-zag path in the lozenge diagram is zero.

We shall prove the result for a typical closed path such as $JKLMNPQRJ$ in Fig. 2-18. This path encloses 3 cells I, II, III.

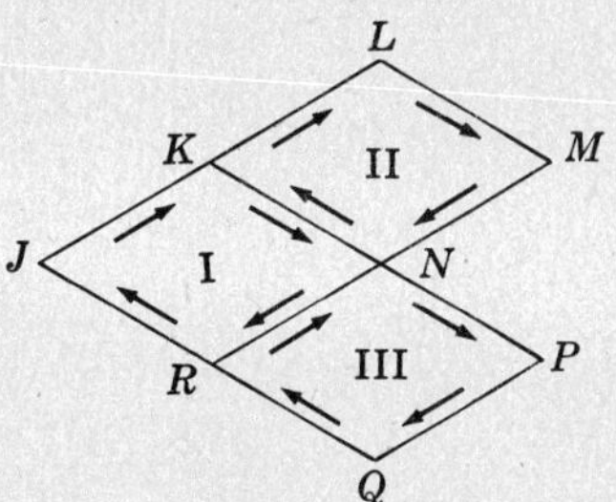

Fig. 2-18

If we denote the contribution corresponding to path JK, for example, by C_{JK} and the total contribution around cell I, for example, by C_{I}, then

$$\begin{aligned} C_{\text{I}} + C_{\text{II}} + C_{\text{III}} &= (C_{JK} + C_{KN} + C_{NR} + C_{RJ}) \\ &\quad + (C_{KL} + C_{LM} + C_{MN} + C_{NK}) \\ &\quad + (C_{RN} + C_{NP} + C_{PQ} + C_{QR}) \\ &= C_{JK} + C_{KL} + C_{LM} + C_{MN} + C_{NP} + C_{PQ} + C_{QR} + C_{RJ} \\ &\quad + (C_{KN} + C_{NK}) + (C_{RN} + C_{NR}) \\ &= C_{JKLMNPQRJ} \end{aligned}$$

since $C_{KN} = -C_{NK}$ and $C_{RN} = -C_{NR}$. Thus since $C_{\text{I}} = 0$, $C_{\text{II}} = 0$, $C_{\text{III}} = 0$ by Problem 2.22, it follows that $C_{JKLMNPQRJ} = 0$.

2.24. Prove that if two zig-zag paths end at the same place in the lozenge diagram then the interpolation formulas corresponding to them are identical.

Consider for example the two zig-zag paths indicated dashed and solid in Fig. 2-19. These two paths end at the same place but begin at different places.

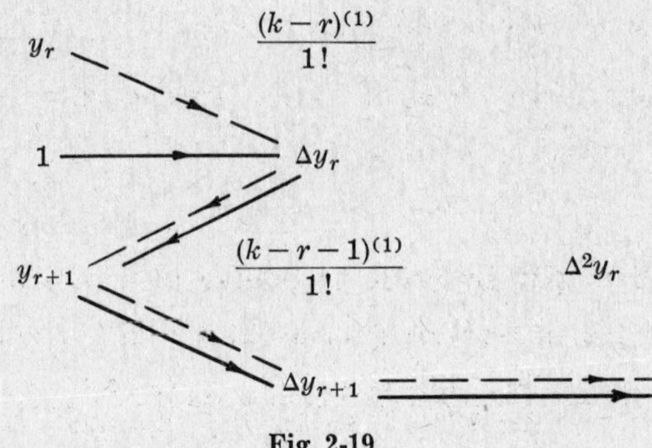

Fig. 2-19

The total contribution to the interpolation formula corresponding to the dashed path is given by

$$y_r + \Delta y_r \frac{(k-r)^{(1)}}{1!} - \Delta y_r \frac{(k-r-1)^{(1)}}{1!} + R \tag{1}$$

when R is the remaining part of the formula.

Similarly the total contribution to the interpolation formula corresponding to the solid path is

$$\tfrac{1}{2}(y_r + y_{r+1}) + \Delta y_r \cdot \frac{1}{2}\left(\frac{(k-r)^{(1)}}{1!} + \frac{(k-r-1)^{(1)}}{1!}\right) - \Delta y_r \frac{(k-r-1)^{(1)}}{1!} + R \tag{2}$$

Note that the minus signs in (*1*) and (*2*) are used because of rule 6, page 37.

To show that the results (*1*) and (*2*) are actually equal we need only show that their difference is zero. We find this difference to be

$$\tfrac{1}{2}y_{r+1} - \tfrac{1}{2}y_r + \tfrac{1}{2}\Delta y_r \frac{(k-r-1)^{(1)}}{1!} - \tfrac{1}{2}\Delta y_r \frac{(k-r)!}{1!} \tag{3}$$

Since $y_{r+1} = y_r + \Delta y_r$, (*3*) can be written

$$\tfrac{1}{2}\Delta y_r \left[1 + \frac{(k-r-1)^{(1)}}{1!} - \frac{(k-r)^{(1)}}{1!}\right] = \tfrac{1}{2}\Delta y_r[1 + k - r - 1 - (k-r)] = 0$$

so that the required result is proved.

2.25. Prove the validity of all the interpolation formulas on page 36.

Since all the formulas can be obtained from any one of them, it follows by Problem 2.24 that they are all valid if we can prove just one of them. However we have already proved one of them, namely the Gregory-Newton formula on page 36. This proves that all of them are valid.

Note that Problems 2.18 and 2.19 illustrate how some of these formulas can be obtained from the Gregory-Newton formula.

LAGRANGE'S INTERPOLATION FORMULA

2.26. Prove Lagrange's interpolation formula on page 38.

Since there are $p+1$ values of y corresponding to $p+1$ values of x in the table of Fig. 2-1 on page 32, we can fit the data by a polynomial of degree p. Choose this polynomial to be

$$y = a_0(x-x_1)(x-x_2)\cdots(x-x_p) + a_1(x-x_0)(x-x_2)\cdots(x-x_p) + \cdots + a_p(x-x_0)(x-x_1)\cdots(x-x_{p-1})$$

Then putting $x = x_0,\ y = y_0$ we have

$$y_0 = a_0(x_0 - x_1)(x_0 - x_2)\cdots(x_0 - x_p) \quad \text{or} \quad a_0 = \frac{y_0}{(x_0 - x_1)(x_0 - x_2)\cdots(x_0 - x_p)}$$

Similarly putting $x = x_1,\ y = y_1$ we find

$$y_1 = a_1(x_1 - x_0)(x_1 - x_2)\cdots(x_1 - x_p) \quad \text{or} \quad a_1 = \frac{y_1}{(x_1 - x_0)(x_1 - x_2)\cdots(x_1 - x_p)}$$

with corresponding values for $a_2, a_3, \ldots, a_p$. Putting these values in the assumed polynomial we obtain Lagrange's formula.

2.27. (*a*) Work Problem 2.10 by using Lagrange's formula and thus (*b*) solve Problem 2.11.

(*a*) Substituting the values from the table of Fig. 2-13 into Lagrange's formula we obtain

$$\begin{aligned} y = {} & 6\,\frac{(x-5)(x-7)(x-9)(x-11)}{(3-5)(3-7)(3-9)(3-11)} + 24\,\frac{(x-3)(x-7)(x-9)(x-11)}{(5-3)(5-7)(5-9)(5-11)} \\ & + 58\,\frac{(x-3)(x-5)(x-9)(x-11)}{(7-3)(7-5)(7-9)(7-11)} + 108\,\frac{(x-3)(x-5)(x-7)(x-11)}{(9-3)(9-5)(9-7)(9-11)} \\ & + 174\,\frac{(x-3)(x-5)(x-7)(x-9)}{(11-3)(11-5)(11-7)(11-9)} \end{aligned}$$

Performing the algebra we find $y = 2x^2 - 7x + 9$ which is a polynomial of degree *two* and *not four* as we might have expected.

(*b*) The values of y for $x = 8, 5.5, 15, 0$ can be obtained by substituting in the Lagrange interpolation formula of part (*a*).

TABLES WITH MISSING ENTRIES

2.28. Find the value of u_3 missing in the following table.

k	0	1	2	3	4
u_k	3	8	15		47

Fig. 2-20

Method 1.

In this problem we cannot construct a difference table. However from the four pairs of values (k, u_k) which are known we would be able to fit a polynomial of degree 3. This leads us to assume that the fourth differences are zero, i.e. $\Delta^4 u_0 = 0$. Thus we have

$$(E-1)^4 u_0 = (E^4 - 4E^3 + 6E^2 - 4E + 1)u_0 = 0$$

which becomes

$$u_4 - 4u_3 + 6u_2 - 4u_1 + u_0 = 0$$

Using the values from the table we then have

$$47 - 4u_3 + 6(15) - 4(8) + 3 = 0 \quad \text{or} \quad u_3 = 27$$

Method 2.

Since there are 4 values of k for which u_k is known, we can fit the data by a polynomial of degree 3 which involves 4 unknown coefficients. Let us therefore assume that

$$u_k = A_0k^3 + A_1k^2 + A_2k + A_3$$

Then putting $k = 0, 1, 2, 4$ in succession we obtain the equations

$$\begin{aligned} A_3 &= 3 \\ A_0 + A_1 + A_2 + A_3 &= 8 \\ 8A_0 + 4A_1 + 2A_2 + A_3 &= 15 \\ 64A_0 + 16A_1 + 4A_2 + A_3 &= 47 \end{aligned}$$

Solving simultaneously we find

$$A_0 = \tfrac{1}{2}, \quad A_1 = -\tfrac{1}{2}, \quad A_2 = 5, \quad A_3 = 3$$

and so

$$u_k = \tfrac{1}{2}k^3 - \tfrac{1}{2}k^2 + 5k + 3$$

Putting $k = 3$ we then find $u_3 = 27$.

Method 3.

By Lagrange's formula we have

$$\begin{aligned} u_k &= u_0 \frac{(k-1)(k-2)(k-4)}{(0-1)(0-2)(0-4)} + u_1 \frac{(k-0)(k-2)(k-4)}{(1-0)(1-2)(1-4)} \\ &\quad + u_2 \frac{(k-0)(k-1)(k-4)}{(2-0)(2-1)(2-4)} + u_4 \frac{(k-0)(k-1)(k-2)}{(4-0)(4-1)(4-2)} \\ &= -\frac{3}{8}(k-1)(k-2)(k-4) + \frac{8}{3}(k)(k-2)(k-4) - \frac{15}{4}(k)(k-1)(k-4) + \frac{47}{24}(k)(k-1)(k-2) \end{aligned}$$

Then putting $k = 3$ we find $u_3 = 27$.

Note that Method 1 has the advantage in that we need not obtain the polynomial.

2.29. Find $\log_{10} 7$ and $\log_{10} 11$ from the following table.

x	6	7	8	9	10	11	12
$f(x) = \log_{10} x$	0.77815		0.90309	0.95424	1.00000		1.07918

Fig. 2-21

Since there are 5 known pairs of values we could fit a polynomial of degree 4 to the data. We are thus led to assume that the 5th differences of $f(x) = \log_{10} x$ are zero.

If we let $x = a + kh$ and $u_k = f(a + kh)$, then taking $a = 6$ and $h = 1$ the given table can be written in the k notation as follows.

k	0	1	2	3	4	5	6
$u_k = \log_{10}(6 + k)$	0.77815		0.90309	0.95424	1.00000		1.07918

Fig. 2-22

From $\Delta^5 u_0 = 0$ and $\Delta^5 u_1 = 0$, i.e. $(E-1)^5 u_0 = 0$ and $(E-1)^5 u_1 = 0$, we are led to the equations

$$(E^5 - 5E^4 + 10E^3 - 10E^2 + 5E - 1)u_0 = 0, \qquad (E^5 - 5E^4 + 10E^3 - 10E^2 + 5E - 1)u_1 = 0$$

or

$$\left.\begin{aligned} u_5 - 5u_4 + 10u_3 - 10u_2 + 5u_1 - u_0 &= 0 \\ u_6 - 5u_5 + 10u_4 - 10u_3 + 5u_2 - u_1 &= 0 \end{aligned}\right\} \qquad (1)$$

Putting in the values of u_0, u_2, u_3, u_4 and u_6 from the table we find

$$\left.\begin{aligned} 5u_1 + u_5 &= 5.26615 \\ u_1 + 5u_5 &= 6.05223 \end{aligned}\right\} \qquad (2)$$

Solving simultaneously we find

$$u_1 = \log_{10} 7 = 0.84494, \qquad u_5 = \log_{10} 11 = 1.04146$$

These should be compared with the exact values to 5 decimal places given by $\log_{10} 7 = 0.84510$, $\log_{10} 11 = 1.04139$ so that the errors made in each case are less than 0.02% and 0.01% respectively.

2.30. The amount A_k of an *annuity* [see Problem 2.144] after k years is given in the following table. Determine the amounts of the annuity for the missing years corresponding to $k = 4$ and 5.

k	3	6	9	12
A_k	3.1216	6.6330	10.5828	15.0258

Fig. 2-23

Let Δ_3 and E_3 respectfully denote the difference and shifting operators for the 3 year intervals while Δ_1 and E_1 denote the corresponding operators for 1 year intervals. Then we have

$$E_3 = 1 + \Delta_3, \qquad E_1 = 1 + \Delta_1$$

Since
$$u_{k+3} = E_3 u_k = E_1^3 u_k$$

we have
$$E_3 = E_1^3$$

or
$$1 + \Delta_3 = (1 + \Delta_1)^3$$

Thus
$$\Delta_1 = (1 + \Delta_3)^{1/3} - 1 \tag{1}$$

Using the binomial theorem we then have

$$\Delta_1 = \frac{1}{3}\Delta_3 + \frac{(1/3)(-2/3)}{2!}\Delta_3^2 + \frac{(1/3)(-2/3)(-5/3)}{3!}\Delta_3^3 + \cdots = \frac{1}{3}\Delta_3 - \frac{1}{9}\Delta_3^2 + \frac{5}{81}\Delta_3^3 + \cdots$$

Since there are only four pairs of values in the table we can assume that third differences Δ are constant. Thus we can omit fourth and higher differences in the expansions.

We thus have
$$\Delta_1 = \frac{1}{3}\Delta_3 - \frac{1}{9}\Delta_3^2 + \frac{5}{81}\Delta_3^3 \tag{2}$$

$$\Delta_1^2 = \frac{1}{9}\Delta_3^2 - \frac{2}{27}\Delta_3^3 \tag{3}$$

$$\Delta_1^3 = \frac{1}{27}\Delta_3^3 \tag{4}$$

The difference table for the 3 year intervals is shown in Fig. 2-24.

k	u_k	$\Delta_3 u_k$	$\Delta_3^2 u_k$	$\Delta_3^3 u_k$
3	3.1216			
		3.5114		
6	6.6330		0.4384	
		3.9498		0.0548
9	10.5828		0.4932	
		4.4430		
12	15.0258			

Fig. 2-24

Using (2), (3) and (4) we then have

$$\Delta_1 u_3 = \frac{1}{3}\Delta_3 u_3 - \frac{1}{9}\Delta_3^2 u_3 + \frac{5}{81}\Delta_3^3 u_3 = \frac{1}{3}(3.5114) - \frac{1}{9}(0.4384) + \frac{5}{81}(0.0548)$$

$$= 1.12513827$$

$$\Delta_1^2 u_3 = \frac{1}{9}\Delta_3^2 u_3 - \frac{2}{27}\Delta_3^3 u_3 = \frac{1}{9}(0.4384) - \frac{2}{27}(0.0548)$$

$$= 0.04465185$$

$$\Delta_1^3 u_3 = \frac{1}{27}\Delta_3^3 u_3 = \frac{1}{27}(0.0548) = 0.00202963$$

By using the above leading differences we can now complete the difference table for 1 year intervals as shown in Fig. 2-25 below.

k	u_k	$\Delta_1 u_k$	$\Delta_1^2 u_k$	$\Delta_1^3 u_k$
3	3.1216			
		1.12513827		
4	4.2467		0.04465185	
		1.16979012		0.00202963
5	5.4165		0.04668148	
		1.21647160		
6	6.6330			

Fig. 2-25

From this table we see that

$$u_4 = 4.2467, \qquad u_5 = 5.4165$$

By proceeding in a similar manner we can find u_7, u_8, u_{10}, u_{11}.

DIVIDED DIFFERENCES

2.31. Construct a divided difference table corresponding to the data in the following table.

x	2	4	5	6	8	10
$f(x)$	10	96	196	350	868	1746

Fig. 2-26

The divided difference table is as follows

x	$f(x)$				
2	10				
		43			
4	96		19		
		100		2	
5	196		27		0
		154		2	
6	350		35		0
		259		2	
8	868		45		
		439			
10	1746				

Fig. 2-27

In the third column the entry 43 is obtained from $\dfrac{96-10}{4-2}$, 100 is obtained from $\dfrac{196-96}{5-4}$, etc.

In the fourth column the entry 19 is obtained from $\dfrac{100-43}{5-2}$, 27 is obtained from $\dfrac{154-100}{6-4}$, etc.

In the fifth column the entries are obtained from $\dfrac{27-19}{6-2}$, $\dfrac{35-27}{8-4}$ and $\dfrac{45-35}{10-5}$ and are all equal to 2.

2.32. Prove that (a) $f(x) = f(x_0) + (x-x_0)f(x, x_0)$

(b) $f(x, x_0) = f(x_0, x_1) + (x - x_1)f(x, x_0, x_1)$

(c) $f(x, x_0, x_1) = f(x_0, x_1, x_2) + (x - x_2)f(x, x_0, x_1, x_2)$

(a) By definition, $f(x, x_0) = \dfrac{f(x)-f(x_0)}{x-x_0}$ so that

$$f(x) = f(x_0) + (x - x_0)f(x, x_0)$$

(b) By definition, $f(x, x_0, x_1) = \dfrac{f(x, x_0) - f(x_0, x_1)}{x - x_1}$ so that

$$f(x, x_0) = f(x_0, x_1) + (x - x_1)f(x, x_0, x_1)$$

(c) By definition, $f(x, x_0, x_1, x_2) = \dfrac{f(x, x_0, x_1) - f(x_0, x_1, x_2)}{x - x_2}$ so that

$$f(x, x_0, x_1) = f(x_0, x_1, x_2) + (x - x_2)f(x, x_0, x_1, x_2)$$

Generalizations can easily be made.

2.33. Prove that

$$f(x) = f(x_0) + (x - x_0)f(x_0, x_1) + (x - x_0)(x - x_1)f(x_0, x_1, x_2) + R$$

where $R = (x - x_0)(x - x_1)(x - x_2)f(x, x_0, x_1, x_2)$

Using the results of Problem 2.32(a), (b) and (c) we have

$$\begin{aligned} f(x) &= f(x_0) + (x - x_0)f(x, x_0) \\ &= f(x_0) + (x - x_0)[f(x_0, x_1) + (x - x_1)f(x, x_0, x_1)] \\ &= f(x_0) + (x - x_0)f(x_0, x_1) + (x - x_0)(x - x_1)f(x, x_0, x_1) \\ &= f(x_0) + (x - x_0)f(x_0, x_1) + (x - x_0)(x - x_1)[f(x_0, x_1, x_2) + (x - x_2)f(x, x_0, x_1, x_2)] \\ &= f(x_0) + (x - x_0)f(x_0, x_1) + (x - x_0)(x - x_1)f(x_0, x_1, x_2) + (x - x_0)(x - x_1)(x - x_2)f(x, x_0, x_1, x_2) \end{aligned}$$

which is equivalent to the required result.

2.34. Prove Newton's divided difference interpolation formula

$$\begin{aligned} f(x) &= f(x_0) + (x - x_0)f(x_0, x_1) + (x - x_0)(x - x_1)f(x_0, x_1, x_2) \\ &\quad + \cdots + (x - x_0)\cdots(x - x_{n-1})f(x_0, \ldots, x_n) + R(x) \end{aligned}$$

where $R(x) = (x - x_0)(x - x_1)\cdots(x - x_n)f(x, x_0, x_1, \ldots, x_n)$

We shall use the principle of mathematical induction. Assume that the formula is true for $n = k$, i.e.

$$\begin{aligned} f(x) &= f(x_0) + (x - x_0)f(x_0, x_1) + \cdots + (x - x_0)\cdots(x - x_{k-1})f(x_0, \ldots, x_k) \\ &\quad + (x - x_0)(x - x_1)\cdots(x - x_k)f(x, x_0, x_1, \ldots, x_k) \end{aligned} \tag{1}$$

Now by definition as in Problem 2.32 we have

$$f(x, x_0, x_1, \ldots, x_{k+1}) = \frac{f(x, x_0, \ldots, x_k) - f(x_0, x_1, \ldots, x_{k+1})}{x - x_{k+1}} \tag{2}$$

so that $$f(x, x_0, x_1, \ldots, x_k) = f(x_0, x_1, \ldots, x_{k+1}) + (x - x_{k+1})f(x, x_0, x_1, \ldots, x_{k+1}) \tag{3}$$

Using this in (1) we find

$$\begin{aligned} f(x) &= f(x_0) + (x - x_0)f(x_0, x_1) + \cdots + (x - x_0)\cdots(x - x_k)f(x_0, \ldots, x_{k+1}) \\ &\quad + (x - x_0)(x - x_1)\cdots(x - x_{k+1})f(x, x_0, x_1, \ldots, x_{k+1}) \end{aligned}$$

Thus assuming that it is true for $n = k$ we have proved it true for $n = k + 1$. But since it is true for $n = 1$ [see Problem 2.32], it is true for $n = 2$ and thus for $n = 3$ and so on for all positive integers.

2.35. Prove that the remainder in Problem 2.34 can be written as

$$R(x) = (x - x_0)(x - x_1)\cdots(x - x_n)\frac{f^{(n+1)}(\eta)}{(n+1)!}$$

where η is a number between the smallest and largest of the values $x, x_0, x_1, \ldots, x_n$.

By Problem 2.34 the remainder is

$$R(x) = (x - x_0)(x - x_1)\cdots(x - x_n)f(x, x_0, x_1, \ldots, x_n) \tag{1}$$

Since $R = 0$ at $x = x_0, x_1, \ldots, x_n$, we know by Problem 1.117, page 31, that the $(n+1)$th derivative $R^{(n+1)}(x)$ is zero for some value η, i.e.

$$R^{(n+1)}(\eta) = 0 \qquad (2)$$

where η is a number between the smallest and largest of the values $x, x_0, x_1, \ldots, x_n$.

Now by taking the $(n+1)$th derivative of $f(x)$ in Newton's divided difference formula, we have

$$f^{(n+1)}(x) = n!\, f(x_0, x_1, \ldots, x_n) + R^{(n+1)}(x) \qquad (3)$$

Then letting $x = \eta$ in (3) and using (2) we find

$$f(x_0, x_1, \ldots, x_n) = \frac{f^{(n+1)}(\eta)}{(n+1)!} \qquad (4)$$

Thus (1) becomes

$$R(x) = (x-x_0)(x-x_1)\cdots(x-x_n)\frac{f^{(n+1)}(\eta)}{(n+1)!} \qquad (5)$$

2.36. Use Newton's divided difference formula to express $f(x)$ as a polynomial in x for the data of Problem 2.31.

From Problem 2.31 we have

$$x_0 = 2, \quad x_1 = 4, \quad x_2 = 5, \quad x_3 = 6, \quad x_4 = 8, \quad x_5 = 10$$

Also,

$$f(x_0) = 10, \quad f(x_0, x_1) = 43, \quad f(x_0, x_1, x_2) = 19, \quad f(x_0, x_1, x_2, x_3) = 2, \quad f(x_0, x_1, x_2, x_3, x_4) = 0$$

Then using Problem 2.33 we have

$$\begin{aligned} f(x) &= 10 + (x-2)(43) + (x-2)(x-4)(19) + (x-2)(x-4)(x-5)(2) + (x-2)(x-4)(x-5)(x-6)(0) \\ &= 2x^3 - 3x^2 + 5x - 4 \end{aligned}$$

2.37. Solve the equation $x^3 - 3x - 4 = 0$.

Consider $y = f(x) = x^3 - 3x - 4$. We are looking for the values of x for which $y = f(x) = 0$. When $x = 2$, $y = -2$ and when $x = 3$, $y = 14$. It follows that x is between 2 and 3 since y changes sign from -2 to $+14$. Also it seems reasonable from these results that the value of x is closer to 2 than to 3. In view of these observations we set up a table of values for x and y as shown in Fig. 2-28. We can think of x as a function of y, i.e. $x = g(y)$.

x	2.0	2.1	2.2	2.3	2.4
y	−2.0000	−1.0390	0.0480	1.2670	2.6240

Fig. 2-28

We now set up the following table of divided differences.

y	$x = g(y)$			
−2.0000	2.0			
		0.10406		
−1.0390	2.1		−0.00589	
		0.09200		0.00048
0.0480	2.2		−0.00432	
		0.08203		0.00029
1.2670	2.3		−0.00324	
		0.07369		
2.6240	2.4			

Fig. 2-29

By Newton's divided difference formula,

$$g(y) = g(y_0) + (y-y_0)g(y_0, y_1) + (y-y_0)(y-y_1)g(y_0, y_1, y_2) + (y-y_0)(y-y_1)(y-y_2)g(y_0, y_1, y_2, y_3) + \cdots \qquad (1)$$

Since the required value of x is that for which $y = 0$, we put $y = 0$ in (1) to obtain

$$x = g(0) = g(y_0) - y_0 g(y_0, y_1) + (y_0)(y_1)g(y_0, y_1, y_2) - y_0 y_1 y_2 g(y_0, y_1, y_2, y_3) + \cdots \qquad (2)$$

Then letting $y_0 = -1.0390$ so that $g(y_0) = 2.1$ and $y_1 = 0.0480$, $y_2 = 1.2670$ we have $g(y_0, y_1) = 0.09200$, $g(y_0, y_1, y_2) = -0.00432$, $g(y_0, y_1, y_2, y_3) = 0.00029$. Thus (2) becomes

$$\begin{aligned} x &= 2.1 + (1.0390)(0.09200) + (-1.0390)(0.0480)(-0.00432) - (-1.0390)(0.0480)(1.2670)(0.00029) \\ &= 2.1 + 0.09559 + 0.00022 + 0.00002 \\ &= 2.19583 \end{aligned}$$

As a check we note that when $x = 2.19583$ we have $x^3 - 3x - 4 = 0.00008$ so that the value is quite accurate.

To find the remaining roots of the equation $x^3 - 3x - 4 = 0$ we use the fact that $x - 2.19583$ must be a factor of $x^3 - 3x - 4$ and find by division that the other factor is $x^2 + 2.19583x + 1.82167$. Setting this factor equal to zero we find that the other two roots are complex numbers given by $-1.09792 \pm 0.78502i$.

INVERSE INTERPOLATION

2.38. Using the data in the table of Problem 2.10 find the value of x corresponding to $y = f(x) = 100$ by using (a) Lagrange's formula, (b) Newton's divided difference formula and (c) the Gregory-Newton forward difference formula.

(a) Interchanging the roles of x and y in Lagrange's formula we have

$$\begin{aligned} x = {} & 3\frac{(y-24)(y-58)(y-108)(y-174)}{(6-24)(6-58)(6-108)(6-174)} + 5\frac{(y-6)(y-58)(y-108)(y-174)}{(24-6)(24-58)(24-108)(24-174)} \\ & + 7\frac{(y-6)(y-24)(y-108)(y-174)}{(58-6)(58-24)(58-108)(58-174)} + 9\frac{(y-6)(y-24)(y-58)(y-174)}{(108-6)(108-24)(108-58)(108-174)} \\ & + 11\frac{(y-6)(y-24)(y-58)(y-108)}{(174-6)(174-24)(174-58)(174-108)} \end{aligned}$$

Then putting $y = 100$ we find $x = 8.656$.

(b) From the table of Problem 2.10 we obtain the following divided difference table.

y	x				
174	11				
		0.030303			
108	9		−0.00008360		
		0.040000		0.000000937	
58	7		−0.00022410		−0.000000040
		0.058824		0.000007661	
24	5		−0.00100552		
		0.111111			
6	3				

Assuming the fourth order difference to be negligible we have by Newton's divided difference formula

$$x = 11 + (y-174)(0.030303) + (y-174)(y-108)(-0.00008360) + (y-174)(y-108)(y-58)(0.000000937)$$

Putting $y = 100$ we find $x = 8.628$.

(c) Since the values of x are equally spaced we can use the Gregory-Newton formula to find an interpolating formula for $f(x)$. As in Problem 2.11 we find $f(x) = 2x^2 - 7x + 9$.

If $f(x) = 100$, i.e. $2x^2 - 7x + 9 = 100$, then $2x^2 - 7x - 91 = 0$ and by solving this equation we find

$$x = \frac{7 \pm \sqrt{777}}{4} = \frac{7 \pm 27.8747}{4}$$

Since only the positive square root is significant for our purposes, we use it to find $x = 8.718$. This can in fact be taken as the true value of x since the table of Problem 2.10 was actually constructed by using the function $f(x) = 2x^2 - 7x + 9$.

2.39. Work Problem 2.37 by using inverse interpolation involving Lagrange's formula.

We use the table of Fig. 2-28, page 58. Then by Lagrange's interpolation formula we have

$$\begin{aligned} x = {} & 2.0 \frac{(y + 1.0390)(y - 0.0480)(y - 1.2670)(y - 2.6240)}{(-2.0000 + 1.0390)(-2.0000 - 0.0480)(-2.0000 - 1.2670)(-2.0000 - 2.6240)} \\ & + 2.1 \frac{(y + 2.0000)(y - 0.0480)(y - 1.2670)(y - 2.6240)}{(-1.0390 + 2.0000)(-1.0390 - 0.0480)(-1.0390 - 1.2670)(-1.0390 - 2.6240)} \\ & + 2.2 \frac{(y + 2.0000)(y + 1.0390)(y - 1.2670)(y - 2.6240)}{(0.0480 + 2.0000)(0.0480 + 1.0390)(0.0480 - 1.2670)(0.0480 - 2.6240)} \\ & + 2.3 \frac{(y + 2.0000)(y + 1.0390)(y - 0.0480)(y - 2.6240)}{(1.2670 + 2.0000)(1.2670 + 1.0390)(1.2670 - 0.0480)(1.2670 - 2.6240)} \\ & + 2.4 \frac{(y + 2.0000)(y + 1.0390)(y - 0.0480)(y - 1.2670)}{(2.6240 + 2.0000)(2.6240 + 1.0390)(2.6240 - 0.0480)(2.6240 - 1.2670)} \end{aligned}$$

Since the required value of x is the one for which $y = 0$, we obtain on putting $y = 0$ in the above result $x = 2.19583$ in agreement with that of Problem 2.37.

APPROXIMATE DIFFERENTIATION

2.40. Establish equation (*16*), page 39, (*a*) by symbolic operator methods and (*b*) by using the Gregory-Newton formula.

(a) In using symbolic operator methods we proceed formally. We start with

$$e^{hD} = 1 + \Delta$$

[equation (*43*), page 8]. Taking logarithms this yields

$$hD = \ln(1 + \Delta) \qquad \text{or} \qquad D = \frac{1}{h}\ln(1 + \Delta)$$

Then using the formal Taylor series on page 8 we find

$$D = \frac{1}{h}\ln(1 + \Delta) = \frac{1}{h}\left(\Delta - \frac{\Delta^2}{2} + \frac{\Delta^3}{3} - \frac{\Delta^4}{4} + \cdots\right)$$

(b) By the Gregory-Newton formula,

$$\begin{aligned} f(a + kh) = {} & f(a) + k\Delta f(a) + \frac{k(k-1)}{2!}\Delta^2 f(a) \\ & + \frac{k(k-1)(k-2)}{3!}\Delta^3 f(a) + \frac{k(k-1)(k-2)(k-3)}{4!}\Delta^4 f(a) + \cdots \end{aligned}$$

Then differentiating with respect to k,

$$\begin{aligned} hf'(a + kh) = {} & \Delta f(a) + \frac{2k-1}{2!}\Delta^2 f(a) + \frac{3k^2 - 6k + 2}{3!}\Delta^3 f(a) \\ & + \frac{4k^3 - 18k^2 + 22k - 6}{4!}\Delta^4 f(a) + \cdots \end{aligned}$$

Putting $k = 0$ we have

$$hf'(a) = \Delta f(a) - \frac{\Delta^2 f(a)}{2} + \frac{\Delta^3 f(a)}{3} - \frac{\Delta^4 f(a)}{4} + \cdots$$

i.e.

$$Df(a) \;=\; \frac{1}{h}\left(\Delta - \frac{\Delta^2}{2} + \frac{\Delta^3}{3} - \frac{\Delta^4}{4} + \cdots\right) f(a)$$

Since $f(a)$ is arbitrary we have

$$D \;=\; \frac{1}{h}\left(\Delta - \frac{\Delta^2}{2} + \frac{\Delta^3}{3} - \frac{\Delta^4}{4} + \cdots\right)$$

If we take into account the remainder term in the Gregory-Newton formula, then we can find a corresponding remainder term for the derivative.

2.41. Given the table of values for the function $f(x)$ in Fig. 2-30, find the approximate value of $f'(0.25)$.

x	0.25	0.35	0.45	0.55	0.65	0.75
$f(x)$	0.24740	0.34290	0.43497	0.52269	0.60518	0.68164

Fig. 2-30

We consider the difference table for $10^5 f(x)$ so as to avoid decimals. This table is as follows. The difference columns are headed $\Delta, \Delta^2, \ldots$ in place of $10^5\Delta f(x), 10^5\Delta^2 f(x), \ldots$ for brevity.

$10^5 f(x)$	Δ	Δ^2	Δ^3	Δ^4	Δ^5
24,740					
	9550				
34,290		−343			
	9207		−92		
43,497		−435		4	
	8772		−88		4
52,269		−523		8	
	8249		−80		
60,518		−603			
	7646				
68,164					

Fig. 2-31

Using the formula of Problem 2.40 we have

$$D[10^5 f(x)] \;=\; \frac{1}{h}\left(\Delta - \frac{\Delta^2}{2} + \frac{\Delta^3}{3} - \frac{\Delta^4}{4} + \cdots\right)[10^5 f(x)]$$

and neglecting differences of order 5 and higher we find for $x = 0.25$

$$10^5 f'(0.25) \;=\; \frac{1}{0.1}\left[9550 - \frac{(-343)}{2} + \frac{(-92)}{3} - \frac{(4)}{4}\right] \;=\; 96{,}898$$

or $f'(0.25) = 0.96898$.

The table of values are those for $f(x) = \sin x$ where x is in radians. From this we see that $f'(x) = \cos x$. By way of comparison the true value is $f'(0.25) = \cos 0.25 = 0.96891$.

2.42. Use the table of Fig. 2-31 to find $f''(0.25)$.

From Problem 2.40,

$$D \;=\; \frac{1}{h}\left(\Delta - \frac{\Delta^2}{2} + \frac{\Delta^3}{3} - \frac{\Delta^4}{4} + \cdots\right)$$

so that on squaring we find

$$D^2 \;=\; \frac{1}{h^2}\left(\Delta^2 - \Delta^3 + \frac{11}{12}\Delta^4 + \cdots\right)$$

Thus

$$D^2f(x) = \frac{1}{h^2}(\Delta^2 - \Delta^3 + \tfrac{11}{12}\Delta^4 + \cdots)f(x)$$

From this it follows that

$$10^5 f''(0.25) = \frac{1}{(0.1)^2}(-343 + 92 + 3.7) = -25{,}470$$

or $f''(0.25) = -0.25470$.

The exact value is $f''(0.25) = -\sin 0.25 = -0.24740$.

2.43. Referring to the table of Fig. 2-31 find $f'(0.40)$.

From the Gregory-Newton formula we have

$$f(a+kh) = f(a) + k\Delta f(a) + \frac{k(k-1)}{2!}\Delta^2 f(a) + \frac{k(k-1)(k-2)}{3!}\Delta^3 f(a) + \cdots$$

Differentiation with respect to k yields

$$hf'(a+kh) = \Delta f(a) + \frac{2k-1}{2!}\Delta^2 f(a) + \frac{3k^2-6k+2}{3!}\Delta^3 f(a) + \cdots + \frac{4k^3-18k^2+22k-6}{4!}$$

Letting $a = 0.25$, $h = 0.1$ and $a + kh = 0.40$ so that $k = 1.5$ we find using the difference table of Problem 2.41

$$10^5 f'(0.40) = \frac{1}{0.1}[9550 + (-343) - \tfrac{1}{24}(-92) + 0(4)]$$

$$f'(0.40) = 0.92108$$

The true value is $f'(0.40) = \cos 0.40 = 0.92106$.

MISCELLANEOUS PROBLEMS

2.44. The population of a particular country during various years is given in the following table. Obtain an estimate for the population in the year (*a*) 1980, (*b*) 1955 and (*c*) 1920.

Year	1930	1940	1950	1960	1970
Population in millions	1.0	1.2	1.6	2.8	5.4

Fig. 2-32

Let us denote the year by x and the population in millions by P. Since the years given are equidistant it is convenient to associate the years 1930, 1940, 1950, 1960, 1970 with the numbers $k = 0, 1, 2, 3, 4$ respectively. This is equivalent to writing $x = a + kh$ where $a = 1930$ and $h = 10$ so that

$$x = 1930 + 10k \quad \text{or} \quad k = \frac{x - 1930}{10} \qquad (1)$$

Using this we can rewrite the given table as follows.

k	0	1	2	3	4
P	1.0	1.2	1.6	2.8	5.4

Fig. 2-33

The difference table corresponding to this is as follows.

k	P	ΔP	$\Delta^2 P$	$\Delta^3 P$	$\Delta^4 P$
0	1.0				
		0.2			
1	1.2		0.2		
		0.4		0.6	
2	1.6		0.8		0
		1.2		0.6	
3	2.8		1.4		
		2.6			
4	5.4				

Fig. 2-34

We shall assume that the third differences are all constant [equal to 0.6] and that the fourth differences are zero. Then using the Gregory-Newton formula we find

$$\begin{aligned} P &= 1.0 + (0.2)k^{(1)} + \frac{0.2k^{(2)}}{2!} + \frac{0.6k^{(3)}}{3!} \\ &= 1.0 + 0.2k + 0.1k(k-1) + 0.1k(k-1)(k-2) \\ &= \tfrac{1}{10}(k^3 - 2k^2 + 3k + 10) \end{aligned}$$

(*a*) From (*1*) we see that the year 1980 corresponds to $k = 5$ so that

$$P = \tfrac{1}{10}[5^3 - 2(5)^2 + 3(5) + 10]$$

Thus we estimate that in 1980 the population will be 10 million.

(*b*) From (*1*) we see that the year 1955 corresponds to

$$k = \frac{1955 - 1930}{10} = 2.5$$

Then

$$P = \tfrac{1}{10}[(2.5)^3 - 2(2.5)^2 + 3(2.5) + 10] = 2.0625$$

Thus to two significant figures we would estimate that the population in 1955 was 2.1 million.

(*c*) The value of k corresponding to the year 1920 is from (*1*)

$$k = \frac{1920 - 1930}{10} = -1$$

Then

$$P = \tfrac{1}{10}[(-1)^3 - 2(-1)^2 + 3(-1) + 10] = 0.4$$

Thus we estimate that the population in 1920 was 0.4 million.

2.45. Suppose that it is necessary to determine some quantity y at the time when it is a maximum or minimum. Due to error we may not be able to do this exactly. However we observe that at times t_1, t_2, t_3 near the time for maximum or minimum, the values are given respectively by y_1, y_2, y_3. Prove that the time at which the quantity is a maximum or minimum is given by

$$\frac{y_1(t_2^2 - t_3^2) + y_2(t_3^2 - t_1^2) + y_3(t_1^2 - t_2^2)}{2[y_1(t_2 - t_3) + y_2(t_3 - t_1) + y_3(t_1 - t_2)]}$$

By Lagrange's formula the value of y at time t is given by

$$y = y_1\frac{(t-t_2)(t-t_3)}{(t_1-t_2)(t_1-t_3)} + y_2\frac{(t-t_1)(t-t_3)}{(t_2-t_1)(t_2-t_3)} + y_3\frac{(t-t_1)(t-t_2)}{(t_3-t_1)(t_3-t_2)} \qquad (1)$$

The maximum or minimum of y occurs where $dy/dt = 0$ [assuming it to be a relative maximum or minimum]. Then taking the derivation of (*1*) with respect to t, setting it equal to zero and solving for t we find the required value given above.

2.46. If an error is made in one of the entries of a table show how the various differences are affected.

Suppose that the value $y_0 + \epsilon$ is given in a table in place of the correct value y_0 so that ϵ is the error. Then the difference table can be constructed as in Fig. 2-35.

y_{-4}				
	Δy_{-4}			
y_{-3}		$\Delta^2 y_{-4}$		
	Δy_{-3}		$\Delta^3 y_{-4}$	
y_{-2}		$\Delta^2 y_{-3}$		$\Delta^4 y_{-4} + \epsilon$
	Δy_{-2}		$\Delta^3 y_{-3} + \epsilon$	
y_{-1}		$\Delta^2 y_{-2} + \epsilon$		$\Delta^4 y_{-3} - 4\epsilon$
	$\Delta y_{-1} + \epsilon$		$\Delta^3 y_{-2} - 3\epsilon$	
$y_0 + \epsilon$		$\Delta^2 y_{-1} - 2\epsilon$		$\Delta^4 y_{-2} + 6\epsilon$
	$\Delta y_0 - \epsilon$		$\Delta^3 y_{-1} + 3\epsilon$	
y_1		$\Delta^2 y_0 + \epsilon$		$\Delta^4 y_{-1} - 4\epsilon$
	Δy_1		$\Delta^3 y_0 - \epsilon$	
y_2		$\Delta^2 y_1$		$\Delta^4 y_0 + \epsilon$
	Δy_2		$\Delta^3 y_1$	
y_3		$\Delta^2 y_2$		
	Δy_3			
y_4				

Fig. 2-35

Note that the error is propagated into the higher ordered differences as indicated by the shading in Fig. 2-35. It is of interest to note that the successive coefficients of ϵ in each column alternate in sign and are given numerically by the binomial coefficients. Also the largest absolute error occurring in an even difference column lies on the same horizontal line in which y_0 occurs.

2.47. In reading a table the entries given in Fig. 2-36 were obtained. (*a*) Show that an error was made in one of the entries and (*b*) correct the error.

x	1	2	3	4	5	6	7	8	9	10	11	12
$f(x)$	.195520	.289418	.379426	.464642	.544281	.617356	.683327	.741471	.791207	.832039	.863558	.885450

Fig. 2-36

(*a*) The difference table omitting decimal points is as follows.

$10^6 f(x)$	Δ	Δ^2	Δ^3	Δ^4	Δ^5	Δ^6
195,520						
	93,898					
289,418		−3890				
	90,008		−902			
379,426		−4792		217		
	85,216		−785		−419	
464,642		−5577		−202		1268
	79,639		−987		649	
544,281		−6564		447		−1279
	73,075		−540		−630	
617,356		−7104		−183		955
	65,971		−723		325	
683,327		−7827		142		−382
	58,144		−581		−57	
741,471		−8408		85		59
	49,736		−496		2	
791,207		−8904		87		6
	40,832		−409		8	
832,039		−9313		95		
	31,519		−314			
863,558		−9627				
	21,892					
885,450						

Fig. 2-37

The violent oscillations of sign which occur near the beginning to the column of fourth differences Δ^4 and the practically constant values toward the end of this column indicate an error near the beginning of the table. This is further supported in the column of sixth differences Δ^6. In view of the fact that the largest absolute error in an even difference column is in the same horizontal line as the entry in which the error occurs, it would seem that the error should occur in the entry corresponding to -1279 which is largest numerically. It thus appears that the error should be in the entry 0.544281 corresponding to $x = 5$.

(b) Denoting the entry which is in possible error, i.e. 0.544281, by $y_0 + \epsilon$ where ϵ is the error, it is clear from comparison with the difference table of Problem 2.46 that

$$\begin{aligned} \Delta^4 y_{-4} + \epsilon &= 217 \\ \Delta^4 y_{-3} - 4\epsilon &= -202 \\ \Delta^4 y_{-2} + 6\epsilon &= 447 \\ \Delta^4 y_{-1} - 4\epsilon &= -183 \\ \Delta^4 y_0 + \epsilon &= 142 \end{aligned}$$

Assuming that the correct fourth differences should all be constant, i.e. $\Delta^4 y_{-4} = \Delta^4 y_{-3} = \Delta^4 y_{-2} = \Delta^4 y_{-1} = \Delta^4 y_{-0}$, we obtain by subtracting the first two equations, the second and third equation, etc. the equations

$$5\epsilon = 419 \text{ or } \epsilon = 84, \quad 10\epsilon = 649 \text{ or } \epsilon = 65, \quad 10\epsilon = 630 \text{ or } \epsilon = 63, \quad 5\epsilon = 325 \text{ or } \epsilon = 65$$

Because most of these indicate the approximate value $\epsilon = 65$ [the value $\epsilon = 84$ seems out of line] we shall consider this as the most probable value.

Taking into account the decimal point, this leads to $\epsilon = 0.000065$ and so since $y_0 + \epsilon = 0.544281$, the true value should be close to $y_0 = 0.544281 - 0.000065 = 0.544216$.

It is of interest to note that the true value was actually 0.544218. It appears that in copying the table the 1 and the 8 in this true value must have been interchanged.

In Fig. 2-38 the difference table taking into account this error is shown.

$10^6 f(x)$	Δ	Δ^2	Δ^3	Δ^4	Δ^5
195,520					
	93,898				
289,418		−3890			
	90,008		−902		
379,426		−4792		54	
	85,216		−848		−4
464,642		−5640		50	
	79,576		−798		19
544,218		−6438		69	
	73,138		−729		0
617,356		−7167		69	
	65,971		−660		10
683,327		−7827		79	
	58,144		−581		6
741,471		−8408		85	
	49,736		−496		2
791,207		−8904		87	
	40,832		−409		−2
832,039		−9313		85	
	31,519		−314		
863,558		−9627			
	21,892				
885,450					

Fig. 2-38

The changes of sign occurring in the fifth differences could be due to *rounding errors* as well as the fact that the function is not actually a polynomial.

2.48. Using the following table find the value of y corresponding to $x = 0.6$.

x	0	0.1	0.2	0.3	0.4	0.5
y	1.250	1.610	2.173	3.069	4.539	7.029

Fig. 2-39

On constructing a difference table from the given data we obtain that shown in Fig. 2-40 where we use 10^3y in place of y to avoid decimals.

x	10^3y	Δ	Δ^2	Δ^3	Δ^4	Δ^5
0	1250					
		360				
0.1	1610		203			
		563		130		
0.2	2173		333		111	
		896		241		94
0.3	3069		574		205	
		1470		446		
0.4	4539		1020			
		2490				
0.5	7029					

Fig. 2-40

Letting $a = 0$, $h = \Delta x = 0.1$ in the Gregory-Newton formula *(6)* on page 34 we have

$$10^3y \;=\; 1250 + \frac{360}{0.1}x^{(1)} + \frac{203}{2!\,(0.1)^2}x^{(2)} + \frac{130}{3!\,(0.1)^3}x^{(3)} + \frac{111}{4!\,(0.1)^4}x^{(4)} + \frac{94}{5!\,(0.1)^5}x^{(5)}$$

or

$$y \;=\; 1.250 + 3.60x^{(1)} + 10.15x^{(2)} + 21.67x^{(3)} + 46.25x^{(4)} + 78.33x^{(5)}$$

Then the value of y corresponding to $x = 0.6$ is

$$\begin{aligned} y &= 1.250 + 3.60(0.6) + 10.15(0.6)(0.5) + 21.67(0.6)(0.5)(0.4) + 46.25(0.6)(0.5)(0.4)(0.3) \\ &\qquad + 78.33(0.6)(0.5)(0.4)(0.3)(0.2) \\ &= 11.283 \end{aligned}$$

2.49. (*a*) Work Problem 2.48 by first obtaining $\log_{10} y$ in place of y. (*b*) Use the result in (*a*) to arrive at a formula for y.

(*a*) From the table of Fig. 2-39 we obtain the following table.

x	0	0.1	0.2	0.3	0.4	0.5
$\log_{10} y$	0.0969	0.2068	0.3371	0.4870	0.6570	0.8469

Fig. 2-41

The difference table corresponding to this table is as follows.

x	$10^4 \log_{10} y$	Δ	Δ^2	Δ^3	Δ^4	Δ^5
0	969					
		1099				
0.1	2068		204			
		1303		−8		
0.2	3371		196		3	
		1499		−5		0
0.3	4870		201		3	
		1700		−2		
0.4	6570		199			
		1899				
0.5	8469					

Fig. 2-42

Then we find by the Gregory-Newton formula

$$10^4 \log_{10} y \;=\; 969 + \frac{1099}{0.1}x^{(1)} + \frac{204}{2!\,(0.1)^2}x^{(2)} - \frac{8}{3!\,(0.1)^3}x^{(3)} + \frac{3}{4!\,(0.1)^4}x^{(4)}$$

or

$$\log_{10} y \;=\; 0.0969 + 1.099x^{(1)} + 1.02x^{(2)} - 0.1333x^{(3)} + 0.1250x^{(4)} \qquad (1)$$

Putting $x = 0.6$ we then obtain

$$\begin{aligned}\log_{10} y &= 0.0969 + 1.099(0.6) + 1.02(0.6)(0.5) - 0.1333(0.6)(0.5)(0.4) + 0.1250(0.6)(0.5)(0.4)(0.3)\\ &= 1.0508\end{aligned}$$

Thus

$$y \;=\; 10^{1.0508} \;=\; 11.24$$

(*b*) The result (*1*) can be written as

$$\begin{aligned}\log_{10} y \;=\; &0.0969 + 1.099x + 1.02x(x-0.1) - 0.1333x(x-0.1)(x-0.2)\\ &+ 0.1250x(x-0.1)(x-0.2)(x-0.3)\end{aligned}$$

To a high degree of approximation this can be written as

$$\log_{10} y \;=\; 0.0969 + 1.099x + 1.02x(x-0.1) \;=\; 0.0969 + 1.02x^2 + 0.997x$$

Thus

$$\begin{aligned}y &= 10^{0.0969+1.02x^2+0.997x} \;=\; 10^{0.0969}\cdot 10^{1.02x^2+0.997x}\\ &= 1.250\cdot 10^{1.02x^2+0.997x}\end{aligned} \qquad (2)$$

The values corresponding to the table of Fig. 2-39 were in fact obtained by using the formula

$$y \;=\; 1.250\cdot 10^{x^2+x} \qquad (3)$$

Thus the exact value corresponding to $x = 0.6$ is 11.40. It is of interest that the value obtained in Problem 2.48 is closer to this true value than that obtained in part (*a*) using logarithms. This can be accounted for by realizing that in the process of taking logarithms and antilogarithms we introduce small errors each time. Further evidence of this is seen by comparing (*2*) and (*3*). By recognizing that such errors do occur however we could be led from (*2*) to (*3*) on replacing 1.02 by 1 and 0.997 by 1. Then (*3*) could be recognized as the true formula by in fact putting $x = 0, 0.1, \ldots, 0.5$ and comparing with the corresponding values of the original table.

Supplementary Problems

SUBSCRIPT NOTATION

2.50. Show that (*a*) $\Delta(y_k - z_k) = \Delta y_k - \Delta z_k$

(*b*) $\Delta(y_k/z_k) = (z_k\Delta y_k - y_k\Delta z_k)/z_k z_{k+1}$

(*c*) $\Delta(ay_k + bz_k) = a\Delta y_k + b\Delta z_k$

where a, b are any constants.

2.51. Show that (*a*) $y_{k+3} = y_k + 3\Delta y_k + 3\Delta^2 y_k + \Delta^3 y_k$

(*b*) $\Delta^3 y_k = y_{k+3} - 3y_{k+2} + 3y_{k+1} - y_k$

2.52. Generalize the results of Problem 2.51.

2.53. Show that (*a*) $\Delta k^{(4)} = 4k^{(3)}$, (*b*) $\Delta^2[2k^{(3)} - 3k^{(5)}] = 12k - 60k^{(3)}$.

2.54. If $y_k = k^3 - 4k^2 + 2k - 3$ (*a*) express y_k as a factorial polynomial and find (*b*) Δy_k, (*c*) $\Delta^2 y_k$, (*d*) $\Delta^3 y_k$, (*e*) $\Delta^4 y_k$.

2.55. If D_k denotes differentiation with respect to k, show that

(a) $\Delta_k = D_k + \frac{1}{2}D_k^2 + \frac{1}{6}D_k^3 + \frac{1}{24}D_k^4 + \cdots$

(b) $\Delta_k^2 = D_k^2 + D_k^3 + \frac{7}{12}D_k^4 + \frac{1}{4}D_k^5 + \cdots$

(c) $\Delta_k^3 = D_k^3 + \frac{3}{2}D_k^4 + \frac{5}{4}D_k^5 + \cdots$

2.56. Illustrate Problem 2.55 by finding $\Delta_k^3(k^5 - 3k^4 + 2k^3)$ and check your answer.

DIFFERENCE TABLES

2.57. Let $f(x) = 5x^2 + 3x - 7$ where $x = 2, 5, 8, 11, 14$. Set up a difference table for $f(x)$.

2.58. By using only the difference table obtained in Problem 2.57, find the values of $y = f(x)$ for $x = 17, 20, 23, 26, 29, 32$.

2.59. Let $f(x) = 2x^3 - 8x^2 + 6x - 10$ where $x = 2, 4, 6, 8, 10$. (a) Set up a difference table for $f(x)$ and (b) deduce the values of $f(x)$ for $x = 12, 14, 16, 18, 20$ from this difference table.

2.60. (a) Suppose that in Problem 2.57 the difference table had been set up for values $x = 2, 5, 8, 11$. Would the resulting table enable us to obtain the values of $y = f(x)$ for $x = 14, 17, 20, 23$? Explain. (b) Answer part (a) if the difference table had been set up for values $x = 2, 5, 8$ and if the values for $x = 11, 14, 17, 20$ were wanted.

2.61. Suppose that it is desired to obtain the values of $f(x) = 3x^4 - 2x^2 + 7x + 5$ for $x = 3, 7, 11, 15, 19, \ldots, 43$. What is the minimum number of values which we must obtain directly from $f(x)$ so that from the resulting difference table we can determine the remaining values of $f(x)$?

2.62. Generalize the results of Problem 2.61 to any polynomial $f(x)$ and any set of equidistant values of x.

2.63. Find the numerical values corresponding to the letters in the following difference table where it is assumed that there is equal spacing of the independent variable x.

y	Δy	$\Delta^2 y$	$\Delta^3 y$	$\Delta^4 y$
A				
	E			
B		H		
	2		-1	
C		J		L
	F		K	
-1		4		
	G			
D				

Fig. 2-43

2.64. A difference table has one entry in the column $\Delta^6 y$. Determine how many entries there would be in the column (a) Δy, (b) y, (c) $\Delta^2 y$.

2.65. Generalize the result of Problem 2.64 by determining how many entries there would be in the column $\Delta^m y$ if there were a entries in the column $\Delta^n y$.

THE GREGORY-NEWTON FORMULA AND APPLICATIONS

2.66. Prove that (a) $(kh)^{(3)} = h^3 k^{(3)}$, (b) $(kh)^{(4)} = h^4 k^{(4)}$.

2.67. Prove that $(kh)^{(n)} = h^n k^{(n)}$ by using mathematical induction and thus complete the proof in Problem 2.6.

2.68. Find a formula for the general term u_k of a sequence whose first few terms are (a) $5, 12, 25, 44$, (b) $1, 5, 14, 30, 55$. What assumptions must you make? Are the results unique? Explain.

2.69. Given a sequence whose first four terms are $3, 7, 14, 30$. (a) Can you obtain a formula for the general term of the sequence? (b) Can you find the 10th term of this sequence? State clearly the assumptions made and whether you feel that results obtained are unique.

2.70. Suppose that in Problem 2.69 the next three terms of the sequence are given as $61, 113, 192$. What effect if any would this additional information have in your answer to that problem?

2.71. Find y as a polynomial function in x which fits the data in the following table and state clearly the assumptions which you make in determining this polynomial.

x	1	2	3	4	5	6
y	3	3	7	21	51	103

Fig. 2-44

2.72. Suppose that the values of x corresponding to those of y in the table of Fig. 2-44 are $1, 3, 5, 7, 9, 11$. Explain how you might obtain y as a polynomial function in x from the answer to Problem 2.71. Can you generalize the result of your conclusions?

INTERPOLATION AND EXTRAPOLATION USING THE GREGORY-NEWTON FORMULA

2.73. Use the table of Problem 2.71 to find the value of y corresponding to (a) $x = 3.5$, (b) $x = 5.6$, (c) $x = 0$, (d) $x = -2$, (e) $x = 8$. What assumptions are you making in these cases?

2.74. Use the following table to find $\sin 45°$ and compare with the true value.

x	0°	30°	60°	90°
$\sin x$	0.00000	0.50000	0.86603	1.00000

Fig. 2-45

2.75. Estimate the error term in Problem 2.74.

2.76. Use the following table to find (a) $\log_{10} 5.2374$, (b) $\log_{10} 5.4933$.

x	5.1	5.2	5.3	5.4	5.5
$\log_{10} x$	0.70757	0.71600	0.72428	0.73239	0.74036

Fig. 2-46

2.77. Use the following table to find (a) $\sin 61°24'$, (b) $\sin 63°48'$.

x	60°	61°	62°	63°	64°	65°
$\sin x$	0.86603	0.87462	0.88295	0.89101	0.89879	0.90631

Fig. 2-47

2.78. Estimate the error terms in (a) Problem 2.76 and (b) Problem 2.77.

2.79. Use the table of Problem 2.74 to find (a) $\sin 68°30'$, (b) $\sin 45°18'$ and compare with the exact values.

2.80. (a) Use the following table to find $\log_{10} \pi$ where we take $\pi = 3.14159$. (b) Give an estimate of the error term.

x	3.12	3.13	3.14	3.15	3.16
$\log_{10} x$	0.49415	0.49554	0.49693	0.49831	0.49969

Fig. 2-48

2.81. Determine if there is any difference to your answer for Problem 2.80 if we add to the table the values of $\log_{10} x$ for $x = 3.11$ and $x = 3.17$ given by 0.49276 and 0.50106 respectively.

2.82. Given the following table find (*a*) $e^{0.243}$, (*b*) $e^{0.411}$.

x	0.1	0.2	0.3	0.4	0.5
e^x	1.10517	1.22140	1.34986	1.49182	1.64872

Fig. 2-49

2.83. The *gamma function* denoted by $\Gamma(x)$ is given for various values of x in the following table. Find $\Gamma(1.1673)$.

x	1.15	1.16	1.17	1.18	1.19
$\Gamma(x)$	0.93304	0.92980	0.92670	0.92373	0.92088

Fig. 2-50

2.84. The *error function*, denoted by erf (x), is given for various values of x in the following table. Find (*a*) erf (0.28), (*b*) erf (0.63), (*c*) erf (1).

x	0.00	0.25	0.50	0.75
erf (x)	0.00000	0.27633	0.52049	0.71116

Fig. 2-51

CENTRAL DIFFERENCE TABLES AND INTERPOLATION FORMULAS

2.85. (*a*) Construct a central difference table corresponding to the data of Problem 2.76 by choosing $a = 5.2$. (*b*) Use Stirling's formula to work Problem 2.76(*a*) and compare with that obtained by use of the Gregory-Newton forward difference formula.

2.86. Work (*a*) Problem 2.77, (*b*) Problem 2.80 by using Stirling's formula.

2.87. Work (*a*) Problem 2.76, (*b*) Problem 2.77, (*c*) Problem 2.80 by using the Gregory-Newton backward difference formula.

2.88. Work (*a*) Problem 2.76, (*b*) Problem 2.77, (*c*) Problem 2.80 by using Bessel's interpolation formula.

2.89. Show how a central difference table can be expressed using the backward difference operator $\nabla, \nabla^2, \nabla^3, \ldots$.

2.90. Write the interpolation formulas on page 36 using the "x notation" in place of the "k notation".

2.91. Show how to write Bessel's interpolation formula using (*a*) central differences and (*b*) backward differences.

2.92. Work Problem 2.83 by using Stirling's formula with $a = 1.17$ and compare with that obtained by using the Gregory-Newton forward difference formula.

2.93. Use Gauss' interpolation formula to work (*a*) Problem 2.76, (*b*) Problem 2.77, (*c*) Problem 2.80, (*d*) Problem 2.83.

2.94. Derive the second Gauss formula and thus complete Problem 2.18.

ZIG-ZAG PATHS AND LOZENGE DIAGRAMS

2.95. Demonstrate the rules for obtaining interpolation formulas [see page 37] by obtaining (*a*) the Gregory-Newton forward difference formula, (*b*) the Gregory-Newton backward difference formula, (*c*) Gauss' first formula on page 36, (*d*) Stirling's formula. Indicate the zig-zag path in the lozenge diagram which corresponds to each formula.

2.96. By choosing an appropriate zig-zag path in Fig. 2-6, page 37, obtain an interpolation formula differing from those on page 36.

2.97. Explain whether you could find the error term in (*a*) the Gauss formulas and (*b*) the Stirling formula by knowing the error in the Gregory-Newton forward difference formula.

LAGRANGE'S INTERPOLATION FORMULA

2.98. Work (*a*) Problem 2.71, (*b*) Problem 2.80 by using Lagrange's interpolation formula.

2.99. (*a*) Find a polynomial in x which fits the data in the following table and use the result to find the values of y corresponding to (*b*) $x = 3$, (*c*) $x = 5$, (*d*) $x = 8$.

x	1	2	4	6	7
y	6	16	102	376	576

Fig. 2-52

2.100. (*a*) The first, third, fourth and sixth terms of a sequence are given by 1, 7, 25 and 211 respectively. Determine a possible formula for the nth term of the sequence and discuss the uniqueness of the result. (*b*) Obtain the second and fifth terms of the sequence.

2.101. (*a*) Obtain a polynomial approximation to $\sin x$ where x is in radians by using the following table.

x	0 (0°)	$\pi/6$ (30°)	$\pi/4$ (45°)	$\pi/3$ (60°)	$\pi/2$ (90°)
$\sin x$	0.00000	0.50000	0.70711	0.86603	1.00000

Fig. 2-53

(*b*) Use the result of (*a*) to find $\sin 15°$, $\sin 75°$ and $\sin 40°20'$ comparing results obtained with the exact values. Give possible reasons for errors obtained.

2.102. Show that Lagrange's interpolation formula reduces to the Gregory-Newton forward difference formula in the case where the x values are equally spaced.

TABLES WITH MISSING ENTRIES

2.103. Find the value u_2 missing from the following table.

k	1	2	3	4	5
u_k	2		40	83	150

Fig. 2-54

2.104. Find the values u_5 and u_8 missing from the following table.

k	4	5	6	7	8	9
u_k	72		146	192		302

Fig. 2-55

2.105. (a) Use the data of Problem 2.101 to find the missing entries for $\sin x$ corresponding to $x = 15°$ and $x = 75°$ and compare with the exact values. (b) Is there any difference between the answers obtained from the method of part (a) and those using the method of Problem 2.101(b)? Explain.

2.106. (a) Find the values of y corresponding to $x = 3$ and $x = 5$ in the table of Problem 2.99 using Method 1 of Problem 2.28. (b) Can you use the method of part (a) to find the value of y corresponding to $x = 8$? If so what is this value?

2.107. Find the values u_1, u_2, u_3, u_4 from the following table.

k	0	5	10	15
u_k	5	27	179	611

Fig. 2-56

2.108. Find the values u_3, u_4, u_5, u_6 from the following table.

k	0	5	10	15	20
u_k	2		92	212	382

Fig. 2-57

2.109. A sequence is given by $u_0, u_1, u_2, u_3, u_4, u_5$ in which all values but u_3 are known. Show that if fourth differences are assumed constant then

$$u_3 = 0.1u_0 - 0.5u_1 + u_2 + 0.5u_4 - 0.1u_5$$

2.110. The following table shows the population of the United States for the years 1930 to 1970 at intervals of 10 years. Use the table to find the population for the years (a) 1980 and (b) 1962, 1964, 1966 and 1968.

Year	1930	1940	1950	1960	1970
Population of United States (millions)	122.8	131.7	151.1	179.3	205.2

Fig. 2-58

2.111. The following table shows the values of e^{-x} for $x = 0.1, 0.2, 0.3, 0.4, 0.5$. Use this table to find the values of e^{-x} for $x = 0.025, 0.050$ and 0.075 and compare with the exact values.

x	0	0.1	0.2	0.3	0.4
e^{-x}	1.00000	0.90484	0.81873	0.74082	0.67032

Fig. 2-59

2.112. Suppose that in the table of Problem 2.82 the entries for e^x are replaced by corresponding ones for e^{-x}. Determine if there is any change in the values obtained and explain.

DIVIDED DIFFERENCES

2.113. (a) Construct a divided difference table corresponding to the data in the following table.

x	3	5	8	10	11
$f(x)$	9	35	119	205	257

Fig. 2-60

(b) Use Newton's divided difference interpolation formula to obtain a polynomial in x which fits the data.

(c) Find values of $f(x)$ for $x = 4$ and 9.

2.114. (*a*) Find a polynomial in x which fits the data in the following table and (*b*) obtain values of $f(x)$ corresponding to $x=5$ and $x=10$.

x	0	2	3	6	7	8
$f(x)$	40	8	7	136	243	392

Fig. 2-61

2.115. Work Problem 2.28 using Newton's divided difference interpolation formula.

2.116. Show how to derive Newton's divided difference formula by assuming an expansion of the form

$$f(x) \;=\; A_0 + A_1(x-x_0) + A_2(x-x_0)(x-x_1) + \cdots + A_n(x-x_0)(x-x_1)\cdots(x-x_n)$$

and then determining the constants $A_0, A_1, A_2, \ldots, A_n$.

2.117. Obtain the Gregory-Newton formula as a special case of Newton's divided difference formula.

2.118. Prove that the value of $f(x_0, x_1, \ldots, x_n)$ remains the same regardless of the arrangement of $x_0, x_1, \ldots, x_n$, i.e. $f(x_0, x_1, \ldots, x_n) = f(x_n, x_0, x_1, \ldots, x_{n-1})$, etc.

2.119. Prove that the divided differences of $f(x)+g(x)$ are equal to the sum of the divided differences of $f(x)$ and $g(x)$.

2.120. Prove that the divided differences of a constant times $f(x)$ is equal to the constant times the divided differences of $f(x)$.

2.121. (*a*) Show that there is a root of $x^3-x-1=0$ between $x=1$ and $x=2$. (*b*) Find the root in (*a*). (*c*) Determine the remaining roots of the equation.

2.122. Solve the equation $x^3+2x^2+10x-20=0$.

2.123 Determine the roots of $x^3-3x^2+2x-5=0$.

2.124. Solve the equation $e^{-x}=x$. [*Hint.* Let $y=f(x)=e^{-x}-x$.]

2.125. Solve the equation $x=5(1-e^{-x})$.

INVERSE INTERPOLATION

2.126. Use the data in Problem 2.71 to find the value of x corresponding to $y=f(x)=15$ by using (*a*) Lagrange's formula, (*b*) Newton's divided difference formula and (*c*) the Gregory-Newton forward difference formula. Compare your results with the exact value.

2.127. The amount of a 10 year annuity in which the semi-annual payment is 1 is given for different interest rates in the following table. Determine the interest rate if the amount of the annuity is 25.

Interest rate	$1\frac{1}{2}\%$	2%	$2\frac{1}{2}\%$	3%
Amount of annuity	23.1237	24.2974	25.5447	26.8704

Fig. 2-62

2.128. Given the data in the following table find the value of x which corresponds to $y = 10$.

x	2	4	6	8	10
y	7	15	11	14	12

Fig. 2-63

2.129. The following table gives values of $f(x)$ for certain values of x. (a) Find the value of x which corresponds to $f(x) = 2.2000$. (b) Using the fact that the entries of the table were actually obtained from $f(x) = 10 \sin \frac{1}{10}(x+2)$ where x is in radians, determine the exact value of x corresponding to $f(x) = 2.2000$ and thus determine the accuracy of the result obtained in (a).

x	0.1	0.2	0.3	0.4	0.5
$f(x)$	2.0846	2.1823	2.2798	2.3770	2.4740

Fig. 2-64

2.130. Work Problem 2.29 by using inverse interpolation with Lagrange's formula.

APPROXIMATE DIFFERENTIATION

2.131. (a) Use the table of Problem 2.76 to find the derivative of $\log_{10} x$ for $x = 5.2$ and (b) compare with the exact value.

2.132. Use the table of Problem 2.77 to find (a) the first derivative and (b) the second derivative of $\sin x$ corresponding to $x = 63°$ and compare with the exact values.

2.133. Find the (a) first derivative and (b) second derivative of e^x at $x = 0.2$ using the table of Problem 2.82 and compare with the exact values.

2.134. Find the derivative of the gamma function of Problem 2.83 for (a) $x = 1.16$, (b) 1.174.

2.135. Find the derivative of the error function of Problem 2.84 for (a) $x = 0.25$, (b) $x = 0.45$.

2.136. Show that (a) $D = \frac{1}{h}\left(\nabla + \frac{1}{2}\nabla^2 + \frac{1}{3}\nabla^3 + \frac{1}{4}\nabla^4 + \cdots\right)$

(b) $D = \frac{1}{h}\left(\delta - \frac{1}{24}\delta^3 + \frac{3}{640}\delta^5 - \frac{5}{7168}\delta^7 + \cdots\right)$

2.137. Show that (a) $D^2 = \frac{1}{h^2}\left(\nabla^2 + \nabla^3 + \frac{11}{12}\nabla^4 + \frac{5}{6}\nabla^5 + \frac{137}{180}\nabla^6 + \cdots\right)$

(b) $D^2 = \frac{1}{h^2}\left(\delta^2 - \frac{1}{12}\delta^4 + \frac{1}{90}\delta^6 - \frac{1}{560}\delta^8 + \cdots\right)$

MISCELLANEOUS PROBLEMS

2.138. (a) Is it possible to determine unique numerical values corresponding to the letters in the following difference table where it is assumed that there is equal spacing of the independent variable x? (b) Does your answer to (a) conflict with the remark made at the end of Problem 2.3?

y	Δy	$\Delta^2 y$	$\Delta^3 y$	$\Delta^4 y$
3				
	D			
A		-1		
	E		K	
B		H		5
	F		2	
C		J		
	G			
2				

Fig. 2-65

2.139. Can you formulate a rule which tells the minimum number of entries in a difference table which must be known before all entries can be uniquely determined?

2.140. A table of cube roots showed the following entries. Find $\sqrt[3]{55}$ from this table and compare with the exact value.

x	50	52	54	56	58	60
$\sqrt[3]{x}$	3.6840	3.7325	3.7798	3.8259	3.8709	3.9149

Fig. 2-66

2.141. If a unit amount of money [the *principal*] is deposited at 6% interest compounded semi-annually and not withdrawn, the total amount A_k accumulated at the end of k years is given in the following table. Determine the amount accumulated after (*a*) 9 years, (*b*) 16 years.

k	6	8	10	12	14
A_k	1.4258	1.6047	1.8061	2.0328	2.2879

Fig. 2-67

2.142. Referring to Problem 2.141 after how many years would one expect the money to double?

2.143. Various values of the *Bessel function of zero order* denoted by $J_0(x)$ are given in the following table. (*a*) Find $J_0(0.5)$. (*b*) Determine the value of x for which $J_0(x) = 0.9000$.

x	0	0.2	0.4	0.6	0.8	1.0
$J_0(x)$	1.0000	0.9900	0.9604	0.9120	0.8463	0.7652

Fig. 2-68

2.144. A unit amount of money deposited at the end of each year for n years at an interest rate of r% compounded annually is called an *annuity*. The following table shows the amount of an annuity accumulated after 20 years at various interest rates. Use the table to determine the amount of an annuity after 20 years if the interest rate is $1\frac{1}{4}$%.

Interest rate r	1	$1\frac{1}{2}$	2	$2\frac{1}{2}$	3
Amount of annuity	22.0190	23.1237	24.2974	25.5447	26.8704

Fig. 2-69

2.145. A man takes out an annuity [see Problem 2.144] with an insurance company who agrees to pay back after 20 years an amount equal to 25 times the yearly payment. Determine the interest rate compounded annually which is involved.

2.146. According to actuarial tables the expectancy of life at age x, given by $f(x)$, is as shown in the following table. Determine the life expectancy of a person aged (*a*) 25 and (*b*) 40.

x	10	20	30	40	50	60	70	80
$f(x)$	50.3	42.8	35.6	29.4	22.3	15.5	10.9	6.8

Fig. 2-70

2.147. (*a*) Give a possible explanation for the fact that the differences in the table of Problem 2.146 tend to decrease and then increase. (*b*) Does the result produce an error in interpolation? Explain. (*c*) How might more accurate results be obtained?

2.148. The following table gives for various values of x corresponding values of $\ln\Gamma(x)$ where $\Gamma(x)$ is the *gamma function* [see page 70]. From the known fact that

$$\Gamma(x)\,\Gamma(1-x) = \frac{\pi}{\sin x\pi}$$

obtain the values of $\ln\Gamma(x)$ corresponding to the missing values.

x	2/12	3/12	4/12	5/12	6/12	7/12	8/12	9/12	10/12
$\ln\Gamma(x)$		0.55938	0.42796		0.24858				

Fig. 2-71

2.149 From the theory of the gamma function we have the result

$$2^{2x-1}\Gamma(x)\,\Gamma(x+\tfrac{1}{2}) = \sqrt{\pi}\,\Gamma(2x)$$

Can this be used in Problem 2.148 to obtain greater accuracy in the values obtained for $\Gamma(x)$? Explain.

2.150. Use Problems 2.148 and 2.149 to find values of $\ln\Gamma(x)$ for $x = 0.1, 0.2, \ldots, 0.9$.

2.151. Find a function $f(x)$ which fits the data in the following table. [*Hint.* Consider $\ln f(x)$.]

x	0.1	0.2	0.3	0.4	0.5	0.6
$f(x)$	0.3423	0.4080	0.5083	0.6616	0.9000	1.2790

Fig. 2-72

2.152. Complete the solution to Problem 2.45 by actually setting the derivative of (*1*) with respect to t equal to zero.

2.153. Suppose that in Problem 2.45 the times in seconds are given by $t_1 = 7.2$, $t_2 = 8.1$, $t_3 = 8.4$ and that the quantity y at these times has values 15.6, 15.9, 15.7 respectively. Find the time at which y is a maximum and the value of this maximum.

2.154. At times $t_1, t_2, \ldots, t_n$ measured from some fixed time as origin an object is observed to be at positions $y_1, y_2, \ldots, y_n$ on a straight line measured from some fixed position as origin. If y is the position of the object at time t show that

$$\frac{dy}{dt} = \frac{(t_2t_3\cdots t_n)(t_2^{-1}+t_3^{-1}+\cdots+t_n^{-1})}{(t_1-t_2)(t_1-t_3)\cdots(t_1-t_n)} + \text{similar terms}$$

2.155. Show that Stirling's interpolation formula can be written in the form

$$\begin{aligned} y_k = {} & y_0 + \tfrac{1}{2}(\Delta y_{-1}+\Delta y_0)k + \Delta^2 y_{-1}\frac{k^2}{2!} + \tfrac{1}{2}(\Delta^3 y_{-2}+\Delta^3 y_{-1})\frac{k(k^2-1)}{3!} \\ & + \Delta^4 y_{-2}\frac{k^2(k^2-1)}{4!} + \cdots + \tfrac{1}{2}(\Delta^{2n-1}y_{-n}+\Delta^{2n-1}y_{-n+1})\frac{k(k^2-1^2)(k^2-2^2)\cdots[k^2-(n-1)^2]}{(2n-1)!} \\ & + \Delta^{2n}y_{-n}\frac{k^2(k^2-1^2)(k^2-2^2)\cdots[k^2-(n-1)^2]}{(2n)!} + \cdots \end{aligned}$$

2.156. Referring to Problem 2.44 what estimate of the population would you give for the year 1910? Explain the significance of your result and give possible limitations on the reliability of interpolation and extrapolation.

2.157. Find the sum of the series

$$1^2 + 2^2 + 3^2 + \cdots + k^2$$

[*Hint.* Find a general term for the sequence $1^2, 1^2+2^2, 1^2+2^2+3^2, \ldots$.]

2.158. Find the sum of the series (*a*) $1+2+3+\cdots+k$, (*b*) $1^3+2^3+3^3+\cdots+k^3$.

2.159. Find the sum of the first k terms of the series

$$2^2 + 5^2 + 8^2 + 11^2 + \cdots$$

2.160. (*a*) Find a formula for the general term of the sequence $2, 5, 14, 31, 60, 109, 194, 347, \ldots$ and (*b*) use the result to find the 10th term.

2.161. Prove that $2^k = 1 + \frac{k^{(1)}}{1!} + \frac{k^{(2)}}{2!} + \frac{k^{(3)}}{3!} + \cdots$ for $k = 0, 1, 2, \ldots$.

2.162. Let $\phi_{k,0} = \begin{cases} 1 & k=0 \\ 0 & k \neq 0 \end{cases}$. Show that for $k = 0, 1, 2, \ldots$,

$$\phi_{k,0} = 1 - \frac{k^{(1)}}{1!} + \frac{k^{(2)}}{2!} - \frac{k^{(3)}}{3!} + \cdots$$

2.163. Prove *Everett's formula*

$$y_k = y_1 \frac{k^{(1)}}{1!} + \Delta^2 y_0 \frac{(k+1)^{(3)}}{3!} + \Delta^2 y_{-1} \frac{(k+2)^{(5)}}{5!} + \cdots$$

$$- y_0 \frac{(k-1)^{(1)}}{1!} - \Delta^2 y_{-1} \frac{k^{(3)}}{3!} - \Delta^4 y_{-2} \frac{(k+1)^{(5)}}{5!} - \cdots$$

An advantage of this formula is that only even order differences are present.

2.164. Prove that

$$y_k = \frac{k(k-1)\cdots(k-n+1)}{(n-1)!}\left[\frac{y_{n-1}}{k-n+1} - \binom{n-1}{1}\frac{y_{n-2}}{k-n+2} + \binom{n-1}{2}\frac{y_{n-3}}{k-n+3} - \cdots + (-1)^{n-1}\binom{n-1}{n-1}\frac{y_0}{k}\right]$$

2.165. Given the following table find the positive value of k such that

$$y_{k+3} = 3(y_{k+2} - y_k) + 18$$

k	4	8	12	16	20
y_k	6	42	110	210	342

Fig. 2-73

2.166. Given the following table find the values of k such that

$$y_{k+2} = 2(y_k - y_{k-4}) + y_{k-6}$$

k	0	5	10	15
y_k	15	25	41	63

Fig. 2-74

2.167. Referring to Problem 2.110(*a*) estimate the population for the year 1920 and compare with the actual population which was 105.7 million. (*b*) Estimate the population for the years 1910, 1900 and 1890 and compare with the actual populations which were 92.0, 76.0 and 62.9 million respectively. (*c*) Estimate the population for the years 1990 and 2000. Discuss the reliability of your predictions.

2.168. In reading a table the entries given in Fig. 2-75 were obtained. (*a*) Show that an error was made in one of the entries and (*b*) correct the error.

x	1	2	3	4	5	6	7	8	9	10
$f(x)$	0.28685	0.36771	0.44824	0.52844	0.60831	0.68748	0.76705	0.84592	0.92444	1.00263

Fig. 2-75

2.169. Suppose that $x = x_0$ is an approximation to the root of the equation $f(x) = 0$. Show that under appropriate conditions a better approximation is given by

$$x_0 - \frac{f(x_0)}{f'(x_0)}$$

and determine these conditions. The method of obtaining roots is called *Newton's method.* [*Hint.* Assume that the exact root is $x_0 + h$ so that $f(x_0 + h) = 0$. Then use the fact that $f(x_0 + h) = f(x_0) + hf'(x_0)$ approximately.]

2.170. Use the method of Problem 2.169 to work (*a*) Problem 2.37, (*b*) Problem 2.122, (*c*) Problem 2.123, (*d*) Problem 2.124, (*e*) Problem 2.125.

Chapter 3

The Sum Calculus

THE INTEGRAL OPERATOR

We have seen on page 2 that if A and B are operators such that $(AB)f = f$, i.e. $AB = 1$, then $B = A^{-1}$ which is called the *inverse operator* corresponding to A or briefly the *inverse* of A.

In order to see the significance of D^{-1} where D is the derivative operator we assume that $f(x)$ and $F(x)$ are such that

$$F(x) = D^{-1}f(x) \tag{1}$$

Then by definition if we apply the operator D to both sides of (1) we obtain

$$D[F(x)] = D[D^{-1}f(x)] = (DD^{-1})f(x) = f(x) \tag{2}$$

Thus $F(x)$ is that function whose derivative is $f(x)$. But from the integral calculus we know that

$$F(x) = \int f(x)\,dx + c \tag{3}$$

where c is the *constant of integration*. Thus we have

$$D^{-1}f(x) = \int f(x)\,dx + c \tag{4}$$

We can interpret D^{-1} or $1/D$ as an *integral operator* or *anti-derivative operator*.

$$D^{-1} = \int (\quad)\,dx \tag{5}$$

The integral in (3) or (4) is called an *indefinite integral* of $f(x)$ and as we have noted all such indefinite integrals can differ only by an arbitrary constant. The process of finding indefinite integrals is called *integration*.

GENERAL RULES OF INTEGRATION

It is assumed that the student is already familiar with the elementary rules for integration of functions. In the following we list the most important ones. In all cases we omit the constant of integration which should be added.

I-1. $\displaystyle\int [f(x) + g(x)]\,dx = \int f(x)\,dx + \int g(x)\,dx$

I-2. $\displaystyle\int \alpha f(x)\,dx = \alpha \int f(x)\,dx \qquad \alpha = \text{constant}$

I-3. $\displaystyle\int f(x)\,Dg(x)\,dx = f(x)\,g(x) - \int g(x)\,Df(x)\,dx$

This is often called *integration by parts.*

I-4. $\displaystyle\int f(x)\,dx = \int f[\phi(t)]\,\frac{d\phi(t)}{dt}\,dt \qquad \text{where } x = \phi(t)$

This is often called *integration by substitution.*

It should be noted that these results could have been written using D^{-1}. Thus, for example, I-1 and I-2 become respectively

$$D^{-1}[f(x)+g(x)] = D^{-1}f(x) + D^{-1}g(x), \qquad D^{-1}[\alpha f(x)] = \alpha D^{-1}f(x)$$

Note that these indicate that D^{-1} is a linear operator [see page 2].

INTEGRALS OF SPECIAL FUNCTIONS

In the following we list integrals of some of the more common functions corresponding to those on page 5. In all cases we omit the constant of integration which should be added.

II-1. $$\int \alpha\, dx = \alpha x$$

II-2. $$\int x^m\, dx = \begin{cases} \dfrac{x^{m+1}}{m+1} & m \neq -1 \\ \ln x & m = -1 \end{cases}$$

II-3. $$\int (ax+b)^m\, dx = \begin{cases} \dfrac{(ax+b)^{m+1}}{(m+1)a} & m \neq -1 \\ \dfrac{\ln(ax+b)}{a} & m = -1 \end{cases}$$

II-4. $$\int b^x\, dx = \frac{b^x}{\ln b} \qquad b > 0,\ b \neq 1$$

II-5. $$\int e^{rx}\, dx = \frac{e^{rx}}{r} \qquad r \neq 0$$

II-6. $$\int \sin rx\, dx = -\frac{\cos rx}{r} \qquad r \neq 0$$

II-7. $$\int \cos rx\, dx = \frac{\sin rx}{r} \qquad r \neq 0$$

DEFINITE INTEGRALS

Let $f(x)$ be defined in an interval $a \leqq x \leqq b$. Divide the interval into n equal parts of length $h = \Delta x = (b-a)/n$. Then the *definite integral* of $f(x)$ between $x = a$ and $x = b$ is defined as

$$\int_a^b f(x)\, dx = \lim_{\substack{h \to 0 \\ \text{or } n \to \infty}} h\{f(a) + f(a+h) + f(a+2h) + \cdots + f[a+(n-1)h]\}$$

if this limit exists.

If the graph of $y = f(x)$ is the curve C of Fig. 3-1, then this limit represents geometrically the *area* bounded by the curve C, the x axis and the ordinates at $x = a$ and $x = b$. In this figure $x_k = a + kh$, $x_{k+1} = a + (k+1)h$ and $y_k = f(x_k)$.

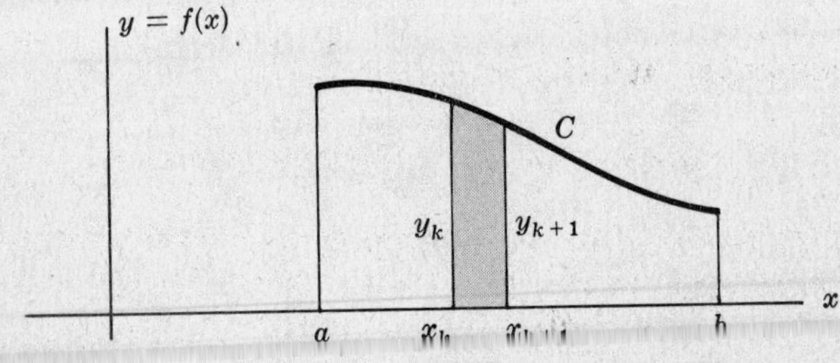

Fig. 3-1

In a definite integral the variable of integration does not matter and so any symbol can be used. Thus we can write

$$\int_a^b f(x)\,dx = \int_a^b f(u)\,du = \int_a^b f(t)\,dt, \quad \text{etc.}$$

For this reason the variable is often called a *dummy variable.*

FUNDAMENTAL THEOREM OF INTEGRAL CALCULUS

If $f(x) = DF(x) = \frac{d}{dx}F(x)$ then the *fundamental theorem of integral calculus* states that

$$\int_a^b f(x)\,dx = \int_a^b \frac{d}{dx}F(x)\,dx = F(x)\Big|_a^b = F(b) - F(a)$$

This can also be written as

$$\int_a^b f(x)\,dx = D^{-1}f(x)\Big|_a^b = F(x)\Big|_a^b = F(b) - F(a)$$

Example 1.

$$\int_1^4 x^2\,dx = \int_1^4 \frac{d}{dx}\left(\frac{x^3}{3}\right)dx = \frac{x^3}{3}\Big|_1^4 = \frac{4^3}{3} - \frac{1^3}{3} = 21$$

SOME IMPORTANT PROPERTIES OF DEFINITE INTEGRALS

1. $\displaystyle\int_a^b f(x)\,dx = -\int_b^a f(x)\,dx$

2. $\displaystyle\int_a^b f(x)\,dx = \int_a^c f(x)\,dx + \int_c^b f(x)\,dx$ where c is between a and b

3. $\displaystyle\left|\int_a^b f(x)\,dx\right| \leqq \int_a^b |f(x)|\,dx$

4. If $m_1 \leqq f(x) \leqq m_2$ then

$$\int_a^b m_1\,dx \leqq \int_a^b f(x)\,dx \leqq \int_a^b m_2\,dx$$

or

$$m_1(b-a) \leqq \int_a^b f(x)\,dx \leqq m_2(b-a)$$

SOME IMPORTANT THEOREMS OF INTEGRAL CALCULUS

Two theorems which will be found useful are the following.

1. **Mean-value theorem for integrals.** If $f(x)$ is continuous in an interval $a \leqq x \leqq b$ and $g(x)$ does not change sign in the interval, then there is a number ξ such that

$$\int_a^b f(x)\,g(x)\,dx = f(\xi)\int_a^b g(x)\,dx \qquad \xi \text{ between } a \text{ and } b$$

2. **Leibnitz's rule for differentiation of an integral.**

$$\frac{d}{d\alpha}\int_{a(\alpha)}^{b(\alpha)} F(x,\alpha)\,dx = \int_{a(\alpha)}^{b(\alpha)} \frac{\partial F}{\partial \alpha}\,dx + F(b(\alpha),\alpha)\frac{db}{d\alpha} - F(a(\alpha),\alpha)\frac{da}{d\alpha}$$

where $a(\alpha)$, $b(\alpha)$ are assumed to be differentiable functions of α and $F(x,\alpha)$ and $\partial F/\partial\alpha$ are assumed to be continuous functions of x. Note that $\partial F/\partial\alpha$ is the partial derivative of F with respect to α, i.e. the ordinary derivative with respect to α, when x is considered to be constant.

THE SUM OPERATOR

Because of the analogy of D and Δ [or more precisely $\Delta/\Delta x = \Delta/h$], it is natural to seek the significance of the operator Δ^{-1} which we conjecture would have properties analogous to those of D^{-1}.

To discover the analogy we suppose that $F(x)$ and $f(x)$ are such that

$$\frac{\Delta}{\Delta x}F(x) = f(x) \qquad \text{or} \qquad \Delta F(x) = hf(x) \tag{6}$$

Then by definition Δ^{-1} is that operator such that

$$F(x) = \Delta^{-1}[hf(x)] \tag{7}$$

Now we can show [see Problem 3.10] that Δ^{-1} is a linear operator and also that if two functions have the same difference, they can differ at most by an arbitrary function $C(x)$ such that $C(x+h) = C(x)$, i.e. an arbitrary periodic function with period h. Since this is analogous to the arbitrary constant of integral calculus we often refer to it as an *arbitrary periodic constant* or briefly *periodic constant.*

From (*7*) we are led by analogy with formulas (*3*) and (*4*) of the integral calculus to define

$$F(x) = \Delta^{-1}[hf(x)] = \sum f(x)h + C(x) \tag{8}$$

or

$$\Delta^{-1}[f(x)] = \sum f(x) + C_1(x) \tag{9}$$

where $C_1(x)$ is also a periodic constant. It is of interest that (*9*) can be written as

$$\left(\frac{\Delta}{\Delta x}\right)^{-1} f(x) = \sum f(x)\,\Delta x + C(x) \tag{10}$$

which displays the remarkable analogy with

$$D^{-1}f(x) = \int f(x)\,dx + c \tag{11}$$

We shall call Σ the *sum operator* and $\Sigma f(x)$ [or $\Sigma f(x)\,\Delta x = h\,\Sigma f(x)$] the *indefinite sum* of $f(x)$. Note that because of (*9*) and (*10*) we can consider Δ^{-1} and Σ or $\Delta/\Delta x$ and $\Sigma(\quad)\,\Delta x$ as inverse operators, i.e. $\Delta\Sigma = 1$ or $\Sigma = \Delta^{-1}$. See Problem 3.11. The process of finding sums is called *summation.*

GENERAL RULES OF SUMMATION

The following formulas bear close resemblance to the rules of integration on page 79. In all cases we omit the periodic constant of summation which should be added.

III-1. $\sum [f(x) + g(x)] = \sum f(x) + \sum g(x)$

III-2. $\sum \alpha f(x) = \alpha \sum f(x) \qquad \alpha = \text{constant}$

III-3. $\sum f(x)\,\Delta g(x) = f(x)\,g(x) - \sum g(x+h)\,\Delta f(x)$

This is often called *summation by parts.*

Note that these formulas could have been written using Δ^{-1} instead of Σ. The first two represent the statement that Σ *or* Δ^{-1} is a linear operator.

SUMMATIONS OF SPECIAL FUNCTIONS

In the following we list sums of some of the more common functions. The results are analogous to those on page 80. In all cases we omit the periodic constant which should be added.

IV-1. $\sum \alpha \quad = \quad \dfrac{\alpha x}{h}$

IV-2. $\sum x^{(m)} \quad = \quad \dfrac{x^{(m+1)}}{(m+1)h} \qquad m \neq -1$

IV-3. $\sum (px+q)^{(m)} \quad = \quad \dfrac{(px+q)^{(m+1)}}{(m+1)ph} \qquad m \neq -1$

For the case where $m = -1$ or is not an integer we use the gamma function [see page 86].

IV-4. $\sum b^x \quad = \quad \dfrac{b^x}{b^h - 1}$

IV-5. $\sum e^{rx} \quad = \quad \dfrac{e^{rx}}{e^{rh} - 1}$

IV-6. $\sum \sin rx \quad = \quad -\dfrac{\cos r(x - \frac{1}{2}h)}{2 \sin \frac{1}{2}rh}$

IV-7. $\sum \cos rx \quad = \quad \dfrac{\sin r(x - \frac{1}{2}h)}{2 \sin \frac{1}{2}rh}$

These results can also be written with Δ^{-1} in place of Σ. The analogy of the above with corresponding integrals is clear if we multiply the sums by $h = \Delta x$. For example, IV-2 becomes

$$\sum x^{(m)} h \quad = \quad \frac{x^{(m+1)}}{m+1} \qquad m \neq -1 \tag{12}$$

which corresponds to

$$\int x^m \, dx \quad = \quad \frac{x^{m+1}}{m+1} \qquad m \neq -1 \tag{13}$$

with $h = \Delta x$, $x^{(m)}$ and Σ corresponding to dx, x^m and $\int$ respectively.

This relationship between the sum calculus and integral calculus is expressed in the following important theorem.

Theorem 3-1.

$$\lim_{h \to 0} \left[\sum f(x)h + C(x) \right] \quad = \quad \int f(x) \, dx + c$$

DEFINITE SUMS AND THE FUNDAMENTAL THEOREM OF SUM CALCULUS

The *definite sum* of $f(x)$ from $x = a$ to $x = a + (n-1)h$ in steps of h is denoted by

$$\sum_{a}^{a+(n-1)h} f(x) \quad = \quad f(a) + f(a+h) + f(a+2h) + \cdots + f[a+(n-1)h] \tag{14}$$

The following theorem analogous to the fundamental theorem of integral calculus on page 81 is of great importance and is called the *fundamental theorem of the sum calculus.*

***Theorem 3-2* [*fundamental theorem of sum calculus*].**

$$\sum_{a}^{a+(n-1)h} f(x) \quad = \quad \Delta^{-1} f(x) \Big|_{a}^{a+nh} \tag{15}$$

where it is emphasized once again that the sum on the left is taken from $x = a$ to $x = a+(n-1)h$ *in steps of h*. It should also be noted that the right side of (*15*) means that the indefinite sum is to be found as a function of x and we are then to substitute the upper limit $a+nh$ and lower limit a and subtract the results. In such case the added arbitrary periodic constant is subtracted out as in the case of definite integrals.

Example 2.

If $a = 1$, $h = 2$, $n = 3$, then from (*15*) and IV-2 we have

$$\sum_1^5 x^{(2)} = \Delta^{-1}x^{(2)}\Big|_1^7 = \frac{x^{(3)}}{3\cdot 2}\Big|_1^7 = \frac{7^{(3)}}{6} - \frac{1^{(3)}}{6} = \frac{7\cdot 5\cdot 3}{6} - \frac{(1)(-1)(-3)}{6} = 17$$

This checks since the left side is $(1)(-1)+(3)(1)+(5)(3) = 17$.

It is of interest to note that if we multiply both sides of (*15*) by h and then let $h \to 0$ and $n \to \infty$ so that $nh = b-a$, we obtain the fundamental theorem of integral calculus.

DIFFERENTIATION AND INTEGRATION OF SUMS

The following theorems are sometimes useful.

Theorem 3-3. If $\Delta^{-1}F(x) = \sum F(x) = G(x)$ then

$$\Delta^{-1}\left(\frac{dF}{dx}\right) = \sum \frac{dF}{dx} = \frac{dG}{dx}$$

Theorem 3-4. If $\Delta^{-1}F(x,\alpha) = \sum F(x,\alpha) = G(x,\alpha)$ then

$$\Delta^{-1}\left(\frac{\partial F}{\partial \alpha}\right) = \sum \frac{\partial F}{\partial \alpha} = \frac{\partial G}{\partial \alpha}$$

Theorem 3-4 is analogous to Leibnitz's rule for differentiating an integral [see page 81].

Similar results hold for integrals. For example we have

Theorem 3-5.

$$\int_{\alpha_1}^{\alpha_2}\left[\sum F(x,\alpha)\right]d\alpha = \sum \int_{\alpha_1}^{\alpha_2} F(x,\alpha)\,d\alpha$$

THEOREMS ON SUMMATION USING THE SUBSCRIPT NOTATION

As in the case of the difference calculus all of the above results for the sum calculus can be expressed in terms of the subscript notation $y_k = f(a+kh)$. For example III-3 on page 82 becomes using $z_k = g(a+kh)$

$$\sum y_k \Delta z_k = y_k z_k - \sum z_{k+1}\Delta y_k \tag{16}$$

The periodic constant of summation in this case becomes an arbitrary constant.

Similarly formulas IV-1 through IV-7 can be written with x replaced by k and $h = 1$. For example, IV-2 becomes

$$\sum k^{(m)} = \frac{k^{(m+1)}}{m+1} \qquad m \neq -1 \tag{17}$$

The fundamental theorem of sum calculus can in subscript notation be written, using the particular value $a = 1$, as

$$\sum_{k=1}^{n} y_k = \Delta^{-1}y_k \Big|_1^{n+1} \tag{18}$$

It is seen that the subscript notation has some obvious advantages since h does not appear.

ABEL'S TRANSFORMATION

The following result, called *Abel's transformation,* is sometimes useful.

$$\sum_{k=1}^{n} u_k v_k = u_{n+1} \sum_{k=1}^{n} v_k - \sum_{k=1}^{n} \left[\Delta u_k \sum_{p=1}^{k} v_p \right] \tag{19}$$

OPERATOR METHODS FOR SUMMATION

Various operator methods are available which can be used to obtain summations quickly. One useful result valid for $\beta \neq 1$ is given by

$$\sum \beta^k P(k) = \Delta^{-1}[\beta^k P(k)] = \frac{1}{\Delta}\beta^k P(k) = \frac{1}{E-1}\beta^k P(k) = \beta^k \frac{1}{\beta E - 1} P(k)$$

$$= \frac{\beta^k}{\beta - 1}\left[1 - \frac{\beta\Delta}{\beta-1} + \frac{\beta^2\Delta^2}{(\beta-1)^2} - \frac{\beta^3\Delta^3}{(\beta-1)^3} + \cdots\right]P(k) + c \tag{20}$$

where c is an arbitrary constant. The series terminates if $P(k)$ is a polynomial.

SUMMATION OF SERIES

The above results are very useful in obtaining sums of various series [see Problems 3.20-3.28]. Further applications involving the important topic of summation of series are given in Chapter 4.

THE GAMMA FUNCTION

In order to generalize the factorial $x^{(m)}$ to the case where m is not an integer we make use of the *gamma function.* This function is defined by

$$\Gamma(p) = \int_0^\infty t^{p-1}e^{-t}\,dt = \lim_{M\to\infty}\int_0^M t^{p-1}e^{-t}\,dt \qquad p > 0 \tag{21}$$

and we can show [see Problem 3.32] that

$$\Gamma(p+1) = p\Gamma(p) \tag{22}$$

The recursion formula *(22)* can be used to define $\Gamma(p)$ for $p \leqq 0$ [see Problem 3.34(*c*)].

Some important properties of the gamma function are as follows.

1. If p is a positive integer, $\Gamma(p+1) = p!$

2. $$\Gamma(\tfrac{1}{2}) = \sqrt{\pi}$$

3. $$\Gamma(r)\,\Gamma(1-r) = \frac{\pi}{\sin r\pi} \qquad 0 < r < 1$$

4. $$\Gamma'(1) = \int_0^\infty e^{-t}\ln t\,dt = -\gamma = -0.5772156649\ldots$$

   where γ is called *Euler's constant.*

If m is an integer we can write [see Problem 3.35]

$$x^{(m)} = \frac{h^m\Gamma\left(\frac{x}{h}+1\right)}{\Gamma\left(\frac{x}{h}-m+1\right)} \tag{23}$$

Since *(23)* has meaning for all values of m we shall take it as a definition of $x^{(m)}$ for all m.

From (*23*) it follows that
$$\Delta x^{(m)} = mx^{(m-1)}h \tag{24}$$

and
$$\sum x^{(m)} = \frac{x^{(m+1)}}{(m+1)h} \qquad m \neq -1 \tag{25}$$

thus generalizing the results (*25*), page 6, and IV-2, page 83.

If $m = -1$, we obtain [see Problem 3.39]
$$\sum x^{(-1)} = \sum \frac{1}{x+h} = \frac{\Gamma'\left(\frac{x}{h}+1\right)}{h\Gamma\left(\frac{x}{h}+1\right)} \tag{26}$$

The function on the right of (*26*) is sometimes called the *digamma function* and is denoted by $\Psi(x)$.

The results (*23*) through (*26*) can also be written in the "k notation" with k replacing x and $h = 1$.

BERNOULLI NUMBERS AND POLYNOMIALS

Various sequences of functions $f_0(x), f_1(x), f_2(x), \ldots$ have the property that
$$Df_n(x) = f_{n-1}(x) \tag{27}$$
where in (*27*), as well as what follows, we shall assume that $n = 1, 2, 3, \ldots$. If we consider the function $f_n(x)$ to be a polynomial of degree n then we can show that
$$f_n(x) = \frac{c_0 x^n}{n!} + \frac{c_1 x^{n-1}}{(n-1)!} + \cdots + c_{n-1}x + c_n \tag{28}$$
where $c_0, c_1, \ldots, c_n$ are constants independent of n [see Problem 3.40].

To specify the polynomials, i.e. to determine $c_0, c_1, \ldots, c_n$, further conditions need to be given. Different conditions will result in different sequences of polynomials.

Let $\beta_n(x)$ be a sequence of polynomials satisfying (*27*), i.e.
$$D\beta_n(x) = \beta_{n-1}(x) \tag{29}$$
and the added condition
$$\Delta\beta_n(x) = \frac{x^{n-1}}{(n-1)!} \tag{30}$$
where we take $h = 1$ and define $0! = 1$. Conditions (*29*) and (*30*) enable us to specify completely the polynomial $\beta_n(x)$. We find in fact that the first few polynomials are
$$\beta_0(x) = 1, \quad \beta_1(x) = x - \tfrac{1}{2}, \quad \beta_2(x) = \tfrac{1}{2}x^2 - \tfrac{1}{2}x + \tfrac{1}{12}, \quad \beta_3(x) = \tfrac{1}{6}x^3 - \tfrac{1}{4}x^2 + \tfrac{1}{12}x \tag{31}$$
See Problem 3.41.

It should be noted that once $\beta_n(x)$ is found, we can use (*29*) to find $\beta_{n-1}(x), \beta_{n-2}(x), \ldots$, by successive differentiations. This is particularly useful since it is possible to obtain $\beta_n(x)$ in terms of Stirling numbers of the second kind by the formula
$$\beta_n(x) = \frac{1}{n!}\frac{d}{dx}\sum_{k=1}^{n} S_k^n \frac{x^{(k+1)}}{k+1} \tag{32}$$
See Problem 3.44.

We define the *Bernoulli polynomials* as
$$B_n(x) = n!\,\beta_n(x) \tag{33}$$

so that from (*32*)
$$B_n(x) = \frac{d}{dx}\sum_{k=1}^{n} S_k^n \frac{x^{(k+1)}}{k+1} \tag{34}$$

The first few Bernoulli polynomials are given by

$$B_0(x) = 1, \quad B_1(x) = x - \tfrac{1}{2}, \quad B_2(x) = x^2 - x + \tfrac{1}{6}, \quad B_3(x) = x^3 - \tfrac{3}{2}x^2 + \tfrac{1}{2}x \tag{35}$$

We define the *Bernoulli numbers*, denoted by B_n, as the values of the Bernoulli polynomials with $x = 0$, i.e.

$$B_n = B_n(0) \tag{36}$$

The first few Bernoulli numbers are given by

$$B_0 = 1, \quad B_1 = -\tfrac{1}{2}, \quad B_2 = \tfrac{1}{6}, \quad B_3 = 0, \quad B_4 = -\tfrac{1}{30} \tag{37}$$

For tables of Bernoulli numbers and polynomials see Appendixes C and D, pages 234 and 235.

IMPORTANT PROPERTIES OF BERNOULLI NUMBERS AND POLYNOMIALS

In the following we list some of the important and interesting properties of Bernoulli numbers and polynomials.

1. $$B_n(x) = x^n + \binom{n}{1}x^{n-1}B_1 + \binom{n}{2}x^{n-2}B_2 + \cdots + \binom{n}{n}B_n$$

   This can be expressed formally by $B_n(x) = (x+B)^n$ where after expanding, using the binomial formula, B^k is replaced by B_k. It is a recurrence formula which can be used to obtain the Bernoulli polynomials.

2. $$B_n(x+1) - B_n(x) = nx^{n-1}$$

3. $$B_n'(x) = nB_{n-1}(x)$$

4. $$B_n = B_n(0) = B_n(1), \qquad B_{2n-1} = 0 \qquad n = 2, 3, 4, \ldots$$

5. $$1 + \binom{n}{1}B_1 + \binom{n}{2}B_2 + \cdots + \binom{n}{n-1}B_{n-1} = 0$$

   This can be expressed formally as $(1+B)^n - B^n = 0$ where after expanding, using the binomial fomula, B^k is replaced by B_k. This recurrence formula can be used to obtain the Bernoulli numbers.

6. $$B_n(x) = (-1)^n B_n(1-x)$$

7. $$\frac{te^{xt}}{e^t - 1} = \sum_{n=0}^{\infty} \frac{B_n(x)\, t^n}{n!} = 1 + B_1(x)\, t + \frac{B_2(x)\, t^2}{2!} + \cdots$$

   This is often called the *generating function for the Bernoulli polynomials* and can be used to define them [see Problems 3.51 and 3.53].

8. $$\frac{t}{e^t - 1} = \sum_{n=0}^{\infty} \frac{B_n t^n}{n!} = 1 + B_1 t + \frac{B_2 t^2}{2!} + \cdots$$

   This is called the *generating function for the Bernoulli numbers* and can be used to obtain these numbers [see Problem 3.52].

9. $$\frac{x}{2}\coth\frac{x}{2} = 1 + \frac{B_2 x^2}{2!} + \frac{B_4 x^4}{4!} + \frac{B_6 x^6}{6!} + \cdots$$

10. $$1^r + 2^r + 3^r + \cdots + (n-1)^r = \frac{(n+B)^{r+1} - B^{r+1}}{r+1}$$

    where $r = 1, 2, 3, \ldots$ and where the right side is to be expanded and then B^k replaced by B_k.

11. $$\frac{1}{1^{2n}} + \frac{1}{2^{2n}} + \frac{1}{3^{2n}} + \cdots = \frac{(-1)^{n-1}B_{2n}(2\pi)^{2n}}{2\cdot(2n)!} \qquad n = 1,2,3,\ldots$$

12. If $n = 1,2,3,\ldots$, then

$$B_{2n-1}(x) = 2(-1)^n(2n-1)!\sum_{k=1}^{\infty}\frac{\sin 2k\pi x}{(2k\pi)^{2n-1}}$$

$$B_{2n}(x) = 2(-1)^{n-1}(2n)!\sum_{k=1}^{\infty}\frac{\cos 2k\pi x}{(2k\pi)^{2n}}$$

EULER NUMBERS AND POLYNOMIALS

The *Euler polynomials* $E_n(x)$ are defined by equation (*27*), i.e.

$$DE_n(x) = E_{n-1}(x) \tag{38}$$

together with the added condition

$$ME_n(x) = \frac{x^n}{n!} \quad \text{or} \quad \tfrac{1}{2}[E_n(x+1) + E_n(x)] = \frac{x^n}{n!} \tag{39}$$

where M is the averaging operator of page 10 and $h = 1$.

The first few Euler polynomials are given by

$$E_0(x) = 1, \quad E_1(x) = x - \tfrac{1}{2}, \quad E_2(x) = \tfrac{1}{2}x^2 - \tfrac{1}{2}x, \quad E_3(x) = \tfrac{1}{6}x^3 - \tfrac{1}{4}x^2 + \tfrac{1}{24} \tag{40}$$

The *Euler numbers* are defined as

$$E_n = 2^n n!\,E_n(\tfrac{1}{2}) \tag{41}$$

and the first few Euler numbers are given by

$$E_0 = 1, \quad E_1 = 0, \quad E_2 = -1, \quad E_3 = 0, \quad E_4 = 5, \quad E_5 = 0 \tag{42}$$

For tables of Euler numbers and polynomials see Appendixes E and F, pages 236 and 237.

IMPORTANT PROPERTIES OF EULER NUMBERS AND POLYNOMIALS

In the following we list some of the important and interesting properties of Euler numbers and polynomials.

1. $$E_n(x) = \frac{e_0x^n}{n!} + \frac{e_1x^{n-1}}{(n-1)!} + \cdots + e_{n-1}x + e_n$$

where $$e_n + \frac{1}{2}\left[\frac{e_0}{n!} + \frac{e_1}{(n-1)!} + \frac{e_2}{(n-2)!} + \cdots + \frac{e_{n-1}}{1!}\right] = 0$$

The first few values of e_k are

$$e_0 = 1, \quad e_1 = -\tfrac{1}{2}, \quad e_2 = 0, \quad e_3 = \tfrac{1}{24}, \quad e_4 = 0, \quad e_5 = -\tfrac{1}{240}$$

2. $$E_n'(x) = E_{n-1}(x)$$

3. $$E_n(x) = (-1)^n E_n(1-x)$$

4. $$E_n(0) + E_n(1) = 0 \qquad \text{for } n = 1,2,3,\ldots$$

5. $$E_n(0) + E_n(-1) = \frac{2(-1)^n}{n!} \qquad \text{for } n = 1,2,3,\ldots$$

6. $$E_{2n} = 2^{2n}(2n)!\,E_{2n}(\tfrac{1}{2}) = 0 \qquad \text{for } n = 1,2,3,\ldots$$

7. $$E_{2n}(0) = E_{2n}(1) = e_{2n} = 0 \qquad \text{for } n = 1,2,3,\ldots$$

8. $$\frac{2e^{xt}}{e^t+1} = \sum_{k=0}^{\infty} E_k(x)t^k = 1 + E_1(x)t + E_2(x)t^2 + \cdots$$

This is often called the *generating function for the Euler polynomials* and can be used to define them.

9. $$\sec x = \sum_{k=0}^{\infty} \frac{|E_{2k}|x^{2k}}{(2k)!} = 1 + \frac{1}{2}x^2 + \frac{5}{24}x^4 + \frac{61}{720}x^6 + \cdots$$

10. $$\tan x = \sum_{k=1}^{\infty} \frac{T_k x^k}{k!} = x + \frac{1}{3}x^3 + \frac{2}{15}x^5 + \frac{17}{315}x^7 + \frac{62}{2835}x^9 + \cdots$$

where $$T_k = |2^k k!\, e_k|$$

are called *tangent numbers* and are integers.

11. $$\frac{1}{1^{2n+1}} - \frac{1}{3^{2n+1}} + \frac{1}{5^{2n+1}} - \cdots = \frac{(-1)^n E_{2n}\pi^{2n+1}}{2^{2n+2}(2n)!} \qquad n = 0, 1, 2, \ldots$$

12. If $n = 1, 2, 3, \ldots$, then for $0 \leqq x \leqq 1$,

$$E_{2n-1}(x) = \frac{4(-1)^n}{\pi^{2n}} \sum_{k=0}^{\infty} \frac{\cos(2k+1)\pi x}{(2k+1)^{2n}}$$

$$E_{2n}(x) = \frac{4(-1)^n}{\pi^{2n+1}} \sum_{k=0}^{\infty} \frac{\sin(2k+1)\pi x}{(2k+1)^{2n+1}}$$

Solved Problems

THE INTEGRAL OPERATOR AND RULES OF INTEGRATION

3.1. Find each of the following:

(*a*) $D^{-1}(2x^2 - 5x + 4)$ (*b*) $D^{-1}(4e^{-3x} - 3\sin 2x)$ (*c*) $D^{-1}\left[\frac{(\sqrt{x}-3)^2}{\sqrt{x^3}}\right]$

(*a*) $$D^{-1}(2x^2 - 5x + 4) = \int (2x^2 - 5x + 4)\,dx = \frac{2x^3}{3} - \frac{5x^2}{2} + 4x + c$$

(*b*) $$\begin{aligned} D^{-1}(4e^{-3x} - 3\sin 2x) &= \int (4e^{-3x} - 3\sin 2x)\,dx \\ &= 4\left(\frac{e^{-3x}}{-3}\right) - 3\left(\frac{-\cos 2x}{2}\right) + c \\ &= -\tfrac{4}{3}e^{-3x} + \tfrac{3}{2}\cos 2x + c \end{aligned}$$

(*c*) $$\begin{aligned} D^{-1}\left[\frac{(\sqrt{x}-3)^2}{\sqrt{x^3}}\right] &= \int \left(\frac{x - 6\sqrt{x} + 9}{\sqrt{x^3}}\right) dx \\ &= \int \frac{x - 6x^{1/2} + 9}{x^{3/2}}\,dx \\ &= \int \left(x^{-1/2} - \frac{6}{x} + 9x^{-3/2}\right) dx \\ &= \frac{x^{1/2}}{1/2} - 6\ln x + 9\frac{x^{-1/2}}{-1/2} + c \\ &= 2x^{1/2} - 6\ln x - 18x^{-1/2} + c \end{aligned}$$

3.2. Prove the formula I-3 on page 79 for integration by parts.

We have by the rules of differentiation

$$D[f(x)\,g(x)] \;=\; f(x)\,Dg(x) + g(x)\,Df(x)$$

or

$$f(x)\,Dg(x) \;=\; D[f(x)\,g(x)] - g(x)\,Df(x)$$

Operating with D^{-1} on both sides and remembering that D^{-1} is a linear operator, we have

$$D^{-1}[f(x)\,Dg(x)] \;=\; D^{-1}D[f(x)\,g(x)] - D^{-1}[g(x)\,Df(x)]$$

or

$$\int f(x)\,Dg(x)\,dx \;=\; f(x)\,g(x) - \int g(x)\,Df(x)\,dx$$

3.3. Find (*a*) $\int x \cos 4x\,dx$, (*b*) $\int \ln x\,dx$.

(*a*) We use integration by parts here. Let

$$f(x) = x, \qquad Dg(x) = \cos 4x$$

so that

$$Df(x) = 1, \qquad g(x) = \frac{\sin 4x}{4}$$

where in finding $g(x)$ we omit the arbitrary constant.

Then using the formula for integration by parts,

$$\int x \cos 4x\,dx \;=\; x\left(\frac{\sin 4x}{4}\right) - \int \frac{\sin 4x}{4}\,dx \;=\; \frac{x \sin 4x}{4} + \frac{\cos 4x}{16} + c$$

(*b*) Let $f(x) = \ln x$, $Dg(x) = 1$ so that

$$Df(x) = \frac{1}{x}, \qquad g(x) = x$$

Then using integration by parts we have

$$\int \ln x\,dx \;=\; x \ln x - \int x \cdot \frac{1}{x}\,dx \;=\; x \ln x - x$$

3.4. Find (*a*) $\int x^2 e^{3x}\,dx$, (*b*) $\int x^4 \sin 2x\,dx$.

(*a*) In this case we must apply integration by parts twice. We obtain

$$\begin{aligned}\int x^2 e^{3x}\,dx &= (x^2)\left(\frac{e^{3x}}{3}\right) - \int (2x)\left(\frac{e^{3x}}{3}\right)dx \\ &= (x^2)\left(\frac{e^{3x}}{3}\right) - \left[(2x)\left(\frac{e^{3x}}{9}\right) - \int (2)\left(\frac{e^{3x}}{9}\right)dx\right] \\ &= (x^2)\left(\frac{e^{3x}}{3}\right) - (2x)\left(\frac{e^{3x}}{9}\right) + (2)\left(\frac{e^{3x}}{27}\right)\end{aligned}$$

apart from the additive constant of integration.

In this result we make the following observations.

(i) The first factors in each of the last two terms, i.e. $2x$ and 2, are obtained by taking successive *derivatives* of x^2 [the first factor of the first term].

(ii) The second factors in each of the last two terms, i.e. $e^{3x}/9$ and $e^{3x}/27$, are obtained by taking successive *integrals* of $e^{3x}/3$ [the second factor of the first term].

(iii) The signs of the terms alternate $+, -, +$. The above observations are quite general and can be used to write down results easily in cases where many integrations by parts might have to be performed. A proof of this method which we shall refer to as *generalized integration by parts* is easily formulated [see Problem 3.143].

(b) Using the generalized integration by parts outlined in part (a) we find the result

$$\int x^4 \sin 2x\, dx = (x^4)\left(\frac{-\cos 2x}{2}\right) - (4x^3)\left(\frac{-\sin 2x}{4}\right)(12x^2)\left(\frac{\cos 2x}{8}\right)$$
$$- (24x)\left(\frac{\sin 2x}{16}\right) + (24)\left(\frac{-\cos 2x}{32}\right) + c$$

3.5. Find (a) $\int \frac{(1+\sin x)\,dx}{\sqrt{x-\cos x}}$, (b) $\int e^{\sqrt[3]{2x-1}}\,dx$, (c) $\int \frac{\ln^2(x+4)}{x+4}\,dx$.

(a) Make the substitution $\sqrt{x-\cos x} = t$ so that $x - \cos x = t^2$ and $(1+\sin x)\,dx = 2t\,dt$ on taking the differential of both sides. Then the given integral becomes

$$\int \frac{(1+\sin x)\,dx}{\sqrt{x-\cos x}} = \int \frac{2t\,dt}{t} = 2t + c = 2\sqrt{x-\cos x} + c$$

(b) Let $\sqrt[3]{2x-1} = t$ so that $2x-1 = t^3$, $x = \frac{1}{2}(t^3+1)$, $dx = \frac{3}{2}t^2\,dt$. Then the given integral becomes

$$\int e^{\sqrt[3]{2x-1}}\,dx = \int e^t \frac{3}{2}t^2\,dt = \frac{3}{2}\int t^2 e^t\,dt$$
$$= \frac{3}{2}[(t^2)(e^t) - (2t)(e^t) + (2)(e^t)] + c$$
$$= \frac{3}{2}e^t(t^2 - 2t + 2) + c = \frac{3}{2}e^{\sqrt[3]{2x-1}}[(2x-1)^{2/3} - 2(2x-1)^{1/3} + 2] + c$$

(c) Let $\ln(x+4) = t$ so that $x+4 = e^t$, $x = e^t - 4$, $dx = e^t\,dt$. Then the given integral becomes

$$\int \frac{\ln^2(x+4)}{x+4}\,dx = \int \frac{t^2(e^t\,dt)}{e^t} = \int t^2\,dt = \frac{t^3}{3} + c = \frac{1}{3}\ln^3(x+4) + c$$

3.6. Prove that
$$\int (ax+b)^m\,dx = \begin{cases} \dfrac{(ax+b)^{m+1}}{(m+1)a} & m \neq -1 \\ \dfrac{\ln(ax+b)}{a} & m = -1 \end{cases}$$

Let $t = ax+b$ so that $dx = dt/a$. Then the required integral becomes

$$\int (ax+b)^m\,dx = \frac{1}{a}\int t^m\,dt = \begin{cases} \dfrac{t^{m+1}}{(m+1)a} & m \neq -1 \\ \dfrac{\ln t}{a} & m = -1 \end{cases}$$

Replacing t by $ax+b$ yields the required result.

DEFINITE INTEGRALS AND THE FUNDAMENTAL THEOREM OF INTEGRAL CALCULUS

3.7. (a) Evaluate $\int_1^2 (x^2+1)^2\,dx$ and (b) give a geometric interpretation.

(a) We first find the corresponding indefinite integral as follows:

$$D^{-1}(x^2+1)^2 = \int (x^2+1)^2\,dx = \int (x^4+2x^2+1)\,dx = \int x^4\,dx + 2\int x^2\,dx + \int dx$$
$$= \frac{x^5}{5} + \frac{2x^3}{3} + x$$

where we can omit the constant of integration. Thus by the fundamental theorem of integral calculus,

$$\int_1^2 (x^2+1)^2\,dx = \frac{x^5}{5} + \frac{2x^3}{3} + x\,\Big|_1^2 = \left(\tfrac{32}{5} + \tfrac{16}{3} + 2\right) - \left(\tfrac{1}{5} + \tfrac{2}{3} + 1\right)$$
$$= \frac{178}{15} = 11.867 \text{ (approx.)}$$

(b) The value $178/15 = 11.867$ (approx.) represents the area [shaded in Fig. 3-2] bounded by the curve $y = (x^2+1)^2$, the x axis and the ordinates at $x = 1$ and $x = 2$.

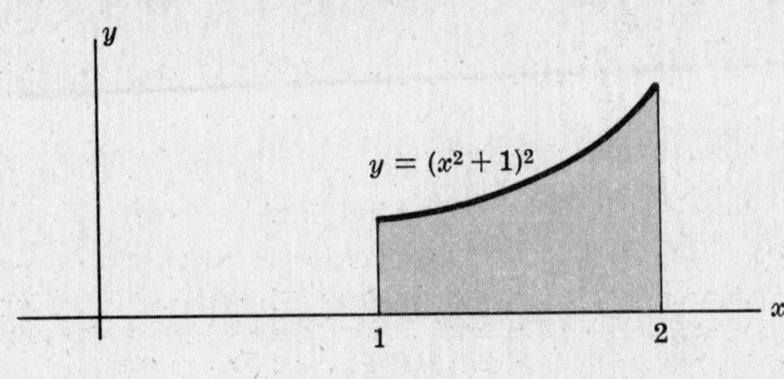

Fig. 3-2

3.8. Find (a) $\int_0^{\pi/2} \sin 6x\,dx$, (b) $\int_1^2 xe^{-3x}\,dx$, (c) $\int_0^2 \frac{dx}{x^2-2x+2}$.

(a)
$$\int_0^{\pi/2} \sin 6x\,dx = -\frac{\cos 6x}{6}\Big|_0^{\pi/2} = -\frac{\cos 3\pi}{6} + \frac{\cos 0}{6} = \frac{1}{6} + \frac{1}{6} = \frac{1}{3}$$

(b)
$$\int_1^2 xe^{-3x}\,dx = (x)\left(\frac{e^{-3x}}{-3}\right) - (1)\left(\frac{e^{-3x}}{9}\right)\Big|_1^2$$
$$= (2)\left(\frac{e^{-6}}{-3}\right) - \left(\frac{e^{-6}}{9}\right) - \left[(1)\left(\frac{e^{-3}}{-3}\right) - \left(\frac{e^{-3}}{9}\right)\right] = \tfrac{4}{9}e^{-3} - \tfrac{7}{9}e^{-6}$$

(c) **Method 1.**
$$\int \frac{dx}{x^2-2x+2} = \int \frac{dx}{(x-1)^2+1}$$

Let $x-1 = \tan t$. Then $dx = \sec^2 t\,dt$ and so
$$\int \frac{dx}{(x-1)^2+1} = \int \frac{\sec^2 t\,dt}{\tan^2 t+1} = \int \frac{\sec^2 t}{\sec^2 t}\,dt$$
$$= \int dt = t + c = \tan^{-1}(x-1) + c.$$

Thus
$$\int_0^2 \frac{dx}{x^2-2x+2} = \tan^{-1}(x-1)\Big|_0^2 = \tan^{-1}(1) - \tan^{-1}(-1) = \frac{\pi}{4} - \left(-\frac{\pi}{4}\right) = \frac{\pi}{2}$$

Method 2.

As in Method 1 we let $x-1 = \tan t$ but we note that when $x = 0$, $\tan t = -1$, $t = -\pi/4$ and when $x = 2$, $\tan t = 1$, $t = \pi/4$. Then
$$\int_{x=0}^2 \frac{dx}{x^2-2x+2} = \int_{x=0}^2 \frac{dx}{(x-1)^2+1} = \int_{t=-\pi/4}^{\pi/4} \frac{\sec^2 t\,dt}{\tan^2 t+1} = \int_{t=-\pi/4}^{\pi/4} dt = \frac{\pi}{2}$$

Note that all of these can be interpreted geometrically as an area.

THE SUM OPERATOR AND RULES OF SUMMATION

3.9. Prove that if $\frac{\Delta}{\Delta x}F_1(x) = f(x)$ and $\frac{\Delta}{\Delta x}F_2(x) = f(x)$, then $F_1(x) - F_2(x) = C(x)$ where $C(x+h) = C(x)$.

By subtraction we have

$$\frac{\Delta}{\Delta x}F_1(x) - \frac{\Delta}{\Delta x}F_2(x) = f(x) - f(x) = 0$$

or

$$\frac{\Delta}{\Delta x}[F_1(x) - F_2(x)] = 0 \quad \text{i.e.} \quad \Delta[F_1(x) - F_2(x)] = 0$$

If we write $F_1(x) - F_2(x) = C(x)$ then

$$\Delta C(x) = 0 \quad \text{or} \quad C(x+h) - C(x) = 0, \quad \text{i.e.} \quad C(x+h) = C(x)$$

We refer to $C(x)$ as a periodic constant with period h. We have proved that if two functions have the same difference, then they can differ by at most an arbitrary periodic constant. The result is analogous to the theorem of integral calculus which states that two functions which have the same derivative can differ at most by an arbitrary constant.

3.10. Prove that Δ^{-1} is a linear operator, i.e. (*a*) $\Delta^{-1}[f(x)+g(x)] = \Delta^{-1}f(x) + \Delta^{-1}g(x)$ and (*b*) $\Delta^{-1}[\alpha f(x)] = \alpha\Delta^{-1}f(x)$ where $f(x)$ and $g(x)$ are any functions and α is any constant.

(*a*) By definition if $\Delta F(x) = f(x)$ and $\Delta G(x) = g(x)$ then

$$F(x) = \Delta^{-1}f(x), \qquad G(x) = \Delta^{-1}g(x) \tag{1}$$

Also by adding $\Delta F(x) = f(x)$ and $\Delta G(x) = g(x)$ we have

$$\Delta[F(x) + G(x)] = f(x) + g(x)$$

so that by definition

$$F(x) + G(x) = \Delta^{-1}[f(x) + g(x)] \tag{2}$$

Using (*1*) in (*2*) we find as required

$$\Delta^{-1}[f(x) + g(x)] = \Delta^{-1}f(x) + \Delta^{-1}g(x) \tag{3}$$

(*b*) For any constant α we have

$$\Delta[\alpha F(x)] = \alpha\Delta F(x) = \alpha f(x)$$

Thus from the first equation in (*1*)

$$\Delta^{-1}[\alpha f(x)] = \alpha F(x) = \alpha\Delta^{-1}[f(x)]$$

3.11. If we define $F(x) = \Delta^{-1}[hf(x)] = \Sigma f(x)h + C(x)$ as in (*8*), page 82, prove that (*a*) $\Delta\Sigma = 1$ or $\Sigma = \Delta^{-1}$, (*b*) $\Delta^{-1}[f(x)] = \Sigma f(x) + C_1(x)$ where $C_1(x)$ is an arbitrary periodic constant, and (*c*) $\left(\frac{\Delta}{\Delta x}\right)^{-1} f(x) = \Sigma f(x)\,\Delta x + C(x)$.

(*a*) By operating with Δ on

$$F(x) = \Delta^{-1}[hf(x)] = \sum f(x)h + C(x) \tag{1}$$

we obtain

$$\Delta F(x) = hf(x) = \Delta \sum f(x)h + \Delta C(x) \tag{2}$$

since Δ is a linear operator. Now by Problem 3.9, $\Delta C(x) = 0$ so that (*2*) becomes

$$hf(x) = \Delta \sum f(x)h \tag{3}$$

But since $f(x)$ is arbitrary and $h \neq 0$ it follows from (*3*) that

$$\Delta \sum = 1 \quad \text{or} \quad \sum = \Delta^{-1}$$

It follows from this and Problem 3.10 that Σ is a linear operator.

(*b*) Dividing equation (*1*) by h using the fact that Σ and Δ^{-1} are linear operators, we have

$$\Delta^{-1}[f(x)] = \sum f(x) + \frac{C(x)}{h} = \sum f(x) + C_1(x)$$

where $C(x)$ and thus $C_1(x) = C(x)/h$ are periodic constants.

(c) This follows at once from (1) on noting that

$$\Delta^{-1}[hf(x)] \quad = \quad \frac{\Delta^{-1}}{h^{-1}}f(x) \quad = \quad \left(\frac{\Delta}{h}\right)^{-1} f(x)$$

and replacing h by Δx.

3.12. Prove that (a) $\Sigma\,[f(x)+g(x)] \;=\; \Sigma\, f(x) + \Sigma\, g(x)$, (b) $\Sigma\, \alpha f(x) \;=\; \alpha\, \Sigma\, f(x)$.

These follow at once from the fact that $\Sigma = \Delta^{-1}$ and Problem 3.10.

3.13. Prove the formula for summation by parts, i.e.

$$\sum f(x)\,\Delta g(x) \quad = \quad f(x)\,g(x) \;-\; \sum g(x+h)\,\Delta f(x)$$

We have

$$\Delta[f(x)\,g(x)] \;=\; f(x)\,\Delta g(x) + g(x+h)\,\Delta f(x)$$

or

$$f(x)\,\Delta g(x) \;=\; \Delta[f(x)\,g(x)] - g(x+h)\,\Delta f(x)$$

Operating on both sides with Δ^{-1} yields

$$\Delta^{-1}[f(x)\,\Delta g(x)] \;=\; f(x)\,g(x) - \Delta^{-1}\,[g(x+h)\,\Delta f(x)]$$

or since $\Delta^{-1} = \Sigma$,

$$\Sigma\, f(x)\,\Delta g(x) \;=\; f(x)\,g(x) \;-\; \Sigma\, g(x+h)\,\Delta f(x)$$

3.14. Prove (a) $\Sigma\, x^{(m)} \;=\; \dfrac{x^{(m+1)}}{(m+1)h}$, $m \neq -1$, (b) $\Sigma\, \sin rx \;=\; \dfrac{-\cos r(x-\frac{1}{2}h)}{2\sin \frac{1}{2}rh}$.

(a) We have from V-2, page 7, or (24), page 86, on replacing m by $m+1$

$$\Delta x^{(m+1)} \;=\; (m+1)x^{(m)}h$$

Then if $m \neq -1$,

$$\Delta\left[\frac{x^{(m+1)}}{(m+1)h}\right] \;=\; x^{(m)}$$

i.e.

$$\Delta^{-1}x^{(m)} \;=\; \frac{x^{(m+1)}}{(m+1)h}$$

or

$$\Sigma\, x^{(m)} \;=\; \frac{x^{(m+1)}}{(m+1)h}$$

apart from an additive periodic constant.

(b) From V-7, page 7,

$$\Delta[\cos rx] \;=\; -2\sin\frac{rh}{2}\sin r\left(x+\frac{h}{2}\right)$$

Replacing x by $x-\frac{1}{2}h$, we have

$$\Delta[\cos r(x-\tfrac{1}{2}h)] \;=\; -2\sin\frac{rh}{2}\sin rx$$

Thus on dividing by $-2\sin\dfrac{rh}{2}$, we have

$$\Delta\left[\frac{\cos r(x-\frac{1}{2}h)}{-2\sin\frac{1}{2}rh}\right] \;=\; \sin rx$$

so that

$$\Delta^{-1} \sin rx = -\frac{\cos r(x - \frac{1}{2}h)}{2 \sin \frac{1}{2}rh}$$

or

$$\sum \sin rx = -\frac{\cos r(x - \frac{1}{2}h)}{2 \sin \frac{1}{2}h}$$

apart from an additive periodic constant.

3.15. Use Problem 3.14(*b*) to obtain $\int \sin rx\, dx$.

From Problem 3.14(*b*) we have on multiplying by h

$$\sum h \sin rx = -\frac{h \cos r(x - \frac{1}{2}h)}{2 \sin \frac{1}{2}rh}$$

Letting $h \to 0$ this yields

$$\begin{aligned}
\int \sin rx\, dx &= \lim_{h\to 0} -\frac{h \cos r(x - \frac{1}{2}h)}{2 \sin \frac{1}{2}rh} \\
&= \lim_{h\to 0} \left[\frac{-\cos r(x - \frac{1}{2}h)}{r}\right]\left[\frac{\frac{1}{2}rh}{\sin \frac{1}{2}rh}\right] \\
&= \left[\lim_{h\to 0} \frac{-\cos r(x - \frac{1}{2}h)}{r}\right]\left[\lim_{h\to 0} \frac{\frac{1}{2}rh}{\sin \frac{1}{2}rh}\right] \\
&= -\frac{\cos rx}{r}
\end{aligned}$$

The arbitrary periodic constant $C(x)$ becomes an arbitrary constant c which should be added to the indefinite integral. The result illustrates Theorem 3-1, page 83.

3.16. (*a*) Find $\Delta^{-1}(xa^x) = \sum xa^x$, $a \neq 1$, and (*b*) check your answer.

(*a*) We have from Problem 3.13

$$\Delta^{-1}[f(x)\,\Delta g(x)] = f(x)\,g(x) - \Delta^{-1}[g(x+h)\,\Delta f(x)]$$

Let $f(x) = x$ and $\Delta g(x) = a^x$. Then $\Delta f(x) = \Delta x = h$ and

$$g(x) = \Delta^{-1} a^x = \frac{a^x}{a^h - 1}$$

Thus

$$\begin{aligned}
\Delta^{-1}[xa^x] &= x \cdot \frac{a^x}{a^h - 1} - \Delta^{-1}\left[\frac{a^{x+h}}{a^h - 1}h\right] \\
&= \frac{xa^x}{a^h - 1} - \frac{ha^h}{a^h - 1}\Delta^{-1}a^x \\
&= \frac{xa^x}{a^h - 1} - \frac{ha^{x+h}}{(a^h - 1)^2}
\end{aligned}$$

(*b*) ***Check:***

$$\begin{aligned}
\Delta\left[\frac{xa^x}{a^h - 1} - \frac{ha^{x+h}}{(a^h - 1)^2}\right] &= \left[\frac{(x+h)a^{x+h}}{a^h - 1} - \frac{ha^{x+2h}}{(a^h - 1)^2}\right] - \left[\frac{xa^x}{a^h - 1} - \frac{ha^{x+h}}{(a^h - 1)^2}\right] \\
&= \frac{(a^h - 1)(x+h)a^{x+h} - ha^{x+2h} - (a^h - 1)(xa^x) + ha^{x+h}}{(a^h - 1)^2} \\
&= xa^x
\end{aligned}$$

DEFINITE SUMS AND THE FUNDAMENTAL THEOREM OF SUM CALCULUS

3.17. Prove Theorem 3-2, the fundamental theorem of sum calculus, i.e.

$$\sum_{a}^{a+(n-1)h} f(x) \;=\; \Delta^{-1}f(x)\Big|_{a}^{a+nh}$$

If we assume that $F(x)$ and $f(x)$ are related by

$$\frac{\Delta F(x)}{\Delta x} = f(x) \qquad \text{or} \qquad \Delta F(x) = hf(x) \tag{1}$$

we have

$$F(x+h) - F(x) = hf(x) \tag{2}$$

Then by putting $x = a, a+h, \ldots, a+(n-1)h$ successively in (2) we obtain the equations

$$\begin{aligned} F(a+h) - F(a) &= hf(a) \\ F(a+2h) - F(a+h) &= hf(a+h) \\ &\cdots\cdots\cdots \\ F(a+nh) - F[a+(n-1)h] &= hf[a+(n-1)h] \end{aligned}$$

Thus by addition we have

$$F(a+nh) - F(a) = h\{f(a) + f(a+h) + \cdots + f[a+(n-1)h]\}$$

or

$$\sum_{a}^{a+(n-1)h} f(x) = \frac{F(a+nh)-F(a)}{h} = \frac{F(x)}{h}\Big|_{a}^{a+nh} \tag{3}$$

where the sum is taken from $x = a$ to $a+(n-1)h$ in steps of h.

But from the second equation in (1) it follows that

$$\Delta\left[\frac{F(x)}{h}\right] = f(x) \qquad \text{or} \qquad \frac{F(x)}{h} = \Delta^{-1}f(x) \tag{4}$$

and the required result is proved on using this in (3).

3.18. Evaluate each of the following using the fundamental theorem of sum calculus and verify directly

(a) $\sum_{1}^{7} 8x^{(3)}$ if $h = 3$

(b) $\sum_{2}^{6} [3x^{(2)} - 8x^{(1)} + 10]$ if $h = 2$

(c) $\sum_{0}^{2\pi} \cos 5x$ if $h = \pi/2$

(d) $\sum_{3}^{19} x^{(-3)}$ if $h = 4$

(a) Since $\sum 8x^{(3)} = 8\sum x^{(3)} = 8\dfrac{x^{(4)}}{4\cdot 3} = \dfrac{2x^{(4)}}{3}$ we have by the fundamental theorem of sum calculus using $a = 1$ and $h = 3$,

$$\begin{aligned} \sum_{1}^{7} 8x^{(3)} &= \frac{2x^{(4)}}{3}\Big|_{1}^{10} = \frac{2\cdot 10^{(4)}}{3} - \frac{2\cdot 1^{(4)}}{3} \\ &= \frac{2(10)(7)(4)(1)}{3} - \frac{2(1)(-2)(-5)(-8)}{3} = 240 \end{aligned}$$

Check: $\sum_{1}^{7} 8x^{(3)} = 8[1^{(3)} + 4^{(3)} + 7^{(3)}] = 8[(1)(-2)(-5) + (4)(1)(-2) + (7)(4)(1)] = 240$

(b)

$$\sum [3x^{(2)} - 8x^{(1)} + 10] = 3\frac{x^{(3)}}{3\cdot 2} - 8\frac{x^{(2)}}{2\cdot 2} + 10\frac{x}{2} = \frac{x^{(3)}}{2} - 2x^{(2)} + 5x$$

Then

$$\begin{aligned} \sum_{2}^{6} [3x^{(2)} - 8x^{(1)} + 10] &= \frac{x^{(3)}}{2} - 2x^{(2)} + 5x\Big|_{2}^{8} \\ &= \left[\frac{(8)(6)(4)}{2} - 2(8)(6) + 5(8)\right] - \left[\frac{(2)(0)(-2)}{2} - 2(2)(0) + 5(2)\right] = 30 \end{aligned}$$

Check:

$$\sum_{2}^{6} [3x^{(2)} - 8x^{(1)} + 10] = [3(2)(0) - 8(2) + 10] + [3(4)(2) - 8(4) + 10] + [3(6)(4) - 8(6) + 10] = 30$$

(*c*) By IV-7, page 83, with $r = 5$, $h = \pi/2$ we have

$$\Delta^{-1} \cos 5x = \frac{\sin 5[x - (\pi/4)]}{2 \sin (5\pi/4)}$$

Thus

$$\sum_{0}^{2\pi} \cos 5x = \left. \frac{\sin 5[x - (\pi/4)]}{2 \sin (5\pi/4)} \right|_{0}^{5\pi/2}$$

$$= \frac{\sin (45\pi/4)}{2 \sin (5\pi/4)} - \frac{\sin (-5\pi/4)}{2 \sin (5\pi/4)} = 1$$

Check:

$$\sum_{0}^{2\pi} \cos 5x = \cos 0 + \cos (5\pi/2) + \cos (5\pi) + \cos (15\pi/2) + \cos (10\pi) = 1$$

(*d*) Since $\Delta^{-1}[x^{(-3)}] = \dfrac{x^{(-2)}}{(-2)(4)} = \dfrac{x^{(-2)}}{-8}$ we have

$$\sum_{3}^{19} x^{(-3)} = \left. \frac{x^{(-2)}}{-8} \right|_{3}^{23} = -\frac{1}{8}[(23)^{(-2)} - (3)^{(-2)}]$$

$$= -\frac{1}{8}\left[\frac{1}{(23)(27)} - \frac{1}{(3)(7)}\right] = \frac{25}{4347}$$

Check:

$$\sum_{3}^{19} x^{(-3)} = \frac{1}{3 \cdot 7 \cdot 11} + \frac{1}{7 \cdot 11 \cdot 15} + \frac{1}{11 \cdot 15 \cdot 19} + \frac{1}{15 \cdot 19 \cdot 23} = \frac{25}{4347}$$

DIFFERENTIATION AND INTEGRATION OF SUMS

3.19. Find $\Delta^{-1}[xe^{rx}]$ by using Theorem 3-4, page 84.

We have from IV-5, page 83,

$$\Delta^{-1}[e^{rx}] = \frac{e^{rx}}{e^{rh} - 1}$$

Then by differentiating both sides with respect to r [which corresponds to α in Theorem 3-4] we find

$$\Delta^{-1}\left[\frac{\partial}{\partial r} e^{rx}\right] = \frac{\partial}{\partial r}\left(\frac{e^{rx}}{e^{rh} - 1}\right)$$

or

$$\Delta^{-1}[xe^{rx}] = \frac{(e^{rh} - 1)(xe^{rx}) - (e^{rx})(he^{rh})}{(e^{rh} - 1)^2}$$

$$= \frac{(x - h)e^{r(x+h)} - xe^{rx}}{(e^{rh} - 1)^2}$$

This can be checked by referring to an alternative method of obtaining the result given in Problem 3.16(*a*). From that problem we find on putting $a = e^r$

$$\Delta^{-1}[xe^{rx}] = \frac{xe^{rx}}{e^{rh} - 1} - \frac{he^{r(x+h)}}{(e^{rh} - 1)^2} = \frac{(x - h)e^{r(x+h)} - xe^{rx}}{(e^{rh} - 1)^2}$$

SUMMATION OF SERIES

3.20. Sum the series $1 \cdot 3 \cdot 5 + 3 \cdot 5 \cdot 7 + 5 \cdot 7 \cdot 9 + \cdots$ to n terms.

Since the nth term of the series is $(2n - 1)(2n + 1)(2n + 3)$, the series can be written as

$$\sum_{5}^{2n+3} x^{(3)} \quad \text{where } h = 2$$

Then by the fundamental theorem of sum calculus,

$$\sum_{5}^{2n+3} x^{(3)} = \Delta^{-1}x^{(3)}\Big|_{5}^{2n+5} = \frac{x^{(4)}}{4\cdot 2}\Big|_{5}^{2n+5}$$

$$= \frac{(2n+5)(2n+3)(2n+1)(2n-1)}{8} - \frac{(5)(3)(1)(-1)}{8}$$

$$= \frac{(2n+5)(2n+3)(2n+1)(2n-1)+15}{8}$$

3.21. Sum the series (*a*) $1^2+2^2+3^2+4^2+\cdots$ and (*b*) $2^2+5^2+8^2+11^2+\cdots$ to n terms.

Since we wish to find $\Sigma\, x^2$ we need to express x^2 as a factorial polynomial. This is found from

$$x^2 = x^{(2)} + hx^{(1)}$$

(*a*) Using $h=1$ the series is given by

$$\sum_{1}^{n} x^2 = \sum_{1}^{n}[x^{(2)}+x^{(1)}] = \Delta^{-1}[x^{(2)}+x^{(1)}]\Big|_{1}^{n+1} = \frac{x^{(3)}}{3}+\frac{x^{(2)}}{2}\Big|_{1}^{n+1}$$

$$= \left[\frac{(n+1)(n)(n-1)}{3}+\frac{(n+1)(n)}{2}\right]-\left[\frac{(1)(0)(-1)}{3}+\frac{(1)(0)}{2}\right]$$

$$= \frac{(n+1)(n)(n-1)}{3}+\frac{(n+1)(n)}{2} = \frac{n(n+1)(2n+1)}{6}$$

As a possible check we can try a particular value of n, say $n=3$. Then we obtain

$$1^2 + 2^2 + 3^2 = \frac{(2)(3)(7)}{6} = 14$$

which is correct.

(*b*) The nth term is $(3n-1)^2$. Thus using $h=3$ the series can be represented by

$$\sum_{2}^{3n-1} x^2 = \sum_{2}^{3n-1}[x^{(2)}+3x^{(1)}] = \Delta^{-1}[x^{(2)}+3x^{(1)}]\Big|_{2}^{3n+2} = \frac{x^{(3)}}{3\cdot 3}+3\,\frac{x^{(2)}}{2\cdot 3}\Big|_{2}^{3n+2}$$

$$= \left[\frac{(3n+2)(3n-1)(3n-4)}{9}+\frac{(3n+2)(3n-1)}{2}\right]-\left[\frac{(2)(-1)(-4)}{9}+\frac{(2)(-1)}{2}\right]$$

$$= \frac{(3n+2)(3n-1)(6n+1)+2}{18} = \frac{n(6n^2+3n-1)}{2}$$

Check: Let $n=3$. Then

$$2^2 + 5^2 + 8^2 = \frac{(3)(6\cdot 3^2+3\cdot 3-1)}{2} = 93$$

which is correct.

3.22. Sum the series $\dfrac{1}{1\cdot 3\cdot 5}+\dfrac{1}{3\cdot 5\cdot 7}+\dfrac{1}{5\cdot 7\cdot 9}+\cdots$ to n terms.

The nth term of the series is $\dfrac{1}{(2n-1)(2n+1)(2n+3)}$.

Now since

$$(x-h)^{(-3)} = \frac{1}{x(x+h)(x+2h)}$$

it follows that if we let $h=2$, the required sum of the series is

$$\sum_{1}^{2n-1} (x-h)^{(-3)} = \sum_{1}^{2n-1} (x-2)^{(-3)} = \Delta^{-1}(x-2)^{(-3)}\Big|_{1}^{2n+1}$$

$$= \frac{(x-2)^{(-2)}}{(-2)(2)}\Big|_{1}^{2n+1} = \left[\frac{(2n-1)^{(-2)}}{-4}\right] - \left[\frac{(-1)^{(-2)}}{-4}\right]$$

$$= \left[-\frac{1}{4}\cdot\frac{1}{(2n+1)(2n+3)}\right] - \left[-\frac{1}{4}\cdot\frac{1}{(1)(3)}\right]$$

$$= \frac{1}{12} - \frac{1}{4(2n+1)(2n+3)}$$

***Check*:** Putting $n=2$ we find $\dfrac{1}{1\cdot3\cdot5} + \dfrac{1}{3\cdot5\cdot7} = \dfrac{1}{12} - \dfrac{1}{4(5)(7)} = \dfrac{8}{105}$ which is correct.

3.23. Sum the series $\dfrac{1}{1\cdot3\cdot5} + \dfrac{1}{3\cdot5\cdot7} + \dfrac{1}{5\cdot7\cdot9} + \cdots$.

In this case we are to sum the series of Problem 3.22 to infinitely many terms. The required sum using the result of Problem 3.22 is

$$\lim_{n\to\infty}\left[\frac{1}{12} - \frac{1}{4(2n+1)(2n+3)}\right] = \frac{1}{12}$$

3.24. Sum the series $\frac{1}{2} + \cos\theta + \cos 2\theta + \cdots + \cos n\theta$.

Using $h=\theta$ the series is given by

$$\frac{1}{2} + \sum_{\theta}^{n\theta} \cos x = \frac{1}{2} + \Delta^{-1}\cos x\Big|_{\theta}^{(n+1)\theta} = \frac{1}{2} + \frac{\sin(x-\frac{1}{2}\theta)}{2\sin\frac{1}{2}\theta}\Big|_{\theta}^{(n+1)\theta}$$

$$= \frac{1}{2} + \frac{\sin(n+\frac{1}{2})\theta}{2\sin\frac{1}{2}\theta} - \frac{\sin\frac{1}{2}\theta}{2\sin\frac{1}{2}\theta} = \frac{\sin(n+\frac{1}{2})\theta}{2\sin\frac{1}{2}\theta}$$

3.25. Sum the series $\dfrac{1}{1\cdot4} + \dfrac{1}{2\cdot5} + \dfrac{1}{3\cdot6} + \cdots$ to n terms.

Using $h=1$ the sum of the series to n terms is given by

$$\sum_{1}^{n} \frac{1}{x(x+3)}$$

To find this we attempt to express the general term as a sum of factorials. We first write

$$\frac{1}{x(x+3)} = \frac{(x+1)(x+2)}{x(x+1)(x+2)(x+3)} = \frac{(x+1)(x+2)}{(x+3)^{(4)}} \qquad (1)$$

We then seek to determine constants $A_0, A_1, A_2, \ldots$ such that the numerator is the sum of factorials in the form

$$(x+1)(x+2) = A_0 + A_1(x+3)^{(1)} + A_2(x+3)^{(2)} + A_3(x+3)^{(3)} + \cdots \qquad (2)$$

Since the left side of (*2*) is of degree 2, it is clear that $A_3, \ldots$ must be zero. Thus

$$(x+1)(x+2) = A_0 + A_1(x+3)^{(1)} + A_2(x+3)^{(2)} = A_0 + A_1(x+3) + A_2(x+3)(x+2)$$

From this identity we find $A_0 = 2$, $A_1 = -2$, $A_2 = 1$. Thus (*1*) becomes

$$\frac{1}{x(x+3)} = \frac{2 - 2(x+3) + (x+3)(x+2)}{x(x+1)(x+2)(x+3)}$$

$$= \frac{2}{x(x+1)(x+2)(x+3)} - \frac{2}{x(x+1)(x+2)} + \frac{1}{x(x+1)}$$

$$= 2(x-1)^{(-4)} - 2(x-1)^{(-3)} + (x-1)^{(-2)}$$

Thus apart from an additive constant,

$$\Delta^{-1}\left[\frac{1}{x(x+3)}\right] = \frac{2(x-1)^{(-3)}}{-3} - \frac{2(x-1)^{(-2)}}{-2} + \frac{(x-1)^{(-1)}}{-1}$$

Then by the fundamental theorem of sum calculus

$$\sum_{1}^{n} \frac{1}{x(x+3)} = \left\{\frac{2(x-1)^{(-3)}}{-3} - \frac{2(x-1)^{(-2)}}{-2} + \frac{(x-1)^{(-1)}}{-1}\right\}\Bigg|_{1}^{n+1}$$

$$= -\frac{2}{3}n^{(-3)} + n^{(-2)} - n^{(-1)} + \frac{2}{3}(0)^{(-3)} - (0)^{(-2)} + (0)^{(-1)}$$

$$= -\frac{2}{3(n+1)(n+2)(n+3)} + \frac{1}{(n+1)(n+2)} - \frac{1}{n+1} + \frac{2}{3(1)(2)(3)} - \frac{1}{(1)(2)} + \frac{1}{1}$$

$$= \frac{11}{18} - \frac{1}{n+1} + \frac{1}{(n+1)(n+2)} - \frac{2}{3(n+1)(n+2)(n+3)}$$

3.26. Sum the series $\frac{1}{1\cdot 4} + \frac{1}{2\cdot 5} + \frac{1}{3\cdot 6} + \cdots$.

From Problem 3.25 by letting $n \to \infty$ it follows that the required sum is 11/18.

SUMMATION USING SUBSCRIPT NOTATION

3.27. Sum the series $1\cdot 3\cdot 5 + 3\cdot 5\cdot 7 + 5\cdot 7\cdot 9 + \cdots$ to n terms by using the subscript or k notation.

The kth term of the series is $y_k = (2k-1)(2k+1)(2k+3) = (2k+3)^{(3)}$ where the difference interval $h = 1$ [as is always the case for the k notation].

Then the sum of the series is

$$\sum_{k=1}^{n} (2k+3)^{(3)} = \frac{(2k+3)^{(4)}}{4\cdot 2}\Bigg|_{1}^{n+1} = \frac{(2n+5)^{(4)}}{8} - \frac{5^{(4)}}{8}$$

$$= \frac{(2n+5)(2n+3)(2n+1)(2n-1)}{8} - \frac{(5)(3)(1)(-1)}{8}$$

$$= \frac{(2n+5)(2n+3)(2n+1)(2n-1)+15}{8}$$

3.28. Sum the series $2^2 + 5^2 + 8^2 + \cdots$ to n terms by using the subscript or k notation.

The kth term of the series is $y_k = (3k-1)^2$. Then the sum of the series is

$$\sum_{k=1}^{n} (3k-1)^2 = \sum_{k=1}^{n} (9k^2 - 6k + 1) = \sum_{k=1}^{n} [9(k^{(2)} + k^{(1)}) - 6k^{(1)} + 1]$$

$$= \sum_{k=1}^{n} [9k^{(2)} + 3k^{(1)} + 1] = 9\frac{k^{(3)}}{3} + 3\frac{k^{(2)}}{2} + k\Bigg|_{1}^{n+1}$$

$$= [3(n+1)^{(3)} + \tfrac{3}{2}(n+1)^{(2)} + n + 1] - [3(1)^{(3)} + \tfrac{3}{2}(1)^{(2)} + 1]$$

$$= [3(n+1)(n)(n-1) + \tfrac{3}{2}(n+1)(n) + (n+1)] - [(3)(1)(0)(-1) + \tfrac{3}{2}(1)(0) + 1]$$

$$= \frac{n(6n^2+3n-1)}{2}$$

ABEL'S TRANSFORMATION

3.29. Prove *Abel's transformation.*

$$\sum_{k=1}^{n} u_k v_k \quad = \quad u_{n+1} \sum_{k=1}^{n} v_k - \sum_{k=1}^{n} \left[\Delta u_k \sum_{p=1}^{k} v_p \right]$$

Using summation by parts and the fundamental theorem of sum calculus we have

$$\sum_{a}^{a+(n-1)h} f(x)\,\Delta g(x) \quad = \quad f(x)\,g(x)\Big|_{a}^{a+nh} - \sum_{a}^{a+(n-1)h} g(x+h)\,\Delta f(x) \qquad (1)$$

where the summation is taken in steps of h.

Putting $a=1,\ h=1,\ x=k$ and writing $f(k)=f_k,\ g(k)=g_k,$ (*1*) becomes

$$\sum_{k=1}^{n} f_k \Delta g_k \quad = \quad f_{n+1}g_{n+1} - f_1 g_1 - \sum_{k=1}^{n} g_{k+1}\Delta f_k \qquad (2)$$

Now let $f_k = u_k$ and $\Delta g_k = v_k$. Then $g_{k+1} - g_k = v_k$ and so on summing from $k=1$ *to* $n-1$ we find

$$\sum_{k=1}^{n-1} (g_{k+1} - g_k) \quad = \quad g_n - g_1 \quad = \quad \sum_{k=1}^{n-1} v_k$$

i.e.

$$g_n \quad = \quad g_1 + \sum_{k=1}^{n-1} v_k$$

From this we have

$$g_{k+1} \quad = \quad g_1 + \sum_{p=1}^{k} v_p$$

Then (*2*) becomes

$$\begin{aligned}
\sum_{k=1}^{n} u_k v_k &= u_{n+1}\left[g_1 + \sum_{k=1}^{n} v_k\right] - u_1 g_1 - \sum_{k=1}^{n}\left\{\Delta u_k \left[g_1 + \sum_{p=1}^{k} v_p\right]\right\} \\
&= u_{n+1}\sum_{k=1}^{n} v_k - \sum_{k=1}^{n}\left[\Delta u_k \sum_{p=1}^{k} v_p\right] + u_{n+1}g_1 - u_1 g_1 - \sum_{k=1}^{n} g_1 \Delta u_k \\
&= u_{n+1}\sum_{k=1}^{n} v_k - \sum_{k=1}^{n}\left[\Delta u_k \sum_{p=1}^{k} v_p\right]
\end{aligned}$$

since

$$\begin{aligned}
u_{n+1}g_1 - u_1 g_1 - \sum_{k=1}^{n} g_1 \Delta u_k &= u_{n+1}g_1 - u_1 g_1 - g_1 \sum_{k=1}^{n} \Delta u_k \\
&= u_{n+1}g_1 - u_1 g_1 - g_1 \sum_{k=1}^{n} (u_{k+1} - u_k) \\
&= u_{n+1}g_1 - u_1 g_1 - g_1(u_{n+1} - u_1) \\
&= 0
\end{aligned}$$

OPERATOR METHODS OF SUMMATION

3.30. Prove that if $\beta \neq 1$

$$\Delta^{-1}[\beta^k P(k)] \quad = \quad \sum \beta^k P(k) \quad = \quad \frac{\beta^k}{\beta - 1}\left[1 - \frac{\beta\Delta}{\beta - 1} + \frac{\beta^2\Delta^2}{(\beta-1)^2} - \frac{\beta^3\Delta^3}{(\beta-1)^3} + \cdots\right]P(k)$$

apart from an arbitrary additive constant.

We have for any $F(k)$

$$\Delta\beta^k F(k) \;=\; \beta^{k+1}F(k+1) - \beta^k F(k) \;=\; \beta^{k+1}EF(k) - \beta^k F(k) \;=\; \beta^k(\beta E - 1)F(k)$$

Let

$$(\beta E - 1)F(k) = P(k) \quad \text{or} \quad F(k) = \frac{1}{\beta E - 1}P(k)$$

Then

$$\Delta \beta^k F(k) = \beta^k P(k)$$

Thus

$$\begin{aligned}
\Delta^{-1}[\beta^k P(k)] &= \beta^k F(k) = \beta^k \frac{1}{\beta E - 1} P(k) \\
&= \beta^k \frac{1}{\beta(1+\Delta) - 1} P(k) = \frac{\beta^k}{\beta - 1} \frac{1}{1 + [\beta\Delta/(\beta - 1)]} P(k) \\
&= \frac{\beta^k}{\beta - 1}\left[1 - \frac{\beta\Delta}{\beta - 1} + \frac{\beta^2\Delta^2}{(\beta-1)^2} - \frac{\beta^3\Delta^3}{(\beta-1)^3} + \cdots\right] P(k)
\end{aligned}$$

apart from an arbitrary additive constant and assuming $\beta \neq 1$. Since $\Delta^{-1}\beta^k P(k) = \sum \beta^k P(k)$ the required result follows. Note that if $P(k)$ is any polynomial the series terminates.

3.31. Use Problem 3.30 to find $\sum_{k=1}^{n} k \cdot 2^k$.

We have by Problem 3.30 with $\beta = 2$, $P(k) = k$

$$\Delta^{-1}[k \cdot 2^k] = \sum k \cdot 2^k = 2^k[1 - 2\Delta + 4\Delta^2 - \cdots]k = 2^k(k-2)$$

Then by the fundamental theorem of sum calculus

$$\sum_{k=1}^{n} k \cdot 2^k = \Delta^{-1}[k \cdot 2^k] \Big|_{k=1}^{n+1} = 2^k(k-2) \Big|_{k=1}^{n+1} = 2^{n+1}(n-1) + 2$$

THE GAMMA FUNCTION

3.32. Prove that $\Gamma(p+1) = p\Gamma(p)$.

Integrating by parts we have for $p > 0$

$$\begin{aligned}
\Gamma(p+1) &= \lim_{M\to\infty} \int_0^M t^p e^{-t}\,dt = \lim_{M\to\infty}\left[(t^p)(-e^{-t})\Big|_0^M - \int_0^M (-e^{-t})(pt^{p-1})\,dt\right] \\
&= \lim_{M\to\infty}\left[-M^p e^{-M} + p\int_0^M t^{p-1}e^{-t}\,dt\right] \\
&= p\int_0^\infty t^{p-1}e^{-t}\,dt \\
&= p\Gamma(p)
\end{aligned}$$

3.33. Prove that $\Gamma(\frac{1}{2}) = \sqrt{\pi}$.

We have

$$\Gamma(\tfrac{1}{2}) = \int_0^\infty t^{-1/2}e^{-t}\,dt = 2\int_0^\infty e^{-x^2}\,dx$$

on letting $t = x^2$. Then

$$\begin{aligned}
\{\Gamma(\tfrac{1}{2})\}^2 &= \left(2\int_0^\infty e^{-x^2}\,dx\right)\left(2\int_0^\infty e^{-y^2}\,dy\right) \\
&= 4\int_0^\infty\int_0^\infty e^{-(x^2+y^2)}\,dx\,dy \qquad (1)
\end{aligned}$$

where the integration is taken over the first quadrant of the xy plane. We can however equivalently perform this integration by using polar coordinates (ρ, ϕ)

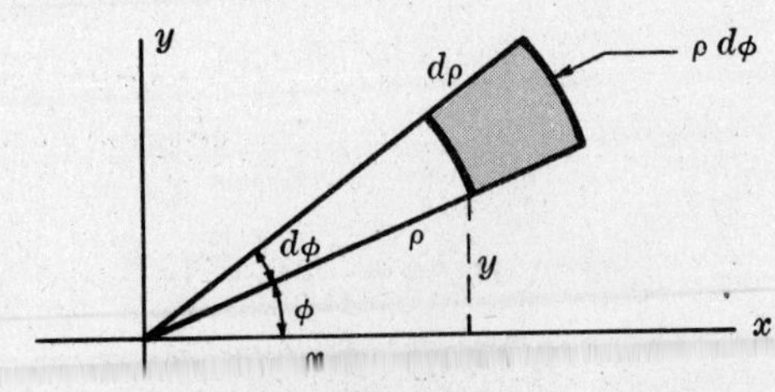

Fig. 3-3

rather than rectangular coordinates (x, y). To do this we note that the element of area in the xy plane shown shaded in Fig. 3-3 is $(\rho\,d\phi)\,d\rho = \rho\,d\rho\,d\phi$. Also since $x = \rho\cos\phi$, $y = \rho\sin\phi$, we have $x^2 + y^2 = \rho^2$ so that the result *(1)* with $dx\,dy$ replaced by $\rho\,d\rho\,d\phi$ becomes

$$\{\Gamma(\tfrac{1}{2})\}^2 = 4\int_{\phi=0}^{\pi/2}\int_{\rho=0}^{\infty} e^{-\rho^2}\rho\,d\rho\,d\phi \qquad (2)$$

The limits for ρ and ϕ in *(2)* are determined from the fact that if we fix ϕ, ρ goes from 0 to ∞ and then we vary ϕ from 0 to $\pi/2$.

The integral in *(2)* is equal to

$$4\int_{\phi=0}^{\pi/2}\left[\int_{\rho=0}^{\infty} e^{-\rho^2}\rho\,d\rho\right]d\phi = 4\int_{\phi=0}^{\pi/2} -\tfrac{1}{2}e^{-\rho^2}\Big|_0^{\infty}\,d\phi = 4\int_{\phi=0}^{\pi/2}\tfrac{1}{2}d\phi = \pi$$

Thus $\{\Gamma(\frac{1}{2})\}^2 = \pi$ and since $\Gamma(\frac{1}{2}) > 0$ we must have $\Gamma(\frac{1}{2}) = \sqrt{\pi}$.

3.34. Find (*a*) $\Gamma(5)$, (*b*) $\Gamma(7/2)$, (*c*) $\Gamma(-1/2)$.

(*a*) Since $\Gamma(p+1) = p\Gamma(p)$ we have on putting $p = 4$, $\Gamma(5) = 4\Gamma(4)$. Similarly on putting $p = 3$, $\Gamma(4) = 3\Gamma(3)$. Continuing, we find $\Gamma(5) = 4\Gamma(4) = 4\cdot 3\Gamma(3) = 4\cdot 3\cdot 2\Gamma(2) = 4\cdot 3\cdot 2\cdot 1\Gamma(1)$. But

$$\Gamma(1) = \int_0^{\infty} e^{-t}\,dt = -e^{-t}\Big|_0^{\infty} = 1$$

Thus $\Gamma(5) = 4\cdot 3\cdot 2\cdot 1 = 4!$

In general if p is any positive integer, $\Gamma(p+1) = p!$

(*b*) Using the recursion formula we find

$$\Gamma(\tfrac{7}{2}) = \tfrac{5}{2}\Gamma(\tfrac{5}{2}) = \tfrac{5}{2}\cdot\tfrac{3}{2}\Gamma(\tfrac{3}{2}) = \tfrac{5}{2}\cdot\tfrac{3}{2}\cdot\tfrac{1}{2}\Gamma(\tfrac{1}{2}) = \tfrac{5}{2}\cdot\tfrac{3}{2}\cdot\tfrac{1}{2}\sqrt{\pi} = \frac{15\sqrt{\pi}}{8}$$

(*c*) Assuming that the gamma function is defined for all values of p by assuming that it satisfies the recursion formula $\Gamma(p+1) = p\Gamma(p)$ for all p, we have on putting $p = -1/2$

$$\Gamma(\tfrac{1}{2}) = (-\tfrac{1}{2})\Gamma(-\tfrac{1}{2}) \quad \text{or} \quad \Gamma(-\tfrac{1}{2}) = -2\Gamma(\tfrac{1}{2}) = -2\sqrt{\pi}$$

3.35. Prove that if m is an integer then

$$x^{(m)} = \frac{h^m\Gamma\left(\dfrac{x}{h}+1\right)}{\Gamma\left(\dfrac{x}{h}-m+1\right)}$$

If m is a positive integer then

$$\begin{aligned} x^{(m)} &= x(x-h)(x-2h)\cdots(x-mh+h) \\ &= h^m\left(\frac{x}{h}\right)\left(\frac{x}{h}-1\right)\left(\frac{x}{h}-2\right)\cdots\left(\frac{x}{h}-m+1\right) \end{aligned} \qquad (1)$$

Now

$$\begin{aligned} \Gamma(p+1) &= p\Gamma(p) = p(p-1)\Gamma(p-1) = p(p-1)(p-2)\Gamma(p-2) = \cdots \\ &= p(p-1)(p-2)\cdots(p-m+1)\Gamma(p-m+1) \end{aligned}$$

Thus

$$\frac{\Gamma(p+1)}{\Gamma(p-m+1)} = p(p-1)(p-2)\cdots(p-m+1) \qquad (2)$$

Using *(2)* with $p = x/h$ together with *(1)* we have

$$x^{(m)} = \frac{h^m\Gamma\left(\dfrac{x}{h}+1\right)}{\Gamma\left(\dfrac{x}{h}-m+1\right)} \qquad (3)$$

Similarly we can prove that if m is a negative integer the result holds. In general we shall take the result as defining $x^{(m)}$ for all values of m. See Problem 3.109.

3.36. Assuming that the result obtained in Problem 3.35 is the defining equation for $x^{(m)}$ for all m, prove that $\Delta x^{(m)} = mx^{(m-1)}h$.

We have

$$\Delta x^{(m)} = \Delta\left[\frac{h^m\Gamma\left(\frac{x}{h}+1\right)}{\Gamma\left(\frac{x}{h}-m+1\right)}\right]$$

$$= \frac{h^m\Gamma\left(\frac{x+h}{h}+1\right)}{\Gamma\left(\frac{x+h}{h}-m+1\right)} - \frac{h^m\Gamma\left(\frac{x}{h}+1\right)}{\Gamma\left(\frac{x}{h}-m+1\right)}$$

$$= \frac{h^m\Gamma\left(\frac{x}{h}+2\right)}{\Gamma\left(\frac{x}{h}-m+2\right)} - \frac{h^m\Gamma\left(\frac{x}{h}+1\right)}{\Gamma\left(\frac{x}{h}-m+1\right)}$$

$$= \frac{h^m\left(\frac{x}{h}+1\right)\Gamma\left(\frac{x}{h}+1\right)}{\left(\frac{x}{h}-m+1\right)\Gamma\left(\frac{x}{h}-m+1\right)} - \frac{h^m\Gamma\left(\frac{x}{h}+1\right)}{\Gamma\left(\frac{x}{h}-m+1\right)}$$

$$= \frac{h^m\Gamma\left(\frac{x}{h}+1\right)}{\Gamma\left(\frac{x}{h}-m+1\right)}\left[\frac{\frac{x}{h}+1}{\frac{x}{h}-m+1}-1\right]$$

$$= \frac{mh^m\Gamma\left(\frac{x}{h}+1\right)}{\Gamma\left(\frac{x}{h}-m+2\right)} = mx^{(m-1)}h$$

3.37. Show that $\sum x^{(m)} = \dfrac{x^{(m+1)}}{(m+1)h}$ for all $m \neq -1$.

From Problem 3.36 $\Delta x^{(m+1)} = (m+1)x^{(m)}h$. Then if $m \neq -1$,

$$\Delta\left[\frac{x^{(m+1)}}{(m+1)h}\right] = x^{(m)} \quad \text{or} \quad \Delta^{-1}x^{(m)} = \frac{x^{(m+1)}}{(m+1)h}; \quad \text{i.e.} \quad \sum x^{(m)} = \frac{x^{(m+1)}}{(m+1)h} \qquad m \neq -1$$

3.38. Let the *digamma function* be given by

$$\Psi(x) = \frac{d}{dx}\ln\Gamma\left(\frac{x}{h}+1\right) = \frac{\Gamma'\left(\frac{x}{h}+1\right)}{h\Gamma\left(\frac{x}{h}+1\right)}$$

Prove that $\Delta\Psi(x) = 1/(x+h)$.

Since Δ and D are commutative with respect to multiplication [see Problem 1.58, page 28],

$$\Delta\Psi(x) = \Delta D\ln\Gamma(x/h+1) = D\Delta\ln\Gamma(x/h+1)$$

$$= D\left[\ln\Gamma\left(\frac{x+h}{h}+1\right) - \ln\Gamma\left(\frac{x}{h}+1\right)\right]$$

$$= D\ln\frac{\Gamma\left(\frac{x}{h}+2\right)}{\Gamma\left(\frac{x}{h}+1\right)} = D\ln\left(\frac{x}{h}+1\right) = \frac{1}{x+h}$$

3.39. Prove that $\sum x^{(-1)} = \sum \frac{1}{x+h} = \frac{\Gamma'\left(\frac{x}{h}+1\right)}{h\Gamma\left(\frac{x}{h}+1\right)}$.

From Problem 3.38 we have since $\frac{1}{x+h} = x^{(-1)}$,

$$\Delta^{-1}\frac{1}{x+h} \quad = \quad \Delta^{-1}x^{(-1)} \quad = \quad \Psi(x) \quad = \quad \frac{\Gamma'\left(\frac{x}{h}+1\right)}{h\Gamma\left(\frac{x}{h}+1\right)}$$

i.e.

$$\sum x^{(-1)} \quad = \quad \sum \frac{1}{x+h} \quad = \quad \frac{\Gamma'\left(\frac{x}{h}+1\right)}{h\Gamma\left(\frac{x}{h}+1\right)}$$

BERNOULLI NUMBERS AND POLYNOMIALS

3.40. Prove that if $f_n(x)$ is a polynomial of degree n having the property that $Df_n(x) = f_{n-1}(x)$, then $f_n(x)$ must have the form

$$f_n(x) \quad = \quad \frac{c_0 x^n}{n!} + \frac{c_1 x^{n-1}}{(n-1)!} + \cdots + c_{n-1}x + c_n$$

where $c_0, c_1, \ldots, c_n$ are constants independent of n.

Suppose that $\quad f_n(x) = A_0(n)x^n + A_1(n)x^{n-1} + \cdots + A_{n-1}(n)x + A_n(n)$

Then by using $Df_n(x) = f_{n-1}(x)$ we have

$$\begin{aligned} nA_0(n)x^{n-1} &+ (n-1)A_1(n)x^{n-2} + \cdots + A_{n-1}(n) \\ &= A_0(n-1)x^{n-1} + A_1(n-1)x^{n-2} + \cdots + A_{n-1}(n-1) \end{aligned}$$

Since this must be an identity, it follows that

$$nA_0(n) = A_0(n-1), \quad (n-1)A_1(n) = A_1(n-1), \quad \ldots, \quad A_{n-1}(n) = A_{n-1}(n-1)$$

From the first of these we have

$$A_0(n) \quad = \quad \frac{A_0(n-1)}{n}$$

Then replacing n by $n-1$ successively we find

$$A_0(n) \quad = \quad \frac{A_0(n-1)}{n} \quad = \quad \frac{A_0(n-2)}{n(n-1)} \quad = \quad \cdots \quad = \quad \frac{A_0(0)}{n(n-1)\cdots 1} \quad = \quad \frac{c_0}{n!}$$

where $c_0 = A_0(0)$ is a constant independent of n.

Similarly from the second we have

$$A_1(n) \quad = \quad \frac{A_1(n-1)}{n-1} \quad = \quad \frac{A_1(n-2)}{(n-1)(n-2)} \quad = \quad \cdots \quad = \quad \frac{A_1(0)}{(n-1)(n-2)\cdots 1} \quad = \quad \frac{c_1}{(n-1)!}$$

where $c_1 = A_1(0)$ is a constant independent of n.

By proceeding in this manner the required result is obtained.

3.41. Find $\beta_n(x)$ for $n = 0, 1, 2, 3$.

By definition [see (*30*), page 86, with $n = 4$] $\Delta\beta_4(x) = x^3/3!$. Thus with $h = 1$

$$\begin{aligned} \beta_4(x) &= \Delta^{-1}\frac{x^3}{3!} = \frac{1}{3!}\Delta^{-1}x^3 \\ &= \frac{1}{6}\Delta^{-1}[x^{(3)} + 3x^{(2)} + x^{(1)}] = \frac{1}{6}\left[\frac{x^{(4)}}{4} + x^{(3)} + \frac{x^{(2)}}{2}\right] + c \\ &= \tfrac{1}{24}(x^4 - 6x^3 + 11x^2 - 6x) + \tfrac{1}{6}(x^3 - 3x^2 + 2x) + \tfrac{1}{12}(x^2 - x) + c \\ &= \tfrac{1}{24}x^4 - \tfrac{1}{12}x^3 + \tfrac{1}{24}x^2 + c_1 \end{aligned}$$

Then since $D\beta_n(x) = \beta_{n-1}(x)$, we find

$$\beta_3(x) = \beta_4'(x) = \tfrac{1}{6}x^3 - \tfrac{1}{4}x^2 + \tfrac{1}{12}x$$

$$\beta_2(x) = \beta_3'(x) = \tfrac{1}{2}x^2 - \tfrac{1}{2}x + \tfrac{1}{12}$$

$$\beta_1(x) = \beta_2'(x) = x - \tfrac{1}{2}$$

$$\beta_0(x) = \beta_1'(x) = 1$$

3.42. Obtain the Bernoulli polynomials $B_n(x)$ for $n = 0, 1, 2, 3$.

Since by definition $B_n(x) = n!\,B_n(x)$, we have using Problem 3.41

$$B_0(x) = 0!\,B_0(x) = 1$$

$$B_1(x) = 1!\,B_1(x) = x - \tfrac{1}{2}$$

$$B_2(x) = 2!\,B_2(x) = x^2 - x + \tfrac{1}{6}$$

$$B_3(x) = 3!\,B_3(x) = x^3 - \tfrac{1}{2}x^2 + \tfrac{1}{2}x$$

3.43. Obtain the Bernoulli numbers B_n for $n = 0, 1, 2, 3$.

Since by definition $B_n = B_n(0)$, we obtain from Problem 3.42

$$B_0 = 1, \quad B_1 = -\tfrac{1}{2}, \quad B_2 = \tfrac{1}{6}, \quad B_3 = 0$$

3.44. (*a*) Prove that

$$\beta_n(x) = \frac{1}{n!}\frac{d}{dx}\sum_{k=1}^{n} S_k^n \frac{x^{(k+1)}}{k+1}$$

where S_k^n are the Stirling numbers of the second kind.

(*b*) Use this result to solve Problems 3.41, 3.42 and 3.43.

(*a*) From definition (*30*), page 86, we have

$$\beta_n(x) = \Delta^{-1}\frac{x^{n-1}}{(n-1)!} = \frac{1}{(n-1)!}\Delta^{-1}x^{n-1}$$

Replacing n by $n+1$ and using (*31*), page 7, with $h = 1$ we have

$$\beta_{n+1}(x) = \frac{1}{n!}\Delta^{-1}\left\{\sum_{k=1}^{n} S_k^n x^{(k)}\right\} = \frac{1}{n!}\sum_{k=1}^{n} S_k^n \Delta^{-1}x^{(k)} = \frac{1}{n!}\sum_{k=1}^{n} S_k^n \frac{x^{(k+1)}}{k+1}$$

Then since $\beta_n(x) = D\beta_{n+1}(x)$, we have

$$\beta_n(x) = \frac{1}{n!}\frac{d}{dx}\sum_{k=1}^{n} S_k^n \frac{x^{(k+1)}}{k+1}$$

(*b*) Putting $n = 3$ in the result of (*a*) we have

$$\beta_3(x) = \frac{1}{3!}\frac{d}{dx}\left[\frac{S_1^3 x^{(2)}}{2} + \frac{S_2^3 x^{(3)}}{3} + \frac{S_3^3 x^{(4)}}{4}\right]$$

$$= \frac{1}{6}\frac{d}{dx}\left[\frac{x^{(2)}}{2} + \frac{3x^{(3)}}{3} + \frac{x^{(4)}}{4}\right]$$

$$= \tfrac{1}{6}x^3 - \tfrac{1}{4}x^2 + \tfrac{1}{12}x$$

in agreement with Problem 3.41. From these we can determine $\beta_2(x)$, $\beta_1(x)$ and $\beta_0(x)$ as in that problem.

We can from these also determine the Bernoulli polynomials and Bernoulli numbers as in Problems 3.42 and 3.43.

3.45. (*a*) If
$$\beta_n(x) = \frac{b_0 x^n}{n!} + \frac{b_1 x^{n-1}}{(n-1)!} + \cdots + b_{n-1}x + b_n$$
prove that for $n = 2, 3, \ldots$
$$\frac{b_0}{n!} + \frac{b_1}{(n-1)!} + \cdots + \frac{b_{n-1}}{1!} = 0$$
where $b_0 = 1$.

(*b*) If B_n are the Bernoulli numbers prove that they must satisfy the recurrence formula $(1+B)^n - B^n = 0$ where, after formally expanding by the binomial theorem, B^k is replaced by B_k.

(*c*) Use (*b*) to obtain the first few Bernoulli numbers.

(*a*) We have for $n = 1, 2, 3, \ldots$
$$\Delta\beta_n(x) = \frac{b_0 \Delta x^n}{n!} + \frac{b_1 \Delta x^{n-1}}{(n-1)!} + \cdots + b_{n-1}\Delta x \tag{1}$$

Then putting $x = 0$ in (*1*) and using the fact that
$$\Delta x^n \,|_{x=0} = [(x+1)^n - x^n]\,|_{x=0} = 1$$
we find
$$\Delta\beta_n(0) = \frac{b_0}{n!} + \frac{b_1}{(n-1)!} + \cdots + b_{n-1} \tag{2}$$

Also since
$$\Delta\beta_n(x) = \frac{x^{n-1}}{(n-1)!}$$
we have on putting $x = 0$
$$\Delta\beta_n(0) = \begin{cases} 0 & n = 2, 3, \ldots \\ 1 & n = 1 \end{cases} \tag{3}$$
and the required result follows from (*2*) and (*3*).

(*b*) The Bernoulli numbers are given by
$$B_n = B_n(0) = n!\,B_n(0) = n!\,b_n$$
Thus
$$b_0 = \frac{B_0}{0!}, \quad b_1 = \frac{B_1}{1!}, \quad b_2 = \frac{B_2}{2!}, \quad \ldots, \quad b_{n-1} = \frac{B_{n-1}}{(n-1)!} \tag{4}$$
where $0! = 1$.

Then using (*4*) in the result of part (*a*) we have
$$\frac{B_0}{n!\,0!} + \frac{B_1}{(n-1)!\,1!} + \frac{B_2}{(n-2)!\,2!} + \cdots + \frac{B_{n-1}}{1!\,(n-1)!} = 0 \tag{5}$$
Multiplying by $n!$ (*5*) becomes since $B_0 = 1$
$$1 + \binom{n}{1}B_1 + \binom{n}{2}B_2 + \cdots + \binom{n}{n-1}B_{n-1} = 0$$
which is the same as expanding $(1+B)^n - B^n = 0$ formally and then replacing B^k by B_k.

(*c*) Putting $n = 2, 3, 4, \ldots$ in (*5*) we find
$$1 + \binom{2}{1}B_1 = 0, \qquad 1 + 2B_1 = 0 \quad \text{or} \quad B_1 = -\tfrac{1}{2}$$
$$1 + \binom{3}{1}B_1 + \binom{3}{2}B_2 = 0, \qquad 1 + 3B_1 + 3B_2 = 0 \quad \text{or} \quad B_2 = \tfrac{1}{6}$$
$$1 + \binom{4}{1}B_1 + \binom{4}{2}B_2 + \binom{4}{3}B_3 = 0, \qquad 1 + 4B_1 + 6B_2 + 4B_3 = 0 \quad \text{or} \quad B_3 = 0$$
etc.

3.46. Prove that for $r = 1, 2, 3, \ldots$

$$1^r + 2^r + 3^r + \cdots + (n-1)^r = \frac{(n+B)^{r+1} - B^{r+1}}{r+1}$$

where the right side is to be expanded formally using the binomial theorem and then replacing B^k by B_k.

From equation *(30)*, page 86, we have on putting $n = r+1$

$$\Delta\beta_{r+1} = \frac{x^r}{r!}$$

Then using $n = r+1$ in equation *(33)*, page 86, and property 1, page 87,

$$\Delta^{-1}x^r = r!\,\beta_{r+1}(x) = \frac{B_{r+1}(x)}{r+1} = \frac{(x+B)^{r+1}}{r+1}$$

Thus by the fundamental theorem of sum calculus,

$$\sum_{x=1}^{n-1} x^r = \left.\frac{(x+B)^{r+1}}{r+1}\right|_1^n = \frac{(n+B)^{r+1} - B^{r+1}}{r+1}$$

3.47. Use Problem 3.46 to evaluate $1^2 + 2^2 + 3^2 + \cdots + n^2$.

Putting $r = 2$ in the result of Problem 3.46 we have

$$\begin{aligned} 1^2 + 2^2 + 3^2 + \cdots + (n-1)^2 &= \frac{(n+B)^3 - B^3}{3} \\ &= \tfrac{1}{3}(n^3 + 3n^2B^1 + 3nB^2 + B^3 - B^3) \\ &= \tfrac{1}{3}(n^3 + 3n^2B_1 + 3nB_2) \\ &= \tfrac{1}{3}n^3 - \tfrac{1}{2}n^2 + \tfrac{1}{6}n \\ &= \tfrac{1}{6}n(n-1)(2n-1) \end{aligned}$$

Then replacing n by $n+1$ we have

$$1^2 + 2^2 + 3^2 + \cdots + n^2 = \tfrac{1}{6}n(n+1)(2n+1)$$

Note that this agrees with the result of Problem 3.21*(a)*.

EULER NUMBERS AND POLYNOMIALS

3.48. Show that the Euler polynomials are given by

$$E_n(x) = \frac{e_0x^n}{n!} + \frac{e_1x^{n-1}}{(n-1)!} + \cdots + e_{n-1}x + e_n$$

where the constants $e_0, \ldots, e_n$ are given for $n = 1, 2, 3, \ldots$ by

$$e_n + \frac{1}{2}\left[\frac{e_0}{n!} + \frac{e_1}{(n+1)!} + \frac{e_2}{(n-2)!} + \cdots + \frac{e_{n-1}}{1!}\right] = 0$$

and $e_0 = 1$.

From Problem 3.40 and definition *(38)* on page 88 it follows that

$$E_n(x) = \frac{e_0x^n}{n!} + \frac{e_1x^{n-1}}{(n-1)!} + \cdots + e_{n-1}x + e_n \tag{1}$$

Then using definition *(39)* on page 88 together with *(1)* it is seen that we must have

$$ME_n(x) = \frac{x^n}{n!} = \frac{e_0}{n!}M(x^n) + \frac{e_1}{(n-1)!}M(x^{n-1}) + \cdots + e_{n-1}M(x) + e_n \tag{2}$$

where M is the averaging operator with $h = 1$ which is a linear operator.

If $n = 0$, (2) becomes

$$ME_0(x) = 1 = e_0 \tag{3}$$

while (1) yields for $n = 0$

$$E_0(x) = e_0 \tag{4}$$

We thus conclude from (3) and (4) that

$$E_0(x) = 1 \tag{5}$$

If $n = 1, 2, 3, \ldots$ we have by definition of M

$$M(x^n) = \tfrac{1}{2}[(x+1)^n + x^n]$$

so that on putting $x = 0$

$$M(x^n)\,|_{x=0} = \tfrac{1}{2}$$

Thus on putting $x = 0$ in (2) we obtain for $n = 1, 2, 3, \ldots$

$$\frac{1}{2}\left[\frac{e_0}{n!} + \frac{e_1}{(n-1)!} + \cdots + e_{n-1}\right] + e_n = 0 \tag{6}$$

as required.

3.49. (a) Find e_0, e_1, e_2, e_3 and (b) obtain the first four Euler polynomials.

(a) We already know from Problem 3.48 that $e_0 = 1$. Then putting $n = 1, 2, 3$ successively in the result (6) of Problem 3.48 we have

$$\frac{1}{2}(1) + e_1 = 0, \qquad \frac{1}{2}\left[\frac{1}{2!} + e_1\right] + e_2 = 0, \qquad \frac{1}{2}\left[\frac{1}{3!} + \frac{e_1}{2!} + e_2\right] + e_3 = 0$$

From this we find

$$e_1 = -\tfrac{1}{2}, \quad e_2 = 0, \quad e_3 = \tfrac{1}{24}$$

(b) Using part (a) we obtain from (1) of Problem 3.48

$$E_0(x) = 1, \quad E_1(x) = x - \frac{1}{2}, \quad E_2(x) = \frac{x^2}{2!} - \frac{x}{2}, \quad E_3(x) = \frac{x^3}{3!} - \frac{x^2}{2\cdot 2!} + \frac{1}{24}$$

For a table of Euler polynomials see page 237.

3.50. Obtain the first four Euler numbers.

By definition the Euler numbers are given by

$$E_n = 2^n n!\, E_n(\tfrac{1}{2})$$

Then from the results of Problem 3.49(b) we have

$$E_0 = 1, \quad E_1 = 0, \quad E_2 = -1, \quad E_3 = 0$$

For a table of Euler numbers see page 236.

MISCELLANEOUS PROBLEMS

3.51. Prove that the generating function for the Bernoulli polynomials is given by

$$\frac{te^{xt}}{e^t - 1} = \sum_{n=0}^{\infty} \frac{B_n(x)t^n}{n!}$$

If we let $G(x, t)$ denote the generating function, then

$$G(x, t) = \sum_{n=0}^{\infty} \beta_n(x)t^n = \sum_{n=0}^{\infty} \frac{B_n(x)}{n!}t^n \tag{1}$$

using the fact that $\beta_n(x) = n!\,\beta_n(x)$.

Now by definition

$$\Delta\beta_n(x) = \frac{x^{n-1}}{(n-1)!}$$

Then multiplying by t^n and summing over n we have

$$\sum_{n=1}^{\infty} \Delta[\beta_n(x)t^n] = \sum_{n=1}^{\infty} \frac{x^{n-1}t^n}{(n-1)!}$$

which can be written

$$\Delta \sum_{n=1}^{\infty} \beta_n(x)t^n = t\sum_{n=1}^{\infty} \frac{(xt)^{n-1}}{(n-1)!} = te^{xt} \qquad (2)$$

Using (*1*), we can write (*2*) as

$$\Delta G(x,t) = te^{xt} \quad \text{or} \quad G(x,t) = \Delta^{-1}(te^{xt})$$

so that

$$G(x,t) = \frac{te^{xt}}{e^t-1} + c \qquad (3)$$

Taking the limit of (*3*) as $t \to 0$ we find since

$$\lim_{t\to 0} G(x,t) = 1, \quad \lim_{t\to 0} \frac{t}{e^t-1} = 1$$

that $c = 0$. Thus we have as required

$$G(x,t) = \frac{te^{xt}}{e^t-1} = \sum_{n=0}^{\infty} \frac{B_n(x)t^n}{n!}$$

This generating function can be used to define the Bernoulli polynomials from which all other properties are obtained. See for example Problem 3.53.

3.52. Prove that the generating function for the Bernoulli numbers is given by

$$\frac{t}{e^t-1} = \sum_{n=0}^{\infty} \frac{B_n t^n}{n!}$$

This follows at once from Problem 3.51 on putting $x = 0$ and noting that $B_n = B_n(0)$. The generating function can be used to obtain all other properties of the Bernoulli numbers.

3.53. Use the generating function of Problem 3.51 to define the Bernoulli polynomials. (*a*) Find the polynomial $B_n(x)$ for $n = 0, 1, 2, 3$. (*b*) Prove that $B_n'(x) = nB_{n-1}(x)$.

(*a*) We have

$$\frac{te^{xt}}{e^t-1} = \sum_{n=0}^{\infty} \frac{B_n(x)t^n}{n!}$$

Then multiplying both sides by

$$e^t - 1 = \frac{t}{1!} + \frac{t^2}{2!} + \frac{t^3}{3!} + \cdots$$

we have

$$te^{xt} = \left(\frac{t}{1!} + \frac{t^2}{2!} + \frac{t^3}{3!} + \cdots\right)\left(B_0(x) + \frac{B_1(x)t}{1!} + \frac{B_2(x)t^2}{2!} + \frac{B_3(x)t^3}{3!} + \cdots\right)$$

Dividing both sides by t and using the expansion

$$e^{xt} = 1 + \frac{xt}{1!} + \frac{x^2t^2}{2!} + \frac{x^3t^3}{3!} + \cdots$$

this becomes

$$1 + \frac{xt}{1!} + \frac{x^2t^2}{2!} + \frac{x^3t^3}{3!} + \cdots = B_0(x) + \left[\frac{B_1(x)}{1!\,1!} + \frac{B_0(x)}{2!}\right]t + \left[\frac{B_2(x)}{2!\,1!} + \frac{B_1(x)}{1!\,2!} + \frac{B_0(x)}{3!}\right]t^2 + \left[\frac{B_3(x)}{3!\,1!} + \frac{B_2(x)}{2!\,2!} + \frac{B_1(x)}{1!\,3!} + \frac{B_0(x)}{4!}\right]t^3 + \cdots$$

Then equating coefficients of corresponding powers of t we have

$$B_0(x) = 1, \quad \frac{B_1(x)}{1!\,1!} + \frac{B_0(x)}{2!} = \frac{x}{1!}, \quad \frac{B_2(x)}{2!\,1!} + \frac{B_1(x)}{1!\,2!} + \frac{B_0(x)}{3!} = \frac{x^2}{2!},$$

$$\frac{B_3(x)}{3!\,1!} + \frac{B_2(x)}{2!\,2!} + \frac{B_1(x)}{1!\,3!} + \frac{B_0(x)}{4!} = \frac{x^3}{3!}, \quad \ldots$$

From these we find

$$B_0(x) = 1, \quad B_1(x) = x - \tfrac{1}{2}, \quad B_2(x) = x^2 - x + \tfrac{1}{6}, \quad B_3(x) = x^3 - \tfrac{3}{2}x^2 + \tfrac{1}{2}x, \quad \ldots$$

(b) Differentiating both sides of

$$\frac{te^{xt}}{e^t - 1} = \sum_{n=0}^{\infty} \frac{B_n(x)t^n}{n!}$$

with respect to x we have

$$\frac{t^2e^{xt}}{e^t - 1} = \sum_{n=0}^{\infty} \frac{B_n'(x)t^n}{n!}$$

But this is the same as

$$t \sum_{n=0}^{\infty} \frac{B_n(x)t^n}{n!} = \sum_{n=0}^{\infty} \frac{B_n'(x)t^n}{n!}$$

i.e.

$$\sum_{n=0}^{\infty} \frac{B_n(x)t^{n+1}}{n!} = \sum_{n=0}^{\infty} \frac{B_n'(x)t^n}{n!}$$

or

$$\sum_{n=1}^{\infty} \frac{B_{n-1}(x)t^n}{(n-1)!} = \sum_{n=1}^{\infty} \frac{B_n'(x)t^n}{n!}$$

since $B_0'(x) = 0$. Equating coefficients of t^n on both sides we thus find

$$\frac{B_{n-1}(x)}{(n-1)!} = \frac{B_n'(x)}{n!}$$

so that

$$B_n'(x) = nB_{n-1}(x)$$

3.54. Find $\displaystyle\frac{d}{d\alpha}\int_{\alpha^2}^{\alpha^3} \frac{\sin x\alpha}{x}\,dx$.

We have by Leibnitz's rule on page 81

$$\frac{d}{d\alpha}\int_{\alpha^2}^{\alpha^3} \frac{\sin x\alpha}{x}\,dx = \int_{\alpha^2}^{\alpha^3} \frac{\partial}{\partial\alpha}\left(\frac{\sin x\alpha}{x}\right)dx + \frac{\sin(\alpha^3\cdot\alpha)}{\alpha^3}\frac{d}{d\alpha}(\alpha^3) - \frac{\sin(\alpha^2\cdot\alpha)}{\alpha^2}\frac{d}{d\alpha}(\alpha^2)$$

$$= \int_{\alpha^2}^{\alpha^3} \cos x\alpha\,dx + \frac{3\sin\alpha^4}{\alpha} - \frac{2\sin\alpha^3}{\alpha}$$

$$= \left.\frac{\sin x\alpha}{\alpha}\right|_{\alpha^2}^{\alpha^3} + \frac{3\sin\alpha^4}{\alpha} - \frac{2\sin\alpha^3}{\alpha}$$

$$= \frac{4\sin\alpha^4 - 3\sin\alpha^3}{\alpha}$$

Supplementary Problems

THE INTEGRAL OPERATOR AND RULES OF INTEGRATION

3.55. Find each of the following:

(a) $D^{-1}(3x^4 - 2x^3 + x - 1)$ (c) $D^{-1}(4\cos 3x - 2\sin 3x)$ (e) $D^{-1}\left(\dfrac{4}{x-2}\right)$

(b) $D^{-1}(2\sqrt{x} - 3\sqrt[3]{x})$ (d) $D^{-1}(4e^{2x} + 3e^{-4x})$ (f) $D^{-1}(2^x)$

3.56. Prove (a) formula I-1 and (b) formula I-2 on page 79.

3.57. Find (a) $\int xe^{-3x}\,dx$, (b) $\int (x\sin x + \cos x)\,dx$, (c) $\int x\ln x\,dx$, (d) $\int x^2\ln^2 x\,dx$.

3.58. Find (a) $\int x^2\sin 2x\,dx$, (b) $\int x^3 e^{2x}\,dx$.

3.59. Find

(a) $\displaystyle\int \frac{e^x + \cos x}{e^x + \sin x}\,dx$ (b) $\int e^{\sqrt{x}}\,dx$ (c) $\int \sin\sqrt[3]{x}\,dx$ (d) $\displaystyle\int \frac{dx}{x^2 - 4x + 8}$ (e) $\displaystyle\int \frac{dx}{(2x-3)^2}$

THE SUM OPERATOR AND RULES OF SUMMATION

3.60. Prove that $\Delta^{-1}[\cos rx] = \sum \cos rx = \dfrac{\sin r(x - \frac{1}{2}h)}{2\sin\frac{1}{2}rh}$.

3.61. Use Problem 3.60 to obtain $\int \cos rx\,dx$.

3.62. Find $\Delta^{-1}[xe^x] = \sum xe^x$.

3.63. (a) Find $\Delta^{-1}[x^2a^x] = \sum x^2a^x$, $a \neq 1$, and (b) check your answer.

3.64. Prove that

$$\Delta^{-1}(x\sin x) = \frac{-x\cos(x - \frac{1}{2}h)}{2\sin\frac{1}{2}h} + \frac{h\sin x}{4\sin^2\frac{1}{2}h}$$

3.65. Find (a) $\Delta^{-1}(x\cos rx)$, (b) $\Delta^{-1}(x\sin rx)$.

DEFINITE SUMS AND THE FUNDAMENTAL THEOREM OF SUM CALCULUS

3.66. Evaluate each of the following using the fundamental theorem of sum calculus and verify directly.

(a) $\displaystyle\sum_{2}^{8} 2x^{(4)}$ if $h = 2$ (c) $\displaystyle\sum_{0}^{\pi} \cos 4x$ if $h = \pi/3$

(b) $\displaystyle\sum_{1}^{4} [3x^{(2)} - 2x^{(3)}]$ if $h = 1$ (d) $\displaystyle\sum_{2}^{10} 6x^{(-3)}$ if $h = 4$

3.67. Find $\displaystyle\sum_{1}^{n} x^{(-3)}$ if $h = 1$.

3.68. Find $\displaystyle\sum_{1}^{n} x \cdot 2^x$ if $h = 1$.

3.69. Show that

$$\sum_{x=a}^{a+(n-1)h} x^2 = na^2 + \tfrac{1}{2}n(n-1)h(2a+h) + \tfrac{1}{3}n(n-1)(n-2)h^2$$

DIFFERENTIATION AND INTEGRATION OF SUMS

3.70. Verify that Theorem 3-3, page 84, is true for

$$\sum x^{(2)} = \frac{x^{(3)}}{3h}$$

3.71. Use Theorem 3-4, page 84, to obtain IV-7 from IV-6, page 83.

3.72. Find (a) $\sum x \cos rx$ (b) $\sum x \sin rx$ by using Theorem 3-4, page 84.

3.73. Prove (a) Theorem 3-3, (b) Theorem 3-4, (c) Theorem 3-5, on page 84.

SUMMATION OF SERIES

3.74. Sum the series (a) $1+2+3+\cdots$ to n terms, (b) $1+4+7+10+\cdots$ to n terms.

3.75. Sum each of the following series to n terms.

(a) $1\cdot 2 + 2\cdot 3 + 3\cdot 4 + \cdots$ (c) $1\cdot 3 + 3\cdot 5 + 5\cdot 7 + \cdots$

(b) $1\cdot 2\cdot 3 + 2\cdot 3\cdot 4 + 3\cdot 4\cdot 5 + \cdots$ (d) $2\cdot 5 + 5\cdot 8 + 8\cdot 11 + \cdots$

3.76. Sum each of the following series to n terms.

(a) $\dfrac{1}{1\cdot 2} + \dfrac{1}{2\cdot 3} + \dfrac{1}{3\cdot 4} + \cdots$ (c) $\dfrac{1}{1\cdot 4\cdot 7} + \dfrac{1}{4\cdot 7\cdot 10} + \dfrac{1}{7\cdot 10\cdot 13} + \cdots$

(b) $\dfrac{1}{1\cdot 3} + \dfrac{1}{3\cdot 5} + \dfrac{1}{5\cdot 7} + \cdots$ (d) $\dfrac{1}{1\cdot 2\cdot 3} + \dfrac{1}{2\cdot 3\cdot 4} + \dfrac{1}{3\cdot 4\cdot 5} + \cdots$

3.77. (a) Show that $\displaystyle\sum_{k=1}^{n} \frac{k}{2^k} = 2 - \frac{n+2}{2^n}$ and thus (b) show that

$$\frac{1}{2} + \frac{2}{2^2} + \frac{3}{2^3} + \frac{4}{2^4} + \cdots = 2$$

3.78. Find the sum of the series $1\cdot 2^1 + 2\cdot 2^2 + 3\cdot 2^3 + 4\cdot 2^4 + \cdots + n\cdot 2^n$.

3.79. Show that $$\sin\theta + \sin 2\theta + \sin 3\theta + \cdots + \sin n\theta = \frac{\cos(\theta/2) - \cos(n+\frac{1}{2})\theta}{2\sin(\theta/2)}$$

3.80. Show that $$1^3 + 2^3 + 3^3 + \cdots + n^3 = (1+2+3+\cdots+n)^2 = \frac{n^2(n+1)^2}{4}$$

3.81. Find the sum to infinity of each of the series in Problem 3.76.

3.82. Sum the series $1\cdot 2^2 + 2\cdot 3^2 + 3\cdot 4^2 + \cdots$ to n terms.

3.83. Sum the series $\displaystyle\sum_{k=1}^{n} k^2\cdot 2^k$.

3.84. Sum the series $1^2 + 3^2 + 5^2 + \cdots$ to n terms.

3.85. (a) Sum the series

$$\frac{1}{1\cdot 2\cdot 3} + \frac{4}{2\cdot 3\cdot 4} + \frac{7}{3\cdot 4\cdot 5} + \frac{10}{4\cdot 5\cdot 6} + \cdots \quad \text{to } n \text{ terms}$$

and (b) find the limit as $n \to \infty$.

3.86. Sum each of the following series to n terms and to infinitely many terms.

(a) $\dfrac{1}{2\cdot 4} + \dfrac{1}{3\cdot 5} + \dfrac{1}{5\cdot 7} + \cdots$ (b) $\dfrac{1}{2\cdot 4\cdot 6} + \dfrac{1}{3\cdot 5\cdot 7} + \dfrac{1}{5\cdot 7\cdot 9} + \cdots$

SUMMATION USING SUBSCRIPT NOTATION

3.87. Use the subscript notation to sum the series of Problem 3.75.

3.88. Use the subscript notation to sum the series of Problem 3.76.

3.89. Sum the series $\sum_{k=1}^{n} k^2(k+1)$.

3.90. Sum the series (a) $\sum_{k=1}^{n} \frac{2k-1}{(k+2)(k+4)}$ (b) $\sum_{k=1}^{\infty} \frac{2k-1}{(k+1)(k+2)(k+3)}$.

3.91. Sum the following series to n terms.

(a) $1\cdot 3\cdot 5\cdot 7 + 3\cdot 5\cdot 7\cdot 9 + 5\cdot 7\cdot 9\cdot 11 + \cdots$

(b) $\frac{1}{1\cdot 3\cdot 5\cdot 7} + \frac{1}{3\cdot 5\cdot 7\cdot 9} + \frac{1}{5\cdot 7\cdot 9\cdot 11} + \cdots$

3.92. Show that

(a) $\sum_{k=0}^{n} ka^k = \frac{(n+1)a^{n+1}}{a-1} - \frac{a^{n+2}-a}{(a-1)^2} \qquad a \neq 1$

(b) $\sum_{k=0}^{\infty} ka^k = \frac{a}{(1-a)^2} \qquad \text{for } |a| < 1$

3.93. Sum the series

$$\frac{5}{1\cdot 3\cdot 5} + \frac{6}{3\cdot 5\cdot 7} + \frac{7}{5\cdot 7\cdot 9} + \cdots$$

(a) to n terms.

(b) to infinity.

ABEL'S TRANSFORMATION

3.94. Use Abel's transformation to sum the series $\sum_{k=1}^{n} \frac{k}{2^k}$.

3.95. Sum the series $\sum_{k=1}^{n} k^2$ by using Abel's transformation.

3.96. Sum the series $\sum_{k=1}^{n} k^2 a^k$, $a \neq 1$, by using Abel's transformation.

OPERATOR METHODS OF SUMMATION

3.97. Work Problem 3.92 by using the result of Problem 3.30.

3.98. Show that

(a) $\Delta^{-1}[r^k \sin \alpha k] = \frac{r^{k+1} \sin \alpha(k-1) - r^k \sin \alpha k}{r^2 - 2r\cos\alpha + 1}$

(b) $\Delta^{-1}[r^k \cos \alpha k] = \frac{r^{k+1} \cos \alpha(k-1) - r^k \cos \alpha k}{r^2 - 2r\cos\alpha + 1}$

3.99. Use Problem 3.98 to show that if $|r| < 1$

(a) $\sum_{k=1}^{\infty} r^k \sin \alpha k = \frac{r\sin\alpha}{r^2 - 2r\cos\alpha + 1}$

(b) $\frac{1}{2} + \sum_{k=1}^{\infty} r^k \cos \alpha k = \frac{1-r^2}{2(r^2 - 2r\cos\alpha + 1)}$

3.100. Show that if $\beta \neq 1$ is a constant and $h \neq 1$,

$$\Delta^{-1}[\beta^x P(x)] = \frac{\beta^x}{\beta^h - 1}\left[1 - \frac{\beta^h\Delta}{\beta^h - 1} + \frac{\beta^{2h}\Delta^2}{(\beta^h-1)^2} - \cdots\right]P(x)$$

thus generalizing the result of Problem 3.30.

3.101. Use Problem 3.100 to find $\Delta^{-1}[xa^x]$. Compare with Problem 3.16.

THE GAMMA FUNCTION

3.102. Find (*a*) $\Gamma(4)$, (*b*) $\Gamma(5/2)$, (*c*) $\Gamma(-5/2)$.

3.103. Evaluate (*a*) $\displaystyle\int_0^\infty x^4 e^{-2x}\,dx$, (*b*) $\displaystyle\int_0^\infty x^2 e^{-x^2}\,dx$.

3.104. Evaluate $\displaystyle\int_0^\infty x e^{-x^3}\,dx$.

3.105. Evaluate $\displaystyle\int_0^1 x^2 \ln^2 x\,dx$.

3.106. Prove that $\displaystyle\Gamma(m) = 2\int_0^\infty x^{2m-1} e^{-x^2}\,dx$.

3.107. Prove that $\displaystyle\frac{1}{x+1} + \frac{1}{x+2} + \cdots + \frac{1}{x+n} = \frac{\Gamma'(x+n+1)}{\Gamma(x+n+1)} - \frac{\Gamma'(x+1)}{\Gamma(x+1)}$.

3.108. Use Problem 3.107 to express in terms of gamma functions

$$\frac{1}{(x+1)^2} + \frac{1}{(x+2)^2} + \cdots + \frac{1}{(x+n)^2}$$

and generalize.

3.109. Show that equation (*3*) of Problem 3.35 holds for all values of m.

BERNOULLI NUMBERS AND POLYNOMIALS

3.110. (*a*) Find $\beta_n(x)$ for $n = 4$ and 5 and thus obtain (*b*) the Bernoulli polynomials and (*c*) the Bernoulli numbers for $n = 4$ and 5.

3.111. Obtain the Bernoulli numbers B_4, B_6, B_8 without first finding the Bernoulli polynomials.

3.112. Prove that $B_n'(x) = nB_{n-1}(x)$ without using the generating function.

3.113. Prove that $B_n = B_n(0) = B_n(1)$.

3.114. Evaluate (*a*) $1^3 + 2^3 + 3^3 + \cdots + n^3$, (*b*) $1^4 + 2^4 + 3^4 + \cdots + n^4$.

3.115. Expand $5x^4 - 8x^3 + 3x^2 - 4x + 5$ in a series of Bernoulli polynomials.

3.116. Prove that $\displaystyle\int_a^x B_n(u)\,du = \frac{B_{n+1}(x) - B_{n+1}(a)}{n+1}$.

3.117. Prove that $\displaystyle\int_0^1 B_n(x)\,dx = \begin{cases} 1 & n = 0 \\ 0 & n > 0 \end{cases}$.

3.118. Prove that $B_{2n+1}(\frac{1}{2}) = 0$.

EULER NUMBERS AND POLYNOMIALS

3.119. Find (*a*) e_4, e_5; (*b*) $E_4(x), E_5(x)$; (*c*) E_4, E_5.

3.120. Prove that (*a*) $E_n(0) = e_n$, (*b*) $E_n(0) + E_n(1) = 0$, (*c*) $E_n(1) = -e_n$.

3.121. Show that (*a*) $e_7 = \dfrac{17}{40{,}320}$, (*b*) $e_9 = \dfrac{-31}{362{,}880}$.

MISCELLANEOUS PROBLEMS

3.122. Determine whether the operators (*a*) D and D^{-1}, (*b*) Δ and Δ^{-1} are commutative.

3.123. Prove that

(a) $$\Delta^{-1}\sin(px+q) = \frac{\sin(px+q-\frac{1}{2}ph-\frac{1}{2}\pi)}{2\sin\frac{1}{2}ph}$$

(b) $$\Delta^{-1}\cos(px+q) = \frac{\cos(px+q-\frac{1}{2}ph-\frac{1}{2}\pi)}{2\sin\frac{1}{2}ph}$$

3.124. Let $\Delta^{-1}f(x) = f_1(x)$, $\Delta^{-1}f_1(x) = f_2(x)$, ... Prove that

$$\begin{aligned}\sum f(x)\,g(x) &= \Delta^{-1}[f(x)\,g(x)]\\ &= f_1(x)\,g(x) - f_2(x+h)\,\Delta g(x) + f_3(x+2h)\,\Delta^2 g(x)\\ &\quad - \cdots + (-1)^{n-1}f_n(x+nh-h)\,\Delta^{n-1}g(x)\\ &\quad + (-1)^n\Delta^{-1}\{f_n(x+nh)\,\Delta^n g(x)\}\end{aligned}$$

which is called *generalized summation by parts.*

3.125. Use Problem 3.124 to find $\Delta^{-1}[x^3 \cdot 2^x]$ if $h = 1$.

3.126. Find $\sum_{k=1}^{n} k^3 \cos 2k$.

3.127. Show that $$\frac{1^2}{2\cdot 3} + \frac{2^2\cdot 4}{3\cdot 4} + \frac{3^2\cdot 4^2}{4\cdot 5} + \cdots + \frac{n^2\cdot 4^{n-1}}{(n+1)(n+2)} = \frac{1}{6} + \frac{(n-1)4^n}{3n+6}$$

3.128. Show that if $\sin\alpha \neq 0$

$$\sin^2\alpha + \sin^2 2\alpha + \cdots + \sin^2 n\alpha = \frac{n}{2} - \frac{\sin n\alpha\cos(n+1)\alpha}{2\sin\alpha}$$

3.129. Show that $$\sum_{k=1}^{n} k^2\cdot a^k = \frac{(a^2n^2-2an^2-2an+n^2+2n+a+1)a^{n+1}-(a^2+a)}{(a-1)^3}$$

3.130. Prove that $$\sum_{k=0}^{n}(-1)^k\binom{n}{k}u_k = (-1)^n\Delta^n u_0$$

3.131. Use Problem 3.130 to show that

(a) $$\sum_{k=0}^{n}(-1)^k\binom{n}{k}\frac{1}{2k+1} = \frac{2^n n!}{1\cdot 3\cdot 5\cdots(2n+1)}$$

(b) $$\sum_{k=0}^{n}(-1)^{k+n}\binom{n}{k}x^n = n!$$

3.132. Evaluate $1^5+2^5+3^5+\cdots+n^5$.

3.133. Show that

$$\frac{1}{\sin\theta} + \frac{1}{\sin 2\theta} + \frac{1}{\sin 4\theta} + \frac{1}{\sin 8\theta} + \cdots \text{ to } n \text{ terms} = \cot(\theta/2) - \cot 2^{n-1}\theta$$

3.134. (a) Show that $\Delta^{-1}\left[\dfrac{\sec^2 x\tan h}{1-\tan x\tan h}\right] = \tan x + c$.

(b) Use (a) to deduce that $\int \sec^2 x\,dx = \tan x + c$.

3.135. (a) Show that $\Delta^{-1}\left[\tan^{-1}\dfrac{h}{x^2+hx+1}\right] = \tan^{-1}x + c$.

(b) Use (a) to deduce that $\int \dfrac{dx}{x^2+1} = \tan^{-1}x + c$.

3.136. (*a*) Show that

$$\sum_{k=1}^{n} \tan^{-1}\left(\frac{1}{k^2+k+1}\right) = \tan^{-1}(n+1) - \frac{\pi}{4}$$

(*b*) Use (*a*) to show that

$$\sum_{k=1}^{\infty} \tan^{-1}\left(\frac{1}{k^2+k+1}\right) = \frac{\pi}{4}$$

3.137. Are there any theorems for integrals corresponding to the Theorems 3-3, 3-4 and 3-5 on page 84? Explain.

3.138. (*a*) Show that formally

$$\Delta^{-1} = \frac{1}{\Delta} = -\frac{1}{1-E} = -(1+E+E^2+\cdots)$$

(*b*) Use (*a*) to obtain the formal result

$$\Delta^{-1} f(x) = -\sum_{u=x}^{\infty} f(u)$$

where the sum on the right is taken for $u = x, x+h, x+2h, \ldots$.

(*c*) Is the result in (*b*) valid? Illustrate by considering the case where $f(x) = \dfrac{1}{x(x+2)}$ with $h=2$.

3.139. Prove Theorem 3-1, page 83.

3.140. Prove (*a*) Theorem 3-3 and (*b*) Theorem 3-4, page 84.

$\left[\textit{Hint.} \quad \text{Let } G(\alpha) = \int_{a(\alpha)}^{b(\alpha)} F(x,\alpha)\,dx \quad \text{and consider } \dfrac{G(\alpha+\Delta\alpha)-G(\alpha)}{\Delta\alpha}.\right]$

3.141. Prove that $\Gamma'(1) = \int_0^\infty e^{-x}\ln x\,dx = -\gamma$ where γ is *Euler's constant* [see page 85].

3.142. Use Problem 3.141 to prove that

$$1+\frac{1}{2}+\frac{1}{3}+\cdots+\frac{1}{n} = \frac{\Gamma'(n+1)}{\Gamma(n+1)}+\gamma$$

3.143. Supply a proof of the method of *generalized integration by parts* [see Problem 3.4].

3.144. The *beta function* is defined as

$$B(m,n) = \int_0^1 x^{m-1}(1-x)^{n-1}\,dx$$

(*a*) Show that $B(m,n) = 2\int_0^{\pi/2} \sin^{2m-1}\theta\,\cos^{2n-1}\theta\,d\theta$.

(*b*) Show that $B(m,n) = B(n,m)$.

3.145. Show that the beta function of Problem 3.144 is related to the gamma function by

$$B(m,n) = \frac{\Gamma(m)\,\Gamma(n)}{\Gamma(m+n)}$$

[*Hint.* Use Problem 3.106 to show that $\Gamma(m)\,\Gamma(n) = 4\int_0^\infty\int_0^\infty x^{2m-1}y^{2n-1}e^{-(x^2+y^2)}\,dx\,dy$. Then transform to polar coordinates and use the result of Problem 3.144.]

3.146. Evaluate (*a*) $\int_0^1 \frac{y^2}{\sqrt{1-y^2}}\,dy$, (*b*) $\int_0^{\pi/2} \sin^4\theta\,\cos^2\theta\,d\theta$, (*c*) $\int_0^{\pi/2} \sqrt{\tan\theta}\,d\theta$.

3.147. Show that (a) $\displaystyle\int_0^{\pi/2} \sin^{2n} x\, dx = \frac{2n-1}{2n}\cdot\frac{2n-3}{2n-2}\cdots\frac{1}{2}\cdot\frac{\pi}{2}$

(b) $\displaystyle\int_0^{\pi/2} \sin^{2n+1} x\, dx = \frac{2n}{2n+1}\cdot\frac{2n-2}{2n-1}\cdots\frac{2}{3}$

3.148. Show that $\Delta^{m-1}\left(\dfrac{1}{n}\right) = (-1)^n B(m,n)$ if $h = 1$.

$\left[\textit{Hint.} \text{ Use the fact that } \dfrac{1}{n} = \displaystyle\int_0^\infty e^{-nx}\, dx.\right]$

3.149. Prove property 6, page 87, i.e. $B_n(x) = (-1)^n B_n(1-x)$.

3.150. Prove that $B_3 = B_5 = B_7 = \cdots = 0$, i.e. all Bernoulli numbers with odd subscript greater than 1 are equal to zero.

3.151. Show that

$$\frac{x}{2}\coth\frac{x}{2} = \frac{x}{2}\left(\frac{e^{x/2}+e^{-x/2}}{e^{x/2}-e^{-x/2}}\right) = 1 + \frac{B_2x^2}{2!} + \frac{B_4x^4}{4!} + \cdots$$

3.152. Show that

(a) $$\cot x = \frac{1}{x} - \frac{2^2}{2!}B_2x + \frac{2^4}{4!}B_4x^3 - \frac{2^6}{6!}B_6x^5 + \cdots$$
$$= \frac{1}{x} - \frac{1}{3}x - \frac{1}{45}x^3 - \frac{2}{945}x^5 - \frac{1}{4725}x^7 - \cdots$$

(b) $$\tan x = \frac{2^2(2^2-1)}{2!}B_2x - \frac{2^4(2^4-1)}{4!}B_4x^3 + \cdots$$
$$= x + \frac{1}{3}x^3 + \frac{2}{15}x^5 + \frac{17}{315}x^7 + \cdots$$

[*Hint.* For (a) use Problem 3.151 with x replaced by ix. For (b) use the identity $\tan x = \cot x - 2\cot 2x$.]

3.153. Show that

$$\csc x = \sum_{k=0}^{\infty} \frac{(-1)^{k-1}(2^{2k}-2)B_{2k}x^{2k-1}}{(2k)!}$$

[*Hint.* Use the identity $\csc x = \cot x + \tan(x/2)$.]

3.154. Let $f(x)$ be periodic with period equal to 1 and suppose that $f(x)$ and $f'(x)$ are bounded and have a finite number of discontinuities. Then $f(x)$ has a *Fourier series* expansion of the form

$$f(x) = \frac{a_0}{2} + \sum_{k=1}^{\infty}(a_k \cos 2k\pi x + b_k \sin 2k\pi x)$$

where $$a_k = 2\int_0^1 f(x)\cos 2k\pi x\, dx, \qquad b_k = 2\int_0^1 f(x)\sin 2k\pi x\, dx$$

(a) If $f(x) = \beta_{2n}(x)$ [see page 88] show that

$$a_0 = 0, \quad a_k = \frac{2(-1)^{n-1}}{(2k\pi)^{2n}}, \quad b_k = 0 \qquad k = 1, 2, \ldots$$

so that $$\beta_{2n}(x) = 2(-1)^{n-1}\sum_{k=1}^{\infty}\frac{\cos 2k\pi x}{(2k\pi)^{2n}}$$

(b) Use the result in (a) to obtain the second result of property 12, page 88.

3.155. Prove the first result of property 12, page 88. [*Hint.* Take the derivative of the second result of property 12 already obtained in Problem 3.154(b).]

3.156. Prove property 11, page 88.

3.157. Show that (*a*) $\frac{1}{1^2}+\frac{1}{2^2}+\frac{1}{3^2}+\cdots = \frac{\pi^2}{6}$

(*b*) $\frac{1}{1^4}+\frac{1}{2^4}+\frac{1}{3^4}+\cdots = \frac{\pi^4}{90}$

3.158. Show that (*a*) $\frac{1}{1^2}-\frac{1}{2^2}+\frac{1}{3^2}-\frac{1}{4^2}+\cdots = \frac{\pi^2}{12}$

(*b*) $\frac{1}{1^4}-\frac{1}{2^4}+\frac{1}{3^4}-\frac{1}{4^4}+\cdots = \frac{7\pi^4}{720}$

3.159. Show that (*a*) $\frac{1}{1^3}-\frac{1}{3^3}+\frac{1}{5^3}-\frac{1}{7^3}+\cdots = \frac{\pi^3}{32}$

(*b*) $\frac{1}{1^5}-\frac{1}{3^5}+\frac{1}{5^5}-\frac{1}{7^5}+\cdots = \frac{5\pi^5}{1536}$

3.160. Prove (*a*) property 4 and (*b*) property 6 on page 88 by using the generating function.

3.161. Show that $B_n(\tfrac{1}{2}) = \left(1-\frac{1}{2^{n-1}}\right)B_n$.

3.162. (*a*) Show that the Bernoulli numbers can be found from

$$B_n = \sum_{k=1}^{n} \frac{(-1)^k k!}{k+1} S_k^n$$

where S_k^n are Stirling numbers of the second kind.

(*b*) Use the result in (*a*) to obtain the first few Bernoulli numbers.

3.163. Prove that

(*a*) $\ln \sin x = \ln x - \frac{x^2}{6} - \frac{x^4}{180} - \frac{x^6}{2835} - \cdots - \frac{(-1)^{k-1}B_{2k}(2x)^{2k}}{2k(2k)!} - \cdots \qquad 0 < x < \pi$

(*b*) $\ln \cos x = -\frac{x^2}{2} - \frac{x^4}{12} - \frac{x^6}{45} - \cdots - \frac{(-1)^{k-1}2^{2k}(2^{2k}-1)B_{2k}x^{2k}}{2k(2k)!} - \cdots \qquad -\pi/2 < x < \pi/2$

3.164. Prove that $1 - \frac{\pi^2}{2!}B_2 + \frac{\pi^4}{4!}B_4 - \cdots = 0$.

3.165. Prove that $B_n = 4n\int_0^\infty \frac{x^{2n-1}}{e^{2\pi x}-1}\,dx$.

3.166. Prove that

$$\frac{1}{e^x+1} = \frac{1}{2} - \frac{(2^2-1)B_2x}{2!} - \frac{(2^4-1)B_4x^3}{4!} - \cdots - \frac{(2^{2k}-1)B_{2k}x^{2k-1}}{(2k)!} - \cdots$$

$\left[\textit{Hint.}\ \frac{1}{e^{2x}-1} = \frac{1}{2}\left(\frac{1}{e^x-1}-\frac{1}{e^x+1}\right).\right]$

3.167. Prove that the only zeros of the polynomial $B_{2n}(x) - B_{2n}$ in the interval $0 \leqq x \leqq 1$ are $x = 0$ and $x = 1$.

3.168. (*a*) Obtain the generating function 8, page 89, for the Euler polynomials and (*b*) use it to find the first four polynomials.

3.169. (*a*) Obtain the generating function for the Euler numbers and (*b*) use it to find the first four numbers.

3.170. Prove properties (*a*) 3 (*b*) 6 (*c*) 7 on page 88.

3.171. Prove formulas (*a*) 9 (*b*) 10 on page 89.

3.172. Derive the results 12 on page 89.

3.173. Obtain property 11 on page 89.

3.174. (*a*) Show that

$$e_n = \frac{1}{n!}\sum_{k=1}^{n} \frac{(-1)^k k!}{2^k} S_k^n$$

where S_k^n are Stirling numbers of the second kind.

(*b*) Use (*a*) to obtain a relationship between the Euler and Stirling numbers and from this find the first four Euler numbers.

3.175. (*a*) Prove that

$$\frac{1}{e^x+1} = \tfrac{1}{2}\{E_0(0) + E_1(0)x + E_2(0)x^2 + \cdots\}$$

(*b*) Use part (*a*) and Problem 3.166 to show that

$$E_0(0) = 1, \quad E_{2k}(0) = 0, \quad E_{2k-1}(0) = \frac{-2(2^{2k}-1)B_{2k}}{(2k)!} \qquad k = 1, 2, \ldots$$

3.176. (*a*) Show that the first few *tangent numbers* [see page 89] are given by

$$T_1 = 1,\; T_2 = 0,\; T_3 = 2,\; T_4 = 0,\; T_5 = 16,\; T_6 = 0,\; T_7 = 272,\; T_8 = 0,\; T_9 = 7936$$

(*b*) Use (*a*) to show that

$$\tan x = x + 2\cdot\frac{x^3}{3!} + 16\cdot\frac{x^5}{5!} + 272\cdot\frac{x^7}{7!} + 7936\cdot\frac{x^9}{9!} + \cdots$$

Chapter 4

Applications of the Sum Calculus

SOME SPECIAL METHODS FOR EXACT SUMMATION OF SERIES

In Chapter 3 we saw how the fundamental theorem of the sum calculus could be used to find the exact sum of certain series. We shall consider, in the following, various special methods which can be used in the exact summation of series.

SERIES OF CONSTANTS

Suppose the series to be summed is

$$u_0 + u_1 + u_2 + \cdots + u_{n-1} = \sum_{k=0}^{n-1} u_k \qquad \text{where} \qquad u_k = f(a+kh) \tag{1}$$

Then we have

$$\sum_{k=0}^{n-1} u_k = nu_0 + \frac{n(n-1)}{2!}\Delta u_0 + \frac{n(n-1)(n-2)}{3!}\Delta^2 u_0 + \cdots \tag{2}$$

In particular the series can be summed exactly if u_k is any polynomial in k. See Problems 4.1 and 4.2.

POWER SERIES

A power series in x is defined as a series having the form

$$a_0 + a_1x + a_2x^2 + \cdots = \sum_{k=0}^{\infty} a_k x^k \tag{3}$$

Suppose that in this series we can write $a_k = u_k v_k$ where u_k can be written as a polynomial in k and where v_k is such that

$$V(x) = \sum_{k=0}^{\infty} v_k x^k \tag{4}$$

i.e. the series on the right of (4) can be summed exactly to $V(x)$. Then we can show that [see Problem 4.3]

$$\begin{aligned}\sum_{k=0}^{\infty} a_k x^k &= V(xE)u_0 = V(x+x\Delta)u_0 \\ &= V(x)u_0 + \frac{xV'(x)}{1!}\Delta u_0 + \frac{x^2V''(x)}{2!}\Delta^2 u_0 + \cdots \end{aligned} \tag{5}$$

where primes denote derivatives of $V(x)$. Since u_k is a polynomial, the series on the right of (5) terminates and yields the exact sum of the series on the left.

The following are some special cases.

1. $v_k = 1$. In this case $V(x) = \sum_{k=0}^{\infty} x^k = \frac{1}{1-x}$ and we find that

$$\sum_{k=0}^{\infty} a_k x^k = \sum_{k=0}^{\infty} u_k x^k = \frac{u_0}{1-x} + \frac{x\Delta u_0}{(1-x)^2} + \frac{x^2\Delta^2 u_0}{(1-x)^3} + \cdots \tag{6}$$

This is sometimes called *Montmort's formula.*

2. $v_k = 1/k!$. In this case $V(x) = \sum_{k=0}^{\infty} \frac{x^k}{k!} = e^x$ and we find that

$$\sum_{k=0}^{\infty} a_k x^k = \sum_{k=0}^{\infty} \frac{u_k x^k}{k!} = e^x \left[u_0 + \frac{x \Delta u_0}{1!} + \frac{x^2 \Delta^2 u_0}{2!} + \cdots \right] \tag{7}$$

Note that we take $0! = 1$ by definition.

For other special cases see Problems 4.37 and 4.47.

The power series method can also be used to sum various series of constants by letting x be equal to particular constants. See for example Problems 4.6 and 4.8.

APPROXIMATE INTEGRATION

Finite difference methods are valuable in obtaining formulas from which we can compute the aproximate value of a definite integral. One important method uses the Gregory-Newton formula [page 9] and yields the result [Problem 4.9]

$$\int_a^{a+nh} f(x)\,dx = nh\left[f(a) + \frac{n}{2}\Delta f(a) + \frac{n(2n-3)}{12}\Delta^2 f(a) + \frac{n(n-2)^2}{24}\Delta^3 f(a) + \frac{n(6n^3 - 45n^2 + 110n - 90)}{120}\Delta^4 f(a) + \cdots \right] \tag{8}$$

The following are some important special cases.

1. **Trapezoidal rule.**

$$\int_a^{a+h} f(x)\,dx = \frac{h}{2}[f(a) + f(a+h)] \tag{9}$$

This is obtained by letting $n = 1$ and neglecting differences of order 2 and higher [see Problem 4.10].

By using (*9*) repeatedly we can arrive at the *extended trapezoidal rule*

$$\int_a^{a+Nh} f(x)\,dx = \frac{h}{2}[f(a) + 2f(a+h) + 2f(a+2h) + \cdots + 2f(a+(N-1)h) + f(a+Nh)] \tag{10}$$

which holds for any positive integer N. If we let $y_k = f(a+kh)$ denote the ordinates, (*10*) can be written as

$$\int_a^{a+Nh} f(x)\,dx = \frac{h}{2}[y_0 + 2y_1 + 2y_2 + \cdots + 2y_{N-1} + y_N] \tag{11}$$

2. **Simpson's one-third rule.**

$$\int_a^{a+2h} f(x)\,dx = \frac{h}{3}[f(a) + 4f(a+h) + f(a+2h)] \tag{12}$$

This is obtained by letting $n = 2$ and neglecting differences of order 4 and higher. It is of interest that third order differences do not enter in this case. See Problem 4.13.

By using (*12*) repeatedly we can arrive at the *extended Simpson's one-third rule*

$$\int_a^{a+Nh} f(x)\,dx = \frac{h}{3}[f(a) + 4f(a+h) + 2f(a+2h) + 4f(a+3h) + \cdots + f(a+Nh)] \tag{13}$$

which holds for any *even* positive integer N. In ordinate notation (*13*) can be written as

$$\int_a^{a+Nh} f(x)\,dx = \frac{h}{3}[y_0 + 4y_1 + 2y_2 + 4y_3 + \cdots + 4y_{N-1} + y_N] \tag{14}$$

3. Simpson's three-eighths rule.

$$\int_a^{a+3h} f(x)\,dx \;=\; \frac{3h}{8}[f(a)+3f(a+h)+3f(a+2h)+f(a+3h)] \tag{15}$$

This is obtained by letting $n=3$ and neglecting differences of order 4 and higher. See Problem 4.16.

By using (*15*) repeatedly we can arrive at an *extended Simpson's three-eighths rule.*

4. Weddle's rule.

$$\int_a^{a+6h} f(x)\,dx \;=\; \frac{3h}{10}[\{f(a)+f(a+2h)+f(a+4h)+f(a+6h)\} + 5\{f(a+h)+f(a+5h)\}+6f(a+3h)] \tag{16}$$

which can also be written as

$$\int_a^{a+6h} f(x)\,dx \;=\; \frac{3h}{10}[(y_0+y_2+y_4+y_6)+5(y_1+y_5)+6y_3] \tag{17}$$

This is obtained by letting $n=6$ and replacing the coefficient 41/140 of $\Delta^6 f(a)$ by $42/140 = 3/10$ and neglecting differences of order 7 and higher. See Problem 4.38.

By using (*16*) or (*17*) repeatedly we can arrive at an *extended Weddle's rule* [see Problem 4.91].

In practice to obtain an approximate value for the definite integral

$$\int_a^b f(x)\,dx \tag{18}$$

we subdivide the interval from a to b into N subintervals where N is some appropriate positive integer and then use one of the above rules.

ERROR TERMS IN APPROXIMATE INTEGRATION FORMULAS

If we let $\mathcal{R}$ denote the error term to be added to the right hand side of any of the approximate integration formulas given above, then we can find $\mathcal{R}$ in terms of h and the derivatives of $f(x)$. The results are given in the following table.

Name of rule	Error term $\mathcal{R}$
Trapezoidal	$-\frac{h^3}{12}f''(\xi)$, ξ between a and $a+h$
Simpson's one-third	$-\frac{h^5}{90}f^{(IV)}(\xi)$, ξ between a and $a+2h$
Simpson's three-eighths	$-\frac{3h^5}{80}f^{(IV)}(\xi)$, ξ between a and $a+3h$
Weddle's	$-\left[\frac{h\Delta^6 f(a)}{140}+\frac{9h^9}{1400}f^{(8)}(\xi)\right]$, ξ between a and $a+6h$

For a specified value of h it is in general true that Simpson's one-third rule and Weddle's rule lead to greatest accuracy followed in order of decreasing accuracy by Simpson's three-eighths rule and trapezoidal rule.

GREGORY'S FORMULA FOR APPROXIMATE INTEGRATION

We can express a definite integral in operator form as

$$\int_a^{a+nh} f(x)\,dx = h\left(\frac{E^n - 1}{\ln E}\right) f(a) \tag{19}$$

(See Problem 4.20.) By using this we obtain *Gregory's formula for approximate integration* given by

$$\begin{aligned}\int_a^{a+nh} f(x)\,dx &= \frac{h}{2}[y_0 + 2y_1 + 2y_2 + \cdots + 2y_{n-1} + y_n] \\ &\quad - \frac{h}{12}(\nabla y_n - \Delta y_0) - \frac{h}{24}(\nabla^2 y_n + \Delta^2 y_0) \\ &\quad - \frac{19h}{720}(\nabla^3 y_n - \Delta^3 y_0) - \frac{3h}{160}(\nabla^4 y_n + \Delta^4 y_0) - \cdots\end{aligned} \tag{20}$$

where the ordinates are given by $y_k = f(a + kh)$, $k = 0, 1, \ldots, n$, and where $\nabla = \Delta E^{-1}$ is the backward difference operator [see page 9]. In practice the series is terminated at or before terms involving the nth differences so that ordinates outside the range of integration are not involved. See Problem 4.23.

THE EULER-MACLAURIN FORMULA

The *Euler-Maclaurin formula,* which is one of the most important formulas of the calculus of finite differences, provides a relationship connecting series and integrals. The formula states that

$$\begin{aligned}\sum_{k=0}^{n-1} f(a+kh) &= \frac{1}{h}\int_a^{a+nh} f(x)\,dx - \frac{1}{2}[f(a+nh) - f(a)] \\ &\quad + \frac{h}{12}[f'(a+nh) - f'(a)] - \frac{h^3}{720}[f'''(a+nh) - f'''(a)] \\ &\quad + \frac{h^5}{30{,}240}[f^{(V)}(a+nh) - f^{(V)}(a)] - \cdots\end{aligned} \tag{21}$$

and in this form is useful in evaluation of series involving $f(x)$ when the integral of $f(x)$ can be found.

Conversely the formula provides an approximate method for evaluating definite integrals in the form

$$\begin{aligned}\int_a^{a+nh} f(x)\,dx &= \frac{h}{2}[y_0 + 2y_1 + 2y_2 + \cdots + 2y_{n-1} + y_n] \\ &\quad - \frac{h^2}{12}[y'_n - y'_0] + \frac{h^4}{720}[y'''_n - y'''_0] \\ &\quad - \frac{h^6}{30{,}240}[y_n^{(V)} - y_0^{(V)}] + \cdots\end{aligned} \tag{22}$$

where we have used the notation $y_k = f(a+kh)$, $y'_0 = f'(a)$, $y'_n = f'(a+nh)$, etc.

We can also express the Euler-Maclaurin formula in terms of differences rather than derivatives [see Problem 4.101].

ERROR TERM IN EULER-MACLAURIN FORMULA

It is possible to express the Euler-Maclaurin formula in terms of Bernoulli numbers and polynomials including an expression for the remainder or error term.

To do this we first show that

$$\int_a^{a+h} f(x)\,dx = \frac{h}{2}[f(a)+f(a+h)] - \sum_{p=1}^{m-1}\frac{B_{2p}h^{2p}}{(2p)!}[f^{(2p-1)}(a+h)-f^{(2p-1)}(a)]$$

$$+ \frac{h^{2m+1}}{(2m)!}\int_0^1 [B_{2m}(u)-B_{2m}]f^{(2m)}(a+hu)\,du \qquad (23)$$

Then replacing a by $a+h,\ a+2h,\ldots,a+(n-1)h$ and adding, we obtain the Euler-Maclaurin formula with a remainder given by

$$\int_a^{a+nh} f(x)\,dx = \frac{h}{2}[f(a)+2f(a+h)+\cdots+2f(a+nh-h)+f(a+nh)]$$

$$- \sum_{p=1}^{m-1}\frac{B_{2p}h^{2p}}{(2p)!}[f^{(2p-1)}(a+nh)-f^{(2p-1)}(a)] + R \qquad (24)$$

where the remainder R is given by

$$R = \frac{h^{2m+1}}{(2m)!}\int_0^1 [B_{2m}(u)-B_{2m}]\left[\sum_{p=0}^{n-1} f^{(2m)}(a+hu+hp)\right]du \qquad (25)$$

This can be used in connection with evaluation of series by writing it as

$$\sum_{k=0}^{n-1} f(a+kh) = \int_a^{a+nh} f(x)\,dx - \frac{h}{2}[f(a+nh)-f(a)]$$

$$+ \sum_{p=1}^{m-1}\frac{B_{2p}h^{2p}}{(2p)!}[f^{(2p-1)}(a+nh)-f^{(2p-1)}(a)] - R \qquad (26)$$

where R is given by (*25*).

The result (*26*) is of course equivalent to (*21*) except that the error term R has been introduced.

STIRLING'S FORMULA FOR $n!$

An important formula which can be obtained by using the Euler-Maclaurin formula is that for $n!$, called *Stirling's formula* and given by

$$n! = \sqrt{2\pi n}\,n^n e^{-n}\left[1+\frac{1}{12n}+\frac{1}{288n^2}+\cdots\right] \qquad (27)$$

When n is not an integer the result can also be used if the left side is replaced by the gamma function $\Gamma(n+1)$ [see page 85]. The series in (*27*) is an *asymptotic series* and is useful for large values of n [actually there is good accuracy even if $n=10$ as shown in Problem 4.78].

Solved Problems

SERIES OF CONSTANTS

4.1. Prove that

$$u_0 + u_1 + u_2 + \cdots + u_{n-1} \;=\; nu_0 + \frac{n(n-1)}{2!}\Delta u_0 + \frac{n(n-1)(n-2)}{3!}\Delta^2 u_0 + \cdots$$

We have from the definition of the operator $E = 1 + \Delta$

$$\begin{aligned} u_0 + u_1 + u_2 + \cdots + u_{n-1} &= u_0 + Eu_0 + E^2u_0 + \cdots + E^{n-1}u_0 \\ &= (1 + E + E^2 + \cdots + E^{n-1})u_0 \\ &= \frac{E^n - 1}{E - 1}u_0 \\ &= \frac{(1+\Delta)^n - 1}{\Delta}u_0 \\ &= \frac{1}{\Delta}\left[n\Delta + \frac{n(n-1)}{2!}\Delta^2 + \frac{n(n-1)(n-2)}{3!}\Delta^3 + \cdots\right]u_0 \\ &= \left[n + \frac{n(n-1)}{2!}\Delta + \frac{n(n-1)(n-2)}{3!}\Delta^2 + \cdots\right]u_0 \\ &= nu_0 + \frac{n(n-1)}{2!}\Delta u_0 + \frac{n(n-1)(n-2)}{3!}\Delta^2 u_0 + \cdots \end{aligned}$$

4.2. Sum the series $1^2 + 2^2 + 3^2 + \cdots$ to n terms.

Comparing with the series of Problem 4.1 we have

$$u_0 = 1^2, \quad u_1 = 2^2, \quad u_2 = 3^2, \quad \ldots, \quad u_{n-1} = n^2$$

Then we can set up the following difference table.

k	u_k	Δu_k	$\Delta^2 u_k$	$\Delta^3 u_k$	$\Delta^4 u_k$
0	1				
		3			
1	4		2		
		5		0	
2	9		2		0
		7		0	
3	16		2		
		9			
4	25				

Fig. 4-1

It is clear from this table that

$$u_0 = 1, \quad \Delta u_0 = 3, \quad \Delta^2 u_0 = 2, \quad \Delta^3 u_0 = 0, \quad \Delta^4 u_0 = 0, \quad \ldots$$

which we could also obtain by observing that $u_k = (k+1)^2$, $\Delta u_k = 2k + 3$, $\Delta^2 u_k = 2$ and putting $k = 0$.

Thus using Problem 4.1 the required sum is

$$\begin{aligned} nu_0 + \frac{n(n-1)}{2!}\Delta u_0 + \frac{n(n-1)(n-2)}{3!}\Delta^2 u_0 + \cdots &= n(1) + \frac{n(n-1)}{2!}(3) + \frac{n(n-1)(n-2)}{3!}(2) \\ &= \frac{n(n+1)(2n+1)}{6} \end{aligned}$$

in agreement with Problem 3.21(*a*), page 98.

Note that the method of Problem 4.1 is especially useful when the kth term can be expressed as a polynomial in k so that differences are ultimately zero.

POWER SERIES

4.3. If $V(x) = \sum_{k=0}^{\infty} v_k x^k$ and $a_k = u_k v_k$ prove that

$$\sum_{k=0}^{\infty} a_k x^k = V(x)u_0 + \frac{xV'(x)}{1!}\Delta u_0 + \frac{x^2V''(x)}{2!}\Delta^2 u_0 + \cdots$$

We have

$$\begin{aligned}
\sum_{k=0}^{\infty} a_k x^k &= \sum_{k=0}^{\infty} u_k v_k x^k \\
&= u_0v_0 + u_1v_1x + u_2v_2x^2 + u_3v_3x^3 + \cdots \\
&= v_0u_0 + v_1xEu_0 + v_2x^2E^2u_0 + v_3x^3E^3u_0 + \cdots \\
&= [v_0 + v_1xE + v_2x^2E^2 + v_3x^3E^3 + \cdots]u_0 \\
&= [v_0 + v_1(xE) + v_2(xE)^2 + v_3(xE)^3 + \cdots]u_0 \\
&= V(xE)u_0 \\
&= V(x + x\Delta)u_0 \\
&= \left[V(x) + V'(x)(x\Delta) + \frac{V''(x)}{2!}(x\Delta)^2 + \cdots\right]u_0 \\
&= V(x)u_0 + xV'(x)\Delta u_0 + \frac{x^2V''(x)}{2!}\Delta^2 u_0 + \cdots
\end{aligned}$$

where we have used the Taylor series expansion

$$V(x+h) = V(x) + V'(x)h + \frac{V''(x)h^2}{2!} + \cdots$$

with $h = x\Delta$.

Note that although the above has been proved without any restriction on u_k an important case where the series can be summed exactly occurs when u_k is a polynomial in k so that higher differences are ultimately zero.

4.4. Prove that

$$\sum_{k=0}^{\infty} u_k x^k = \frac{u_0}{1-x} + \frac{x\Delta u_0}{(1-x)^2} + \frac{x^2\Delta^2 u_0}{(1-x)^3} + \cdots$$

Put $v_k = 1$ in the results of Problem 4.3. Then

$$V(x) = \sum_{k=0}^{\infty} v_k x^k = \sum_{k=0}^{\infty} x^k = 1 + x + x^2 + x^3 + \cdots = \frac{1}{1-x}$$

Thus

$$V'(x) = \frac{1}{(1-x)^2}, \qquad V''(x) = \frac{2}{(1-x)^3}, \qquad \ldots$$

and so from Problem 4.3 we have

$$\begin{aligned}
\sum_{k=0}^{\infty} u_k x^k &= V(x)u_0 + \frac{xV'(x)}{1!}\Delta u_0 + \frac{x^2V''(x)}{2!}\Delta^2 u_0 + \cdots \\
&= \frac{1}{1-x}u_0 + \frac{x}{(1-x)^2}\Delta u_0 + \frac{x^2}{(1-x)^3}\Delta^2 u_0 + \cdots
\end{aligned}$$

4.5. Sum the series $\sum_{k=0}^{\infty} (k+1)(k+2)x^k = 1\cdot 2 + 2\cdot 3x + 3\cdot 4x^2 + \cdots$.

Comparing with Problem 4.4 we have $u_k = (k+1)(k+2)$. Then we can construct the following difference table.

k	u_k	Δu_k	$\Delta^2 u_k$	$\Delta^3 u_k$	$\Delta^4 u_k$
0	2				
		4			
1	6		2		
		6		0	
2	12		2		0
		8		0	
3	20		2		
		10			
4	30				

Fig. 4-2

From this table it is clear that

$$u_0 = 2, \quad \Delta u_0 = 4, \quad \Delta^2 u_0 = 2, \quad \Delta^3 u_0 = 0, \quad \Delta^4 u_0 = 0, \quad \ldots$$

These could also have been obtained by letting $k = 0$ in the results $u_k = (k+1)(k+2)$, $\Delta u_k = 2k+4$, $\Delta^2 u_k = 2$, ...

Thus we have from Problem 4.4

$$\begin{aligned} \sum_{k=0}^{\infty} (k+1)(k+2)x^k &= \frac{2}{1-x} + \frac{4x}{(1-x)^2} + \frac{2x^2}{(1-x)^3} \\ &= \frac{2(1-x)^2 + 4x(1-x) + 2x^2}{(1-x)^3} \\ &= \frac{2}{(1-x)^3} \end{aligned}$$

It should be noted that the series converges only if $|x| < 1$.

4.6. Sum the series $1 \cdot 2 - \dfrac{2 \cdot 3}{5} + \dfrac{3 \cdot 4}{5^2} - \dfrac{4 \cdot 5}{5^3} + \dfrac{5 \cdot 6}{5^4} - \cdots$.

This series is a special case of the series of Problem 4.5 in which we let $x = -1/5$. Then the sum of the series is

$$\frac{2}{(1+1/5)^3} = \frac{125}{108}$$

4.7. Prove that

$$\sum_{k=0}^{\infty} \frac{u_k x^k}{k!} = e^x \left[u_0 + \frac{x \Delta u_0}{1!} + \frac{x^2 \Delta^2 u_0}{2!} + \cdots \right]$$

Put $v_k = 1/k!$ in the results of Problem 4.3. Then

$$V(x) = \sum_{k=0}^{\infty} v_k x^k = \sum_{k=0}^{\infty} \frac{x^k}{k!} = 1 + x + \frac{x^2}{2!} + \frac{x^3}{3!} + \cdots = e^x$$

Thus

$$V'(x) = e^x, \qquad V''(x) = e^x, \qquad \ldots$$

and so from Problem 4.3 we have

$$\begin{aligned} \sum_{k=0}^{\infty} u_k x^k &= V(x)u_0 + \frac{xV'(x)}{1!} \Delta u_0 + \frac{x^2 V''(x)}{2!} \Delta^2 u_0 + \cdots \\ &= e^x u_0 + x e^x \Delta u_0 + \frac{x^2 e^x}{2!} \Delta^2 u_0 + \cdots \\ &= e^x \left[u_0 + x \Delta u_0 + \frac{x^2 \Delta^2 u_0}{2!} + \cdots \right] \end{aligned}$$

4.8. Sum the series $2^2 + \dfrac{5^2}{1!4} + \dfrac{8^2}{2!4^2} + \dfrac{11^2}{3!4^3} + \cdots$.

The series is a special case with $x = 1/4$ of

$$2^2 + \frac{5^2}{1!}x + \frac{8^2}{2!}x^2 + \frac{11^2}{3!}x^3 + \cdots$$

or

$$\sum_{k=0}^{\infty} \frac{(3k+2)^2 x^k}{k!}$$

Thus we can obtain the sum by putting $u_k = (3k+2)^2$ in the result of Problem 4.7. From the difference table in Fig. 4-3 we find

$$u_0 = 4, \quad \Delta u_0 = 21, \quad \Delta^2 u_0 = 18, \quad \Delta^3 u_0 = 0, \quad \ldots$$

k	u_k	Δu_k	$\Delta^2 u_k$	$\Delta^3 u_k$
0	4			
		21		
1	25		18	
		39		0
2	64		18	
		57		0
3	121		18	
		75		
4	196			

Fig. 4-3

Thus

$$\sum_{k=0}^{\infty} \frac{(3k+2)^2 x^k}{k!} = e^x[4 + 21x + 9x^2]$$

and putting $x = 1/4$ the required sum is seen to be $157\sqrt[4]{e}/16$.

APPROXIMATE INTEGRATION

4.9. Prove that

$$\int_a^{a+nh} f(x)\,dx = nh\Big[f(a) + \frac{n}{2}\Delta f(a) + \frac{n(2n-3)}{12}\Delta^2 f(a) + \frac{n(n-2)^2}{24}\Delta^3 f(a) + \frac{n(6n^3 - 45n^2 + 110n - 90)}{120}\Delta^4 f(a) + \cdots\Big]$$

By the Gregory-Newton formula (*44*) on page 9 we have

$$f(x) = f(a) + \frac{\Delta f(a)}{\Delta x}(x-a)^{(1)} + \frac{\Delta^2 f(a)}{\Delta x^2}\frac{(x-a)^{(2)}}{2!} + \frac{\Delta^3 f(a)}{\Delta x^3}\frac{(x-a)^{(3)}}{3!} + \cdots$$

Using $\Delta x = h$ and the definition of $(x-a)^{(m)}$ this can be written

$$f(x) = f(a) + \frac{\Delta f(a)}{h}(x-a) + \frac{\Delta^2 f(a)}{2!\,h^2}(x-a)(x-a-h) + \frac{\Delta^3 f(a)(x-a)(x-a-h)(x-a-2h)}{3!\,h^3} + \cdots$$

Integrating from $x = a$ to $a + nh$ we then have

$$\begin{aligned}\int_a^{a+nh} f(x)\,dx &= \int_a^{a+nh} f(a)\,dx + \frac{\Delta f(a)}{h}\int_a^{a+nh}(x-a)\,dx \\ &\quad + \frac{\Delta^2 f(a)}{2!\,h^2}\int_a^{a+nh}(x-a)(x-a-h)\,dx \\ &\quad + \frac{\Delta^3 f(a)}{3!\,h^3}\int_a^{a+nh}(x-a)(x-a-h)(x-a-2h)\,dx \\ &\quad + \frac{\Delta^4 f(a)}{4!\,h^4}\int_a^{a+nh}(x-a)(x-a-h)(x-a-2h)(x-a-3h)\,dx + \cdots \\ &= nh\left[f(a) + \frac{n}{2}\Delta f(a) + \frac{n(2n-3)}{12}\Delta^2 f(a)\right. \\ &\quad \left. + \frac{n(n-2)^2}{24}\Delta^3 f(a) + \frac{n(6n^3-45n^2+110n-90)}{120}\Delta^4 f(a) + \cdots\right]\end{aligned}$$

4.10. Prove the *trapezoidal rule*

$$\int_a^{a+h} f(x)\,dx = \frac{h}{2}[f(a) + f(a+h)]$$

Put $n = 1$ in the result of Problem 4.9. Then assuming that differences of order higher than two are zero we have

$$\begin{aligned}\int_a^{a+h} f(x)\,dx &= h[f(a) + \tfrac{1}{2}\Delta f(a)] \\ &= h[f(a) + \tfrac{1}{2}\{f(a+h) - f(a)\}] \\ &= \frac{h}{2}[f(a) + f(a+h)]\end{aligned}$$

4.11. Use Problem 4.10 to obtain the *extended trapezoidal rule*

$$\int_a^{a+Nh} f(x)\,dx = \frac{h}{2}[f(a) + 2f(a+h) + 2f(a+2h) + \cdots + 2f(a+(N-1)h) + f(a+Nh)]$$

From the results of Problem 4.10 we have

$$\begin{aligned}\int_a^{a+h} f(x)\,dx &= \frac{h}{2}[f(a) + f(a+h)] \\ \int_{a+h}^{a+2h} f(x)\,dx &= \frac{h}{2}[f(a+h) + f(a+2h)] \\ &\cdots\cdots\cdots\cdots \\ \int_{a+(N-1)h}^{a+Nh} f(x)\,dx &= \frac{h}{2}[f(a+(N-1)h) + f(a+Nh)]\end{aligned}$$

Then by addition

$$\int_a^{a+Nh} f(x)\,dx = \frac{h}{2}[f(a) + 2f(a+h) + \cdots + 2f(a+(N-1)h) + f(a+Nh)]$$

or using ordinates $y_k = f(a+kh)$,

$$\int_a^{a+Nh} f(x)\,dx = \frac{h}{2}[y_0 + 2y_1 + 2y_2 + \cdots + 2y_{N-1} + y_N]$$

4.12. (*a*) Find $\int_0^1 \frac{dx}{1+x^2}$ by the extended trapezoidal rule where the interval of integration is subdivided into 6 equal parts.

(*b*) Compare the result with the exact value.

(*a*) We have $a = 0$, $a + Nh = 1$ so that since $N = 6$, $h = 1/6$. The work of computation can be arranged in the following table.

x	0	1/6	2/6	3/6	4/6	5/6	1
$f(x) = 1/(1+x^2)$	1.00000000	0.97297297	0.90000000	0.80000000	0.69230769	0.59016393	0.50000000

Fig. 4-4

Then applying the extended trapezoidal rule of Problem 4.11 we have

$$\begin{aligned}\int_0^1 \frac{dx}{1+x^2} &= \frac{1/6}{2}[1 + 2(0.97297297 + 0.90000000 + 0.80000000 + 0.69230769 + 0.59016393) \\ &\qquad + 0.50000000] \\ &= 0.78424077\end{aligned}$$

(*b*) The exact value is

$$\int_0^1 \frac{dx}{1+x^2} = \tan^{-1} x \Big|_0^1 = \tan^{-1} 1 = \frac{\pi}{4} = 0.785398163\ldots$$

Thus the trapezoidal rule for this number of subdivisions yields a result which is accurate to two decimal places. The percent error is 0.147%.

4.13. Prove Simpson's one-third rule

$$\int_a^{a+2h} f(x)\,dx = \frac{h}{3}[f(a) + 4f(a+h) + f(a+2h)]$$

Put $n = 2$ in the result of Problem 4.9. Then assuming that differences of order 4 and higher are zero we obtain

$$\begin{aligned}\int_a^{a+2h} f(x)\,dx &= 2h[f(a) + \Delta f(a) + \tfrac{1}{6}\Delta^2 f(a)] \\ &= 2h[f(a) + \{f(a+h) - f(a)\} + \tfrac{1}{6}\{f(a+2h) - 2f(a+h) + f(a)\}] \\ &= \frac{h}{3}[f(a) + 4f(a+h) + f(a+2h)]\end{aligned}$$

4.14. Use Problem 4.13 to obtain the *extended Simpson's one-third rule*

$$\int_a^{a+Nh} f(x)\,dx = \frac{h}{3}[f(a) + 4f(a+h) + 2f(a+2h) + 4f(a+3h) + \cdots + f(a+Nh)]$$

where it is assumed that N is even.

From the results of Problem 4.13 we have

$$\int_a^{a+2h} f(x)\,dx = \frac{h}{3}[f(a) + 4f(a+h) + f(a+2h)]$$

$$\int_{a+2h}^{a+4h} f(x)\,dx = \frac{h}{3}[f(a+2h) + 4f(a+3h) + f(a+4h)]$$

. . . .

$$\int_{a+(N-2)h}^{a+Nh} f(x)\,dx = \frac{h}{3}[f(a+(N-2)h) + 4f(a+(N-1)h) + f(a+Nh)]$$

Then by addition

$$\int_a^{a+Nh} f(x)\,dx = \frac{h}{3}[f(a) + 4f(a+h) + 2f(a+2h) + 4f(a+3h) + \cdots + f(a+Nh)]$$

Using ordinates $y_k = f(a+kh)$ this can be written

$$\int_a^{a+Nh} f(x)\,dx = \frac{h}{3}[y_0 + 4y_1 + 2y_2 + 4y_3 + \cdots + 4y_{N-1} + y_N]$$

4.15. (*a*) Work Problem 4.12 by using the extended Simpson's one-third rule and (*b*) compare your result with the exact value and that obtained from the trapezoidal rule.

(*a*) We can apply the rule since $N = 6$ [which is a multiple of 3]. It is convenient to write the rule in the form

$$\begin{aligned}\int_0^1 \frac{dx}{1+x^2} &= \frac{h}{3}[y_0 + y_6 + 4(y_1 + y_3 + y_5) + 2(y_2 + y_4)] \\ &= \frac{1/6}{3}[1.00000000 + 0.50000000 + 4(0.97297297 + 0.80000000 + 0.59016393) \\ &\qquad + 2(0.90000000 + 0.69230769)] \\ &= 0.78539794\end{aligned}$$

where we have used the value $h = 1/6$ and the table of values obtained in Problem 4.12 which is applicable here.

(*b*) The result is seen to be accurate to 6 decimal places on comparison with the exact value 0.78539816... as obtained in Problem 4.12(*b*). The percent error is 0.00028%. It is clear that the result is far more accurate than that obtained using the trapezoidal rule.

4.16. Prove *Simpson's three-eighths rule*

$$\int_a^{a+3h} f(x)\,dx = \frac{3h}{8}[f(a) + 3f(a+h) + 3f(a+2h) + f(a+3h)]$$

Put $n = 3$ in Problem 4.9 and neglect differences of order 4 and higher. Then

$$\begin{aligned}\int_a^{a+3h} f(x)\,dx &= 3h\,[f(a) + \tfrac{3}{2}\Delta f(a) + \tfrac{3}{4}\Delta^2 f(a) + \tfrac{1}{8}\Delta^3 f(a)] \\ &= 3h\,[f(a) + \tfrac{3}{2}\{f(a+h) - f(a)\} + \tfrac{3}{4}\{f(a+2h) - 2f(a+h) + f(a)\} \\ &\qquad + \tfrac{1}{8}\{f(a+3h) - 3f(a+2h) + 3f(a+h) - f(a)\}] \\ &= \frac{3h}{8}[f(a) + 3f(a+h) + 3f(a+2h) + f(a+3h)]\end{aligned}$$

We can obtain an *extended Simpson's three-eighths rule* for the evaluation of $\int_a^{a+Nh} f(x)\,dx$ when N is a multiple of 3 [see Problem 4.57].

4.17. (*a*) Work Problem 4.12 by the extended Simpson's three-eighths rule and (*b*) compare with the result obtained from the one-third rule.

(*a*) Using Problem 4.1 and also the table of Problem 4.12 we have

$$\begin{aligned}\int_0^1 \frac{dx}{1+x^2} &= \frac{3h}{8}[(y_0 + y_6) + 3(y_1 + y_2 + y_4 + y_5) + 2y_3] \\ &= \frac{3(1/6)}{8}[(1.00000000 + 0.50000000) \\ &\qquad + 3(0.97297297 + 0.90000000 + 0.69230769 + 0.59016393) + 2(0.80000000)] \\ &= 0.78539586\end{aligned}$$

(*b*) The percent error is 0.00293% which is about 10 times the error made using Simpson's one-third rule.

ERROR TERMS IN APPROXIMATE INTEGRATION FORMULAS

4.18. Find the error term in the trapezoidal rule.

The Gregory-Newton formula with a remainder after 2 terms is

$$f(x) = f(a) + \frac{\Delta f(a)}{\Delta x}(x-a)^{(1)} + R$$

where

$$R = \frac{f''(\eta)(x-a)^{(2)}}{2!} \qquad \eta \text{ between } a \text{ and } x$$

Putting $\Delta x = h$ and integrating from $x = a$ to $x = a+h$ we have

$$\int_a^{a+h} f(x)\,dx = \int_a^{a+h} f(a)\,dx + \frac{\Delta f(a)}{h}\int_a^{a+h}(x-a)\,dx + \mathcal{R} \tag{1}$$

where

$$\mathcal{R} = \frac{1}{2!}\int_a^{a+h} f''(\eta)(x-a)^{(2)}\,dx \tag{2}$$

Performing the integrations in (1) we find

$$\int_a^{a+h} f(x)\,dx = \frac{h}{2}[f(a)+f(a+h)] + \mathcal{R}$$

Now to estimate the remainder or error $\mathcal{R}$ we note that $(x-a)^{(2)} = (x-a)(x-a-h)$ does not change sign for x between a and $a+h$. We can thus apply the mean-value theorem [see page 81] to (2) to obtain

$$\mathcal{R} = \frac{1}{2!}f''(\xi)\int_a^{a+h}(x-a)(x-a-h)\,dx = -\frac{h^3}{12}f''(\xi)$$

where ξ is also between a and $a+h$. This yields the required error term.

4.19. Find the error term in Simpson's one-third rule.

If we proceed as in Problem 4.18 the error term will be given by

$$\frac{1}{3!}\int_a^{a+2h} f'''(\eta)(x-a)^{(3)}\,dx$$

However $(x-a)^{(3)} = (x-a)(x-a-h)(x-a-2h)$ changes sign between $x = a$ and $x = a+2h$ so that we cannot apply the mean-value theorem. We must thus proceed in another manner.

One approach is to write the error as

$$\int_a^{a+2h} f(x)\,dx - \frac{h}{3}[f(a)+4f(a+h)+f(a+2h)] \tag{1}$$

To simplify details by introducing a symmetry to this we shall choose $a = -h$ so that the error (1) is given by

$$\mathcal{R}(h) = \int_{-h}^{h} f(x)\,dx - \frac{h}{3}[f(-h)+4f(0)+f(h)] \tag{2}$$

We now consider h as a parameter and differentiate with respect to it. We find using Leibnitz's rule on page 81

$$\begin{aligned}
\mathcal{R}'(h) &= f(h) + f(-h) - \frac{h}{3}[-f'(-h)+f'(h)] - \tfrac{1}{3}[f(-h)+4f(0)+f(h)] \\
&= \tfrac{2}{3}f(h) + \tfrac{2}{3}f(-h) - \tfrac{4}{3}f(0) + \frac{h}{3}[f'(-h)-f'(h)] \\[1ex]
\mathcal{R}''(h) &= \tfrac{2}{3}f'(h) - \tfrac{2}{3}f'(-h) + \frac{h}{3}[-f''(-h)-f''(h)] + \tfrac{1}{3}[f'(-h)-f'(h)] \\
&= \tfrac{1}{3}f'(h) - \tfrac{1}{3}f'(-h) - \frac{h}{3}[f''(-h)+f''(h)] \\[1ex]
\mathcal{R}'''(h) &= \tfrac{1}{3}f''(h) + \tfrac{1}{3}f''(-h) - \frac{h}{3}[-f'''(-h)+f'''(h)] - \tfrac{1}{3}[f''(-h)+f''(h)] \\
&= -\frac{h}{3}[f'''(h)-f'''(-h)]
\end{aligned}$$

Thus

$$\mathcal{R}(0) = \mathcal{R}'(0) = \mathcal{R}''(0) = \mathcal{R}'''(0) = 0 \tag{3}$$

If we now assume that $f^{(IV)}(x)$ is continuous we can apply the mean-value theorem for derivatives [see page 8] to obtain

$$\mathcal{R}'''(h) = -\frac{h}{3}[f'''(h)-f'''(-h)] = -\frac{2h^2}{3}[f^{(IV)}(\eta)] \qquad \eta \text{ between } -h \text{ and } h$$

Now considering

$$\mathcal{R}'''(h) \quad = \quad G(h) \quad = \quad -\frac{2h^2}{3} f^{(\mathrm{IV})}(\eta) \tag{4}$$

we can show by integrating three times using conditions (3) that [see Problem 4.89]

$$\mathcal{R}(h) \quad = \quad -\frac{1}{3}\int_0^h u^2(h-u)^2 f^{(\mathrm{IV})}(\eta)\, du \tag{5}$$

Since $u^2(h-u)^2$ does not change sign between $u=0$ and $u=h$ we can now apply the mean-value theorem for integrals in (5) to obtain

$$\mathcal{R}(h) \quad = \quad -\tfrac{1}{3} f^{(\mathrm{IV})}(\xi)\int_0^h u^2(h-u)^2\, du \quad = \quad -\frac{h^5 f^{(\mathrm{IV})}(\xi)}{90}$$

which is the required error term.

GREGORY'S FORMULA FOR APPROXIMATE INTEGRATION

4.20. Show that $\displaystyle\int_a^{a+nh} f(x)\, dx \;=\; h\left(\frac{E^n-1}{\ln E}\right) f(a)$.

We have by definition

$$D^{-1}f(x) \quad = \quad \int f(x)\, dx \tag{1}$$

so that

$$\int_a^{a+nh} f(x)\, dx \quad = \quad D^{-1}f(x)\Big|_a^{a+nh} \tag{2}$$

But since $E = e^{hD}$ we formally have $D = \dfrac{1}{h}\ln E$ so that

$$D^{-1}f(x) \quad = \quad \frac{1}{D} f(x) \quad = \quad \frac{h}{\ln E} f(x)$$

Then (2) can be written as

$$\begin{aligned}\int_a^{a+nh} f(x)\, dx \quad = \quad \frac{h}{\ln E} f(x)\Big|_a^{a+nh} \quad &= \quad \frac{h}{\ln E}[f(a+nh) - f(a)] \\ &= \quad \frac{h}{\ln E}[E^n f(a) - f(a)] \\ &= \quad h\left(\frac{E^n-1}{\ln E}\right) f(a)\end{aligned}$$

4.21. Show that

$$\int_a^{a+nh} f(x)\, dx \quad = \quad \frac{h}{\ln E}[f(a+nh) - f(a)] \quad = \quad \frac{h}{\ln E}(y_n - y_0)$$

where $y_k = f(a+kh)$.

From Problem 4.20 we have, using $y_k = f(a+kh)$ so that $y_0 = f(a)$, $y_n = f(a+nh)$,

$$\int_a^{a+nh} f(x)\, dx \quad = \quad h\left(\frac{E^n-1}{\ln E}\right) f(a) \quad = \quad \frac{h}{\ln E}[f(a+nh) - f(a)] \quad = \quad \frac{h}{\ln E}(y_n - y_0)$$

4.22. Show that

$$\frac{1}{\ln(1+x)} \quad = \quad \frac{1}{x}\left(1 + \frac{1}{2}x - \frac{1}{12}x^2 + \frac{1}{24}x^3 - \frac{19}{720}x^4 + \frac{3}{160}x^5 - \cdots\right)$$

We have [see page 8]

$$\ln(1+x) \quad = \quad x - \frac{x^2}{2} + \frac{x^3}{3} - \frac{x^4}{4} + \frac{x^5}{5} - \cdots$$

Then by ordinary long division we obtain

$$\frac{1}{\ln(1+x)} = \frac{1}{x - \dfrac{x^2}{2} + \dfrac{x^3}{3} - \dfrac{x^4}{4} + \cdots}$$

$$= \frac{1}{x} + \frac{1}{2} - \frac{1}{12}x + \frac{1}{24}x^2 - \frac{19}{720}x^3 + \frac{3}{160}x^4 - \cdots$$

$$= \frac{1}{x}\left(1 + \frac{1}{2}x - \frac{1}{12}x^2 + \frac{1}{24}x^3 - \frac{19}{720}x^4 + \frac{3}{160}x^5 - \cdots\right)$$

4.23. Obtain Gregory's formula (*20*), page 124, for approximate integration.

From Problem 4.21 we have

$$\int_a^{a+nh} f(x)\,dx = h\left(\frac{E^n-1}{\ln E}\right)f(a) = \frac{h}{\ln E}[f(a+nh)-f(a)] = \frac{h}{\ln E}(y_n - y_0) \tag{1}$$

Now $$E = 1+\Delta, \quad E^{-1} = 1-\nabla \tag{2}$$

so that $$\ln E = \ln(1+\Delta), \quad \ln E = -\ln(1-\nabla) \tag{3}$$

Then (*1*) can be written

$$\int_a^{a+nh} f(x)\,dx = h\left[\frac{-1}{\ln(1-\nabla)}y_n - \frac{1}{\ln(1+\Delta)}y_0\right] \tag{4}$$

Now using Problem 4.22 we find

$$\frac{-1}{\ln(1-\nabla)}y_n = \left(\frac{1}{\nabla} - \frac{1}{2} - \frac{1}{12}\nabla - \frac{1}{24}\nabla^2 - \frac{19}{720}\nabla^3 - \frac{3}{160}\nabla^4 - \cdots\right)y_n$$

$$\frac{1}{\ln(1+\Delta)}y_0 = \left(\frac{1}{\Delta} + \frac{1}{2} - \frac{1}{12}\Delta + \frac{1}{24}\Delta^2 - \frac{19}{720}\Delta^3 + \frac{3}{160}\Delta^4 - \cdots\right)y_0$$

By subtraction we then have

$$\frac{-1}{\ln(1-\nabla)}y_n - \frac{1}{\ln(1+\Delta)}y_0 = \left(\frac{1}{\nabla}y_n - \frac{1}{\Delta}y_0\right) - \frac{1}{2}(y_n+y_0) - \frac{1}{12}(\nabla y_n - \Delta y_0)$$
$$-\frac{1}{24}(\nabla^2 y_n + \Delta^2 y_0) - \frac{19}{720}(\nabla^3 y_n - \Delta^3 y_0) - \frac{3}{160}(\nabla^4 y_n + \Delta^4 y_0) - \cdots$$

Also since $\nabla = E^{-1}\Delta = \Delta E^{-1}$ [see Problem 1.39, page 25] we have

$$\frac{1}{\nabla}y_n - \frac{1}{\Delta}y_0 = \frac{1}{E^{-1}\Delta}y_n - \frac{1}{\Delta}y_0 = \frac{Ey_n - y_0}{\Delta} = \frac{(E^{n+1}-1)y_0}{\Delta} = \left(\frac{E^{n+1}-1}{E-1}\right)y_0$$
$$= (1+E+E^2+\cdots+E^n)y_0 = y_0 + y_1 + y_2 + \cdots + y_n$$

Thus (*4*) becomes

$$\int_a^{a+nh} f(x)\,dx = \frac{h}{2}[y_0 + 2y_1 + 2y_2 + \cdots + 2y_{n-1} + y_n]$$
$$-\frac{h}{12}(\nabla y_n - \Delta y_0) - \frac{h}{24}(\nabla^2 y_n + \Delta^2 y_0)$$
$$-\frac{19h}{720}(\nabla^3 y_n - \Delta^3 y_0) - \frac{3h}{160}(\nabla^4 y_n + \Delta^4 y_0) - \cdots$$

4.24. Work Problem 4.12 by using Gregory's formula.

Putting $a = 0$, $n = 6$, $h = 1/6$ and $f(x) = 1/(1+x^2)$ in Gregory's formula it becomes

$$\int_0^1 \frac{dx}{1+x^2} = \frac{1/6}{2}[y_0 + 2y_1 + 2y_2 + \cdots + 2y_5 + y_6]$$

$$- \frac{1/6}{12}(\nabla y_6 - \Delta y_0) - \frac{1/6}{24}(\nabla^2 y_6 + \Delta^2 y_0)$$

$$- \frac{19(1/6)}{720}(\nabla^3 y_6 - \Delta^3 y_0) - \frac{3(1/6)}{160}(\nabla^4 y_6 + \Delta^4 y_0)$$

The ordinates $y_0, y_1, \ldots, y_6$ can be taken from the table of Problem 4.12. To obtain the various differences we form the following difference table where the factor 10^8 has been used so as to avoid decimals.

k	$y_k \times 10^8$	$\Delta y_k \times 10^8$	$\Delta^2 y_k \times 10^8$	$\Delta^3 y_k \times 10^8$	$\Delta^4 y_k \times 10^8$
0	100000000				
		−2702703			
1	97297297		−4594594		
		−7297297		1891891	
2	90000000		−2702703		41581
		−10000000		1933472	
3	80000000		−769231		−609386
		−10769231		1324086	
4	69230769		+554855		−680958
		−10214376		643128	
5	59016393		+1197983		
		−9016393			
6	50000000				

Fig. 4-5

From this table we see that

$$\begin{aligned}
\nabla y_6 &= \Delta E^{-1} y_6 = \Delta y_5 = -0.09016393, & \Delta y_0 &= -0.02702703\\
\nabla^2 y_6 &= \Delta^2 E^{-2} y_6 = \Delta^2 y_4 = 0.01197983, & \Delta^2 y_0 &= -0.04594594\\
\nabla^3 y_6 &= \Delta^3 E^{-3} y_6 = \Delta^3 y_3 = 0.00643128, & \Delta^3 y_0 &= 0.01933472\\
\nabla^4 y_6 &= \Delta^4 E^{-4} y_6 = \Delta^4 y_2 = -0.00680958, & \Delta^4 y_0 &= 0.00041581
\end{aligned}$$

Thus we have

$$\int_0^1 \frac{dx}{1+x^2} = 0.78424077 - \frac{(1/6)}{12}(-0.0631369)$$

$$- \frac{(1/6)}{24}(-0.03396611) - \frac{19(1/6)}{720}(-0.01290344) - \frac{3(1/6)}{160}(-0.00639377)$$

$$= 0.78424077 + 0.00087690 + 0.00023588 + 0.00005675 + 0.00001998$$

$$= 0.78543028$$

It should be noted that the value 0.78424077 in the above is that obtained using the trapezoidal rule. Thus we can think of the remaining terms in Gregory's formula as supplying correction terms to the trapezoidal rule. It is seen that these correction terms provide a considerable improvement since the value obtained above has only an error of 0.00423% whereas that obtained without the correction terms is 0.147%. It should be noted however that Simpson's one-third rule is still the victor.

THE EULER-MACLAURIN FORMULA

4.25. Show that

$$\Delta^{-1} f(x) = \frac{1}{e^{hD} - 1} f(x) = \frac{1}{hD}\left[\sum_{k=0}^{\infty} \frac{B_k (hD)^k}{k!}\right] f(x)$$

$$= \frac{1}{hD} f(x) - \frac{1}{2} f(x) + \frac{1}{12} hDf(x) - \frac{1}{720} h^3 D^3 f(x) + \cdots$$

We have since $\Delta = E - 1 = e^{hD} - 1$ [see page 8]

$$\Delta^{-1}f(x) = \frac{1}{\Delta}f(x) = \frac{1}{e^{hD}-1}f(x) = \frac{1}{hD}\frac{hD}{e^{hD}-1}f(x)$$

Then using property 8, page 87, with $t = hD$ this becomes

$$\begin{aligned}\Delta^{-1}f(x) &= \frac{1}{hD}\left[\sum_{k=0}^{\infty}\frac{B_k(hD)^k}{k!}\right]f(x)\\ &= \frac{1}{hD}\left[B_0 + \frac{B_1 hD}{1!} + \frac{B_2h^2D^2}{2!} + \frac{B_3h^3D^3}{3!} + \frac{B_4h^4D^4}{4!} + \cdots\right]f(x)\\ &= \frac{1}{hD}f(x) - \frac{1}{2}f(x) + \frac{1}{12}hDf(x) - \frac{1}{720}h^3D^3f(x) + \cdots\end{aligned}$$

on using the values for $B_0, B_1, B_2, B_3, \ldots$.

4.26. Derive the Euler-Maclaurin formula (*21*) on page 124.

By the fundamental theorem of sum calculus we have [see page 83]

$$\sum_{a}^{a+(n-1)h} f(x) = \Delta^{-1}f(x)\Big|_a^{a+nh} \tag{1}$$

where the summation in (*1*) is to be taken from a to $a+(n-1)h$ *in steps of h*. Using the results of Problem 4.25 we then have

$$\sum_{a}^{a+(n-1)h} f(x) = \frac{1}{hD}f(x)\Big|_a^{a+nh} - \frac{1}{2}f(x)\Big|_a^{a+nh} + \frac{1}{12}hf'(x)\Big|_a^{a+nh} - \frac{1}{720}h^3f'''(x)\Big|_a^{a+nh} + \cdots$$

which can be written as

$$\begin{aligned}f(a) + f(a+h) + \cdots + f(a+(n-1)h) &= \frac{1}{h}\int_a^{a+nh} f(x)\,dx - \frac{1}{2}[f(a+nh) - f(a)]\\ &\quad + \frac{h}{12}[f'(a+nh) - f'(a)] - \frac{h^3}{720}[f'''(a+nh) - f'''(a)] + \cdots\end{aligned}$$

4.27. Find $\sum_{k=100}^{199}\frac{1}{k}$ by using the Euler-Maclaurin formula.

Let $f(x) = 1/x$, $a = 100$, $a+(n-1)h = 199$, $h = 1$ in the result of Problem 4.26 so that $n = 100$. Then $f'(x) = -1/x^2$, $f''(x) = 2x^{-3}$, $f'''(x) = -6x^{-4}$ so that

$$\begin{aligned}\sum_{k=100}^{199}\frac{1}{k} &= \int_{100}^{200}\frac{dx}{x} - \frac{1}{2}\left[\frac{1}{200} - \frac{1}{100}\right] + \frac{1}{12}\left[-\frac{1}{(200)^2} + \frac{1}{(100)^2}\right]\\ &\quad - \frac{1}{720}\left[-\frac{6}{(200)^4} + \frac{6}{(100)^4}\right] + \cdots\\ &= \ln 2 + \tfrac{1}{2}(0.005) + \tfrac{1}{12}(0.000075) + \cdots\\ &= 0.693147 + 0.002500 + 0.00000625 + \cdots\\ &= 0.695653\end{aligned}$$

to 6 decimal places. In Problem 4.32 we shall discover how accurate the result is.

ERROR TERM IN THE EULER-MACLAURIN SERIES

4.28. Let $\phi_p(x) = B_p(x) - B_p$, $p = 0, 1, 2, 3, \ldots$, where $B_p(x)$ are the Bernoulli polynomials and B_p are the Bernoulli numbers. Show that

(*a*) $\phi_p'(x) = pB_{p-1}(x)$

(*b*) $\phi_p(x) = \dfrac{\phi_{p+1}'(x)}{p+1} - B_p$

(*c*) $\phi_p(0) = 0$

(*d*) $\phi_p(1) = \begin{cases}1 & p = 1\\ 0 & p > 1\end{cases}$

(*a*) We have using property 3, page 87,

$$\phi_p'(x) \;=\; B_p'(x) \;=\; pB_{p-1}(x)$$

(*b*) We have since $\phi_{p+1}(x) = B_{p+1}(x) - B_{p+1}$

$$\phi_{p+1}'(x) \;=\; B_{p+1}'(x) \;=\; (p+1)B_p(x) \;=\; (p+1)[\phi_p(x) + B_p]$$

Then

$$\phi_p(x) \;=\; \frac{\phi_{p+1}'(x)}{p+1} - B_p$$

(*c*) Since by definition $B_p(0) = B_p$ we have

$$\phi_p(0) \;=\; B_p(0) - B_p \;=\; 0$$

(*d*) We have

$$\phi_1(x) \;=\; B_1(x) - B_1 \;=\; x - \tfrac{1}{2} - (-\tfrac{1}{2}) \;=\; x$$

Thus

$$\phi_1(1) \;=\; 1 \tag{1}$$

If $p > 1$ then by property 4, page 87,

$$\phi_p(1) \;=\; B_p(1) - B_p \;=\; 0 \tag{2}$$

Then from (*1*) and (*2*) we see that

$$\phi_p(1) \;=\; \begin{cases} 1 & p = 1 \\ 0 & p > 1 \end{cases}$$

4.29. Show that

$$\int_a^{a+h} f(x)\,dx \;=\; \frac{h}{2}[f(a) + f(a+h)] + \frac{h^3}{2!}\int_0^1 \phi_2(u)\,f''(a+hu)\,du$$

Letting $x = a + hu$, using the fact that $\phi_2(u) = u^2 - u$, and integrating by parts we have

$$\begin{aligned}
\int_a^{a+h} f(x)\,dx \;&=\; h\int_0^1 f(a+hu)\,du \\
&=\; \frac{h}{2}\int_0^1 f(a+hu)\,d(2u-1) \\
&=\; \frac{h}{2}\left[(2u-1)f(a+hu)\Big|_0^1 - h\int_0^1 (2u-1)f'(a+hu)\,du\right] \\
&=\; \frac{h}{2}[f(a) + f(a+h)] - \frac{h^2}{2}\int_0^1 (2u-1)f'(a+hu)\,du \\
&=\; \frac{h}{2}[f(a) + f(a+h)] - \frac{h^2}{2}\int_0^1 f'(a+hu)\,d\phi_2(u) \\
&=\; \frac{h}{2}[f(a) + f(a+h)] - \frac{h^2}{2}f'(a+hu)\phi_2(u)\Big|_0^1 + \frac{h^3}{2!}\int_0^1 \phi_2(u)f''(a+hu)\,du \\
&=\; \frac{h}{2}[f(a) + f(a+h)] + \frac{h^3}{2!}\int_0^1 \phi_2(u)f''(a+hu)\,du
\end{aligned}$$

4.30. Show that

(*a*)
$$\frac{h^3}{2!}\int_0^1 \phi_2(u)f''(a+hu)\,du \;=\; -\frac{B_2h^2}{2!}[f'(a+h) - f'(a)] + \frac{h^5}{4!}\int_0^1 \phi_4(u)f^{(\mathrm{IV})}(a+hu)\,du$$

(*b*)
$$\frac{h^5}{4!}\int_0^1 \phi_4(u)f^{(\mathrm{IV})}(a+hu)\,du \;=\; -\frac{B_4h^4}{4!}[f'''(a+h) - f'''(a)] + \frac{h^7}{6!}\int_0^1 \phi_6(u)f^{(\mathrm{VI})}(a+hu)\,du$$

(*a*) From Problem 4.28(*b*) with $p = 2$ we have

$$\phi_2(u) \;=\; -B_2 + \tfrac{1}{3}\phi_3'(u) \qquad \text{and} \qquad \phi_3(u) \;=\; \tfrac{1}{4}\phi_4'(u)$$

Then

$$\begin{aligned}
\frac{h^3}{2!}\int_0^1 \phi_2(u)f''(a+hu)\,du &= \frac{h^3}{2!}\int_0^1 [-B_2+\tfrac{1}{3}\phi_3'(u)]f''(a+hu)\,du \\
&= -\frac{B_2h^3}{2!}\int_0^1 f''(a+hu)\,du + \frac{h^3}{3!}\int_0^1 \phi_3'(u)f''(a+hu)\,du \\
&= -\frac{B_2h^2}{2!}f'(a+hu)\Big|_0^1 + \frac{h^3}{3!}\left[\phi_3(u)f''(a+hu)\Big|_0^1\right. \\
&\qquad\qquad \left. - h\int_0^1 \phi_3(u)f'''(a+hu)\,du\right] \\
&= -\frac{B_2h^2}{2!}[f'(a+h)-f'(a)] - \frac{h^4}{3!}\int_0^1 \phi_3(u)f'''(a+hu)\,du \\
&= -\frac{B_2h^2}{2!}[f'(a+h)-f'(a)] - \frac{h^4}{4!}\int_0^1 \phi_4'(u)f'''(a+hu)\,du \\
&= -\frac{B_2h^2}{2!}[f'(a+h)-f'(a)] - \frac{h^4}{4!}\left[\phi_4(u)f'''(a+hu)\Big|_0^1\right. \\
&\qquad\qquad \left. - h\int_0^1 \phi_4(u)f^{(IV)}(a+hu)\,du\right] \\
&= -\frac{B_2h^2}{2!}[f'(a+h)-f'(a)] + \frac{h^5}{4!}\int_0^1 \phi_4(u)f^{(IV)}(a+hu)\,du
\end{aligned}$$

(*b*) From Problem 4.28(*b*) with $p = 4$ we have

$$\phi_4(u) = -B_4 + \tfrac{1}{5}\phi_5'(u) \quad \text{and} \quad \phi_5(u) = \tfrac{1}{6}\phi_6'(u)$$

Then

$$\begin{aligned}
\frac{h^5}{4!}\int_0^1 \phi_4(u)f^{(IV)}(a+hu)\,du &= \frac{h^5}{4!}\int_0^1 [-B_4+\tfrac{1}{5}\phi_5'(u)]f^{(IV)}(a+hu)\,du \\
&= -\frac{B_4h^5}{4!}\int_0^1 f^{(IV)}(a+hu)\,du + \frac{h^5}{5!}\int_0^1 \phi_5'(u)f^{(IV)}(a+hu)\,du \\
&= -\frac{B_4h^4}{4!}f'''(a+hu)\Big|_0^1 + \frac{h^5}{5!}\left[\phi_5(u)f^{(IV)}(a+hu)\Big|_0^1\right. \\
&\qquad\qquad \left. - h\int_0^1 \phi_5(u)f^{(V)}(a+hu)\,du\right] \\
&= -\frac{B_4h^4}{4!}[f'''(a+h)-f'''(a)] - \frac{h^6}{5!}\int_0^1 \phi_5(u)f^{(V)}(a+hu)\,du \\
&= -\frac{B_4h^4}{4!}[f'''(a+h)-f'''(a)] - \frac{h^6}{6!}\int_0^1 \phi_6'(u)f^{(V)}(a+hu)\,du \\
&= -\frac{B_4h^4}{4!}[f'''(a+h)-f'''(a)] - \frac{h^6}{6!}\left[\phi_6(u)f^{(V)}(a+hu)\Big|_0^1\right. \\
&\qquad\qquad \left. - h\int_0^1 \phi_6(u)f^{(VI)}(a+hu)\,du\right] \\
&= -\frac{B_4h^4}{4!}[f'''(a+h)-f'''(a)] + \frac{h^7}{6!}\int_0^1 \phi_6(u)f^{(VI)}(a+hu)\,du
\end{aligned}$$

4.31. Obtain the Euler-Maclaurin series with a remainder for the case where $n = 1$ [see equation (*24*), page 125].

From Problems 4.29 and 4.30 we have

$$\begin{aligned}
\int_a^{a+h} f(x)\,dx &= \frac{h}{2}[f(a)+f(a+h)] - \frac{B_2h^2}{2!}[f'(a+h)-f'(a)] - \frac{B_4h^4}{4!}[f'''(a+h)-f'''(a)] \\
&\qquad + \frac{h^7}{6!}\int_0^1 \phi_6(u)f^{(VI)}(a+hu)\,du
\end{aligned}$$

which is the required result. This result can of course be generalized and leads to that on page 125.

4.32. Estimate the error made in the approximation to the sum in Problem 4.27.

Using $f(x) = 1/x$, $a = 100$, $h = 1$, $n = 100$, $m = 3$ in the result of Problem 4.27 we can write it as

$$\sum_{k=100}^{199} \frac{1}{k} = \int_{100}^{200} \frac{dx}{x} - \frac{1}{2}\left[\frac{1}{200} - \frac{1}{100}\right]$$
$$+ \frac{1}{12}\left[-\frac{1}{(200)^2} + \frac{1}{(100)^2}\right] - \frac{1}{720}\left[-\frac{6}{(200)^4} + \frac{6}{(100)^4}\right] - R$$

where

$$R = \frac{1}{6!}\int_0^1 [B_6(u) - B_6]\left[\sum_{p=0}^{99} \frac{720}{(p+100+u)^7}\right] du$$
$$= \int_0^1 [u^6 - 3u^5 + \tfrac{5}{2}u^4 - \tfrac{1}{2}u^2]\left[\sum_{p=0}^{99} \frac{1}{(p+100+u)^7}\right] du$$

Now by the mean-value theorem for integrals we have

$$R = \left[\sum_{p=0}^{99} \frac{1}{(p+100+\xi)^7}\right] \int_0^1 (u^6 - 3u^5 + \tfrac{5}{2}u^4 - \tfrac{1}{2}u^2)\, du$$
$$= -\frac{1}{42}\sum_{p=0}^{99} \frac{1}{(p+100+\xi)^7}$$

Thus the absolute value of the error is

$$|R| = \frac{1}{42}\sum_{p=0}^{99} \frac{1}{(p+100+\xi)^7}$$

But since $0 < \xi < 1$, it follows that each term in the series on the right is less than $1/(100)^7$ and so the sum of the first 100 terms of the series is less than $100/(100)^7$ so that

$$|R| < \frac{1}{42}\frac{100}{(100)^7} = 2.38 \times 10^{-14}$$

The fact that the error is so small shows that not only is the result 0.695653 accurate to 6 decimal places but that we could have obtained accuracy to 12 decimal places at least by simply obtaining more decimal places in Problem 4.27. If we do this we find

$$\sum_{k=100}^{199} \frac{1}{k} = 0.693147180559945 + 0.0025 + 0.00000625 + 0.000000000078125$$
$$= 0.695653430638$$

accurate to *all 12 decimal places.*

By taking only one more term in the Euler-Maclaurin formula it is possible to obtain even more decimal place accuracy [see Problem 4.75].

This problem serves to illustrate further the importance of being able to find an error term since one then can determine how many figures in a calculation are accurate.

STIRLING'S FORMULA FOR $n!$

4.33. Prove that

$$\ln n! = c + (n + \tfrac{1}{2}) \ln n - n + \frac{1}{12n} - \frac{1}{360n^2} + \cdots$$

where c is a constant independent of n.

Let $f(x) = \ln x$, $h = 1$ in the Euler-Maclaurin formula. Then

$$\int_a^{a+n} \ln x\, dx = \tfrac{1}{2}\ln a + \ln(a+1) + \cdots + \ln(a+n-1) + \tfrac{1}{2}\ln(a+n)$$
$$-\frac{1}{12}\left(\frac{1}{a+n} - \frac{1}{a}\right) + \frac{1}{360}\left(\frac{1}{(a+n)^3} - \frac{1}{a^3}\right) - \cdots \quad (1)$$

From Problem 3.3(*b*), page 90,

$$\int_a^{a+n} \ln x\,dx \;=\; x\ln x - x\Big|_a^{a+n} \;=\; (a+n)\ln(a+n) - (a+n) \qquad (2)$$

Thus we have on adding $\frac{1}{2}\ln a + \frac{1}{2}\ln(a+n)$ to both sides of equation (*1*) and using equation (*2*)

$$(a+n+\tfrac{1}{2})\ln(a+n) - n + a - (a+\tfrac{1}{2})\ln a$$
$$= \ln[(a)(a+1)(a+2)\cdots(a+n)] - \frac{1}{12}\left(\frac{1}{a+n} - \frac{1}{a}\right) + \frac{1}{360}\left(\frac{1}{(a+n)^3} - \frac{1}{a^3}\right) - \cdots$$

Assuming that a and n are both positive integers and replacing $a+n$ by n, i.e. n by $n-a$, in this result we have

$$(n+\tfrac{1}{2})\ln n - n + 2a - (a+\tfrac{1}{2})\ln a \;=\; \ln\left[\frac{n!}{a!}\right] - \frac{1}{12}\left(\frac{1}{n} - \frac{1}{a}\right) + \frac{1}{360}\left(\frac{1}{n^3} - \frac{1}{a^3}\right) - \cdots$$

which can be written

$$\ln n! \;=\; c + (n+\tfrac{1}{2})\ln n - n + \frac{1}{12n} - \frac{1}{360n^2} + \cdots$$

where c is a constant independent of n.

4.34. Prove that in the result of Problem 4.33 the constant $c = \frac{1}{2}\ln 2\pi$.

We make use of *Wallis' product* [see Problem 4.100] which states that

$$\lim_{n\to\infty} \frac{2\cdot 2\cdot 4\cdot 4\cdot 6\cdot 6\cdot \cdots \cdot (2n)\cdot(2n)}{1\cdot 3\cdot 3\cdot 5\cdot 5\cdot 7\cdot \cdots \cdot (2n-1)\cdot(2n+1)} \;=\; \frac{\pi}{2} \qquad (1)$$

Now we can write

$$\frac{2\cdot 2\cdot 4\cdot 4\cdot 6\cdot 6\cdot \cdots \cdot (2n)\cdot(2n)}{1\cdot 3\cdot 3\cdot 5\cdot 5\cdot 7\cdot \cdots \cdot (2n-1)\cdot(2n+1)} \;=\; \frac{2^{4n}(n!)^4}{(2n+1)[(2n)!]^2}$$

It follows that

$$\lim_{n\to\infty} \ln \frac{2^{4n}(n!)^4}{(2n+1)[(2n)!]^2} \;=\; \ln\frac{\pi}{2}$$

or

$$\lim_{n\to\infty} [4n\ln 2 + 4\ln n! - \ln(2n+1) - 2\ln(2n)!] \;=\; \ln\frac{\pi}{2} \qquad (2)$$

Using the result of Problem 4.33 to find $\ln n!$ and $\ln(2n)!$ and substituting in (*2*) we have

$$\lim_{n\to\infty}\Bigg[4n\ln 2 + 4c + (4n+2)\ln n - 4n + \frac{1}{3n} - \frac{1}{90n^2} + \cdots$$
$$- \ln(2n+1) - 2c - (4n+1)\ln(2n) + 4n - \frac{1}{12n} + \frac{1}{720n^2} + \cdots\Bigg] \;=\; \ln\frac{\pi}{2}$$

Thus

$$2c - 2\ln 2 \;=\; \ln\frac{\pi}{2} \qquad \text{or} \qquad c \;=\; \tfrac{1}{2}\ln 2\pi$$

4.35. Prove Stirling's formula for $n!$.

Using $c = \frac{1}{2}\ln 2\pi$ in Problem 4.33 we have

$$\ln n! \;=\; \tfrac{1}{2}\ln 2\pi + (n+\tfrac{1}{2})\ln n - n + \frac{1}{12n} - \frac{1}{360n^2} + \cdots \qquad (1)$$

or

$$n! \;=\; e^{\frac{1}{2}\ln 2\pi} e^{(n+\frac{1}{2})\ln n} e^{-n} e^{1/12n - \cdots}$$

i.e.

$$n! \;=\; \sqrt{2\pi}\, n^{n+\frac{1}{2}} e^{-n}\left(1 + \frac{1}{12n} + \cdots\right)$$
$$=\; \sqrt{2\pi n}\, n^n e^{-n}\left(1 + \frac{1}{12n} + \cdots\right) \qquad (2)$$

4.36. (*a*) Evaluate 20! by using Stirling's formula and (*b*) compare with the true value.

(*a*) **Method 1.**

From Problem 4.35, equation (*2*),

$$\begin{aligned} 20! &= \sqrt{40\pi}\,(20)^{20}e^{-20}\left(1+\frac{1}{240}+\cdots\right) \\ &= \sqrt{125.6636}\,(2^{20}\cdot 10^{20})(e^{-20})(1.004167) \\ &= (11.21000)(1.048576\times 10^{26})(2.061096\times 10^{-9})(1.004167) \\ &= 2.43282\times 10^{18} \end{aligned}$$

Method 2.

From Problem 4.35, equation (*1*),

$$\begin{aligned} \ln 20! &= \frac{1}{2}\ln 2\pi + \frac{41}{2}\ln 20 - 20 + \frac{1}{240} - \cdots \\ &= 0.91894 + \frac{41}{2}(2.99574) - 20 + 0.0041667 - \cdots \\ &= 0.91894 + 61.41267 - 20 + 0.0041667 \\ &= 42.33578 \end{aligned}$$

Then

$$\log_{10} 20! = 0.434294\ln 20! = 18.38618$$

and so

$$20! = 2.433\times 10^{18}$$

(*b*) The exact result is 2,432,902,008,176,640,000 and thus the percent error of the result in Method 1 of part (*a*) is less than 0.0033%.

MISCELLANEOUS PROBLEMS

4.37. Sum the series

$$x + \frac{2^2x^2}{1} + \frac{3^2x^3}{2} + \frac{4^2x^4}{3} + \cdots$$

Consider the series

$$u_0x + \frac{u_1x^2}{1} + \frac{u_2x^3}{2} + \frac{u_3x^4}{3} + \cdots$$

This can be written

$$\begin{aligned} u_0x + \frac{Eu_0x^2}{1} + \frac{E^2u_0x^3}{2} + \frac{E^3u_0x^4}{3} + \cdots \\ = \left[1 + \frac{xE}{1} + \frac{x^2E^2}{2} + \frac{x^3E^3}{3} + \cdots\right]u_0x \end{aligned}$$

Using series 4, page 8, this can be written as

$$\begin{aligned} [-\ln(1-xE)]u_0x &= \{-\ln(1-x[1+\Delta])\}u_0x = \{-\ln(1-x-x\Delta)\}u_0x \\ &= \left\{-\ln\left[(1-x)\left(1-\frac{x\Delta}{1-x}\right)\right]\right\}u_0x \\ &= \left[-\ln(1-x) - \ln\left(1-\frac{x\Delta}{1-x}\right)\right]u_0x \\ &= -u_0x\ln(1-x) + \left[\frac{x\Delta}{1-x} + \frac{1}{2}\left(\frac{x\Delta}{1-x}\right)^2 + \frac{1}{3}\left(\frac{x\Delta}{1-x}\right)^3 + \cdots\right]u_0x \\ &= -u_0x\ln(1-x) + \frac{x^2\Delta u_0}{1-x} + \frac{1}{2}\frac{x^3\Delta^2u_0}{(1-x)^2} + \frac{1}{3}\frac{x^4\Delta^3u_0}{(1-x)^3} + \cdots \end{aligned}$$

Since $u_n = (n+1)^2$ we have the following difference table:

	u_n	Δu_n	$\Delta^2 u_n$	$\Delta^3 u_n$
$n=0$	1			
		3		
$n=1$	4		2	
		5		0
$n=2$	9		2	
		7		0
$n=3$	16		2	
		9		
$n=4$	25			

Fig. 4-6

Thus

$$u_0 = 1, \quad \Delta u_0 = 3, \quad \Delta^2 u_0 = 2, \quad \Delta^3 u_0 = 0, \quad \Delta^4 u_0 = 0, \quad \ldots$$

Then the sum of the series is

$$-x \ln(1-x) + \frac{3x^2}{1-x} + \frac{x^3}{(1-x)^2} = \frac{3x^2 - 2x^3}{(1-x)^2} - x\ln(1-x)$$

4.38. Prove Weddle's rule [equation (*16*), page 123].

Put $n = 6$ in Problem 4.9 and neglect differences of order 6 and higher. In doing this we need the terms involving $\Delta^5 f(a)$ and $\Delta^6 f(a)$ which are not given in Problem 4.9 but are presented in Problem 4.60. We then find

$$\begin{aligned}\int_a^{a+6h} f(x)\,dx &= 6h\left[f(a) + 3\Delta f(a) + \frac{9}{2}\Delta^2 f(a) + 4\Delta^3 f(a)\right.\\ &\qquad \left. + \frac{41}{20}\Delta^4 f(a) + \frac{11}{20}\Delta^5 f(a) + \frac{41}{840}\Delta^6 f(a)\right]\\ &= h\left[6f(a) + 18\Delta f(a) + 27\Delta^2 f(a) + 24\Delta^3 f(a)\right.\\ &\qquad \left. + \frac{123}{10}\Delta^4 f(a) + \frac{33}{10}\Delta^5 f(a) + \frac{41}{140}\Delta^6 f(a)\right]\end{aligned}$$

Now if we change the coefficient $41/140$ to $42/140 = 3/10$ the error made is quite small. If we make this change we obtain the approximation

$$\begin{aligned}\int_a^{a+6h} f(x)\,dx &= h\left[6f(a) + 18\Delta f(a) + 27\Delta^2 f(a) + 24\Delta^3 f(a)\right.\\ &\qquad \left. + \frac{123}{10}\Delta^4 f(a) + \frac{33}{10}\Delta^5 f(a) + \frac{3}{10}\Delta^6 f(a)\right]\\ &= \frac{3h}{10}[\{f(a) + f(a+2h) + f(a+4h) + f(a+6h)\}\\ &\qquad + 5\{f(a+h) + f(a+5h)\} + 6f(a+3h)]\end{aligned}$$

The result can be extended to evaluate $\int_a^{a+Nh} f(x)\,dx$ when N is a multiple of 6 [see Problem 4.91].

4.39. (*a*) Work Problem 4.12 by Weddle's rule and (*b*) compare the result with the exact value and that obtained by Simpson's one-third rule and Simpson's three-eighths rule.

(*a*) We can use the table of Problem 4.12 since $N = 6$. Then using Problem 4.38 we have

$$\begin{aligned}\int_0^1 \frac{dx}{1+x^2} &= \frac{3(1/6)}{10}[(1.00000000 + 0.90000000 + 0.69230769 + 0.50000000)\\ &\qquad + 5(0.97297297 + 0.59016393) + 6(0.80000000)]\\ &= 0.78539961\end{aligned}$$

(*b*) The percent error from the exact value [Problem 4.12(*b*)] is 0.00185%. Although the accuracy is very good it is not as good as that obtained from Simpson's one-third rule but it is better than that obtained from Simpson's three-eighths rule.

The following table shows for comparison purposes the percent errors for the various rules.

Rule	Percent error
Trapezoidal	0.147%
Simpson's one-third	0.00028%
Simpson's three-eighths	0.00293%
Weddle's	0.00185%

Fig. 4-7

Supplementary Problems

SERIES OF CONSTANTS

4.40. Sum each of the following series to n terms by using the method of Problem 4.1, page 126.

(*a*) $1 + 3 + 5 + 7 + \cdots$

(*b*) $1^2 + 3^2 + 5^2 + 7^2 + \cdots$

(*c*) $2^2 + 5^2 + 8^2 + 11^2 + \cdots$

(*d*) $2 \cdot 4 + 4 \cdot 6 + 6 \cdot 8 + \cdots$

(*e*) $1^3 + 2^3 + 3^3 + 4^3 + \cdots$

(*f*) $1 \cdot 3 \cdot 5 + 3 \cdot 5 \cdot 7 + 5 \cdot 7 \cdot 9 + \cdots$

(*g*) $1 \cdot 2 \cdot 5 + 2 \cdot 3 \cdot 6 + 3 \cdot 4 \cdot 7 + \cdots$

(*h*) $3 \cdot 5 \cdot 10 + 4 \cdot 6 \cdot 12 + 5 \cdot 7 \cdot 14 + \cdots$

4.41. Sum the series $1^2 + 2^2x + 3^2x^2 + 4^2x^3 + \cdots$.

4.42. Can the series of Problem 4.41 be summed to n terms? Explain.

4.43. Sum the series $1^2 - \frac{2^2}{3} + \frac{3^2}{3^2} - \frac{4^2}{3^3} + \frac{5^2}{3^4} - \frac{6^2}{3^5} + \cdots$.

POWER SERIES

4.44. Sum the series

(*a*) $x + 2x^2 + 3x^3 + 4x^4 + \cdots$

(*b*) $x^2 + 2^2x^2 + 3^2x^3 + 4^2x^4 + \cdots$

(*c*) $1 \cdot 3 + 3 \cdot 5x + 5 \cdot 7x^2 + 7 \cdot 9x^3 + \cdots$

4.45. Sum the series $\frac{2 \cdot 5}{3^2} - \frac{5 \cdot 8}{3^4} + \frac{8 \cdot 11}{3^6} - \frac{11 \cdot 14}{3^8} + \cdots$.

4.46. Sum the series $\frac{1^2}{1!} + \frac{3^2}{2!} + \frac{5^2}{3!} + \frac{7^2}{4!} + \cdots$.

4.47. If

$$A(x) = u_0 - \frac{x^2\Delta^2 u_0}{2!} + \frac{x^4\Delta^4 u_0}{4!} - \cdots$$

and

$$B(x) = x\Delta u_0 - \frac{x^3\Delta^3 u_0}{3!} + \frac{x^5\Delta^5 u_0}{5!} - \cdots$$

show that

(*a*) $$\sum_{k=0}^{\infty} \frac{(-1)^k u_k x^k}{(2k+1)!} = A(x) \sin x + B(x) \cos x$$

(*b*) $$\sum_{k=0}^{\infty} \frac{(-1)^k u_k x^k}{(2k)!} = A(x) \cos x - B(x) \sin x$$

4.48. Sum the series $1^3x - \frac{3^3x^3}{2!} + \frac{5^3x^5}{4!} + \frac{7^3x^7}{6!} + \cdots$.

4.49. Sum the series $\frac{1^2x}{1!} + \frac{2^2x^2}{3!} + \frac{3^2x^3}{5!} + \frac{4^2x^4}{7!} + \cdots$.

4.50. Can the power series method of page 121 be used for the finite series

$$\frac{1^2x}{3} + \frac{2^2x^2}{3^2} + \frac{3^2x^3}{3^3} + \frac{4^2x^4}{3^4} + \cdots + \frac{n^2x^n}{3^n}$$

4.51. Can the sum of the first n terms of the series of Problem 4.48 be found by using the power series method of page 121? Explain.

APPROXIMATE INTEGRATION

4.52. Find $\int_1^3 \frac{dx}{x}$ by using the extended trapezoidal rule where the interval of integration is subdivided into (*a*) 6 (*b*) 10 equal parts. Show that the true value is $\ln 3 = 1.09861$ approx. and compare this with the values obtained in (*a*) and (*b*).

4.53. Discuss how you would improve the accuracy in Problem 4.52.

4.54. Work Problem 4.52 by using Simpson's one-third rule and discuss the accuracy.

4.55. Show that Simpson's one-third rule is exact if $f(x)$ is a polynomial of degree 3. What is a geometrical significance of this?

4.56. (*a*) Work Problem 4.52(*a*) by using Simpson's three-eighths rule and discuss the accuracy. (*b*) Can you work Problem 4.52(*b*) by Simpson's rule? Explain.

4.57. Derive the extended Simpson's three-eighths rule.

4.58. (*a*) Explain the difficulty involved in attempting to find the approximate value of

$$\int_0^1 \frac{dx}{\sqrt{1-x^2}}$$

without actually evaluating the integral.

(*b*) Can you resolve the difficulty in part (*a*)?

4.59. (*a*) Prove the formula

$$\int_a^{a+4h} f(x)\,dx = \frac{h}{45}[14(y_0+y_4) + 64(y_1+y_3) + 24y_2]$$

where $y_k = f(a+kh)$.

(*b*) Use (*a*) to obtain an approximate value for

$$\int_1^3 \frac{dx}{x}$$

and compare with the exact value.

4.60. Show that the next two terms in the series for

$$\int_a^{a+nh} f(x)\,dx$$

given in equation (*8*), page 122, are

$$\frac{n(2n^4 - 24n^3 + 105n^2 - 200n + 144)}{1440}\Delta^5 f(a)$$
$$+ \frac{n(12n^5 - 210n^4 + 1428n^3 - 4725n^2 + 7672n - 5040)}{60{,}480}\Delta^6 f(a)$$

and use this to complete the proof in Problem 4.38.

4.61. Find an approximate value of $\int_0^{\pi/2} \sin x\, dx$ and compare with the exact value.

4.62. Use the following table to find an approximate value for

$$\int_0^1 f(x)\, dx$$

x	0	0.2	0.4	0.6	0.8	1.0
$f(x)$	1.00000	0.81873	0.67032	0.54881	0.44933	0.36788

Fig. 4-8

4.63. Discuss what steps you would use to work Problem 4.62 if the values corresponding to (*a*) $x = 0.4$ (*b*) $x = 0.4$ and $x = 0.8$ were not present in the table.

THE EULER-MACLAURIN FORMULA

4.64. Use the Euler-Maclaurin formula to find the sums of the series

(*a*) $1^2 + 2^2 + 3^2 + \cdots + n^2$

(*b*) $1^3 + 2^3 + 3^3 + \cdots + n^3$

(*c*) $1\cdot 2 + 2\cdot 3 + 3\cdot 4 + \cdots + n(n+1)$

(*d*) $1^4 + 2^4 + 3^4 + \cdots$

4.65. Use the Euler-Maclaurin formula to work Problem 3.46, page 108.

4.66. Show that $\dfrac{1}{1} + \dfrac{1}{2} + \dfrac{1}{3} + \cdots + \dfrac{1}{n} = \gamma + \ln n + \dfrac{1}{2n} - \dfrac{1}{12n^2} + \dfrac{1}{120n^4} - \cdots$

where γ is *Euler's constant.*

4.67. Use Problem 4.66 to obtain an approximate value for γ.

4.68. Show that $\dfrac{1}{1^2} + \dfrac{1}{2^2} + \dfrac{1}{3^2} + \cdots + \dfrac{1}{(n-1)^2} = \dfrac{\pi^2}{6} - \dfrac{1}{n} - \dfrac{1}{2n^2} - \dfrac{1}{6n^3} - \dfrac{1}{30n^5} - \cdots$

4.69. Show that $1 + \dfrac{1}{3} + \dfrac{1}{5} + \cdots + \dfrac{1}{2n-1} \approx \dfrac{1}{2}\gamma + \ln 2 + \dfrac{1}{2}\ln n$

4.70. Sum the series $\dfrac{1}{1^3} + \dfrac{1}{2^3} + \dfrac{1}{3^3} + \cdots$ to 5 decimal place accuracy by using the Euler-Maclaurin formula.

ERROR TERM IN THE EULER-MACLAURIN FORMULA

4.71. Show that

(*a*) $$\frac{h^7}{6!}\int_0^1 \phi_6(u) f^{(\text{VI})}(a+hu)\, du = -\frac{B_6 h^6}{6!}[f^{(\text{V})}(a+h) - f^{(\text{V})}(a)] + \frac{h^9}{8!}\int_0^1 \phi_8(u) f^{(\text{VIII})}(a+hu)\, du$$

(*b*) $$\frac{h^9}{8!}\int_0^1 \phi_8(u) f^{(\text{VIII})}(a+hu)\, du = -\frac{B_8 h^8}{8!}[f^{(\text{VII})}(a+h) - f^{(\text{VII})}(a)] + \frac{h^{11}}{10!}\int_0^1 \phi_{10}(u) f^{(10)}(a+hu)\, du$$

4.72. Obtain the Euler-Maclaurin formula with a remainder for the case where $m = 5$.

4.73. (*a*) Use mathematical induction to extend the results of Problems 4.71 and 4.72 and thus (*b*) prove result (*23*), page 125.

4.74. Prove the result (*25*), page 125.

4.75. (*a*) Show how to improve the accuracy of the result in Problem 4.32 by taking one more term in the Euler-Maclaurin formula. (*b*) Determine the accuracy of the result obtained in part (*a*).

4.76. Prove that if $f^{(2p-1)}(x)$ does not change sign in the interval from a to $a+nh$ then the remainder R given in the Euler-Maclaurin formula (*24*), page 125, has the property that

$$|R| \leqq \frac{B_{2m}h^{2m+1}}{(2m)!}\sum_{p=0}^{n-1}|f^{(2m)}(a+hu+hp)|$$

4.77. Work Problem 4.70 by taking into account the error term in the Euler-Maclaurin formula.

STIRLING'S FORMULA FOR $n!$

4.78. Obtain 10! from Stirling's formula and compare with the exact value.

4.79. Calculate the approximate value of the binomial coefficient $\binom{100}{50}$.

4.80. Use Stirling's formula to show that

$$\sqrt{\pi} = \lim_{n\to\infty}\frac{2^{2n}(n!)^2}{(2n)!\sqrt{n}}$$

and show that the result is equivalent to Wallis' product.

4.81. Show that for large values of n

$$\binom{-1/2}{n} \approx \frac{1}{\sqrt{\pi n}}$$

4.82. Generalize the result of Problem 4.35 to obtain

$$n! = \sqrt{2\pi n}\, n^n e^{-n}\left(1+\frac{1}{12n}+\frac{1}{288n^2}-\frac{139}{51{,}840n^3}+\cdots\right)$$

MISCELLANEOUS PROBLEMS

4.83. Sum the series $\dfrac{(1^2+1)x}{1!} - \dfrac{(2^2+2)x^2}{2!} + \dfrac{(3^2+3)x^3}{3!} - \cdots$.

4.84. Sum the series $\dfrac{1}{1\cdot3\cdot7}+\dfrac{1}{3\cdot5\cdot9}+\dfrac{1}{5\cdot7\cdot11}+\cdots$ (*a*) to n terms and (*b*) to infinity.

4.85. Show that

$$1+x+x^2+\cdots+x^n = n+\frac{n(n-1)}{2!}(x-1)+\frac{n(n-1)(n-2)}{3!}(x-1)^2+\cdots+(x-1)^n$$

4.86. Find an approximate value for $\displaystyle\int_0^2 \frac{dx}{1+x^3}$.

4.87. Obtain an approximate value for the integral

$$\int_0^{\pi/2} x\cot x\, dx$$

and compare with the exact value given by $-\frac{1}{2}\pi\ln 2 = -1.0888\ldots$.

4.88. (*a*) Use integration by parts and the result of Problem 4.87 to find

$$\int_0^{\pi/2} \ln\sin\theta\, d\theta$$

(*b*) Explain why numerical methods are not immediately applicable to the integral in (*a*).

4.89. Obtain equation (5) of Problem 4.19 and thus complete the proof in that problem.

4.90. Show that the seventh differences in Weddle's rule drop out and explain the significance of this.

4.91. Derive Weddle's extended rule.

4.92. Show that

(a) $$\sum_{n=0}^{\infty} (-1)^n f(n) \quad = \quad \tfrac{1}{2}(-1)^{n-1}\left[1 - \frac{\Delta}{2} + \frac{\Delta^2}{4} - \frac{\Delta^3}{8} + \cdots\right] f(n)$$

(b) $$\sum_{n=0}^{\infty} (-1)^n f(n) \quad = \quad \left[\frac{1}{2} - \frac{B_1(2^2-1)}{2!} D + \frac{B_2(2^4-1)}{4!} D^3 - \cdots\right] f(n)$$

These are sometimes called *Euler transformations* and are often useful for speeding up the convergence of slowly convergent series.

4.93. Show how Problem 4.92 can be used to find an approximate value for the series

(a) $\frac{1}{1} - \frac{1}{2} + \frac{1}{3} - \frac{1}{4} + \frac{1}{5} - \cdots$

(b) $\frac{1}{1} - \frac{1}{3} + \frac{1}{5} - \frac{1}{7} + \frac{1}{9} - \cdots$

and compare with the exact values given by $\ln 2$ and $\pi/4$ respectively.

4.94. Sum the series $1^2 - 2^2 + 3^2 - 4^2 + \cdots + (-1)^{n-1}n^2$.

4.95. Sum the series $1^2 + 2 \cdot 2^2 + 4 \cdot 3^2 + 8 \cdot 4^2 + 16 \cdot 5^2 + \cdots$ to n terms.

4.96. If $x \neq -1, -2, -3, \ldots$ show that

$$\frac{1}{x+1} + \frac{1!}{(x+1)(x+2)} + \frac{2!}{(x+1)(x+2)(x+3)} + \cdots \quad = \quad \frac{1}{x}$$

4.97. Show that $$\frac{1}{(n+1)^2} + \frac{1}{(n+2)^2} + \cdots \quad = \quad \frac{1}{n+1} + \frac{1}{2(n+1)^2} + R$$

where $$0 \leqq R \leqq \frac{1}{6(n+1)^3}$$

4.98. Let $f(x)$ have the values y_0, y_1, y_2 at $x = a, a+h, a+2h$ respectively.

(a) Obtain an interpolating polynomial for $f(x)$.

(b) Use the interpolating polynomial in (a) to obtain the approximate result

$$\int_a^{a+h} f(x)\,dx \quad = \quad \frac{h}{12}[5f(a) + 8f(a+h) - f(a+2h)]$$

(c) Use the interpolating polynomial in (a) to obtain the approximate result

$$\int_a^{a+4h} f(x)\,dx \quad = \quad \frac{4h}{3}[2f(a+h) - f(a+2h) + 2f(a+3h)]$$

(d) Show that the formula in (c) is exact when $f(x)$ is a polynomial of degree 3 but that the formula in (b) is not exact in such case.

4.99. Use the results of Problem 4.98(b) and (c) to obtain approximate values of

(a) $\displaystyle\int_1^3 \frac{dx}{x}$ (b) $\displaystyle\int_0^1 \frac{dx}{1+x^2}$ (c) $\displaystyle\int_0^1 e^{-x}\,dx$

and discuss the accuracy.

4.100. (*a*) Show that

$$\frac{\pi}{2} = \frac{2\cdot 2}{1\cdot 3}\cdot\frac{4\cdot 4}{3\cdot 5}\cdot\frac{6\cdot 6}{5\cdot 7}\cdots\frac{2n\cdot 2n}{(2n-1)(2n+1)}\,\frac{\int_0^{\pi/2}\sin^{2n}x\,dx}{\int_0^{\pi/2}\sin^{2n+1}x\,dx}$$

(*b*) By taking the limit as $n\to\infty$ in (*a*) obtain *Wallis' product formula* of Problem 4.34.

[*Hint.* For (*a*) see Problem 3.147, page 118, while for (*b*) first prove that $1 \leqq \dfrac{\int_0^{\pi/2}\sin^{2n}x\,dx}{\int_0^{\pi/2}\sin^{2n+1}x\,dx} \leqq 1+\dfrac{1}{2n}$.]

4.101. Obtain the Euler-Maclaurin formula in terms of differences instead of derivatives and give an example of its use.

4.102. Let $S_n(x) = \sum_{k=0}^{n-1} a_k x^k$.

(*a*) Show that on using summation by parts

$$S_n(x) = \frac{a_0 - a_n x^n}{1-x} + \frac{x}{1-x}\sum_{k=0}^{n-1}(\Delta a_k)x^k$$

(*b*) Generalize the result of (*a*) by showing that after p summations by parts

$$S_n(x) = \frac{1}{1-x}\sum_{k=0}^{p-1}\left(\frac{x}{1-x}\right)^k \Delta^k a_0 + \left(\frac{x}{1-x}\right)^p \sum_{k=0}^{n-1}(\Delta^p a_k)x^k - \frac{x^n}{1-x}\sum_{k=0}^{p-1}\left(\frac{x}{1-x}\right)^k \Delta^k a_n$$

4.103. (*a*) By letting $n\to\infty$ in Problem 4.102(*b*) show that

$$S(x) = \lim_{n\to\infty} S_n(x) = \frac{1}{1-x}\sum_{k=0}^{p-1}\left(\frac{x}{1-x}\right)^k \Delta^k a_0 + \left(\frac{x}{1-x}\right)^k \sum_{k=0}^{\infty}(\Delta^p a_k)x^k$$

(*b*) Discuss the case where $p\to\infty$ in part (*a*). Under what conditions will it be true that

$$S(x) = \frac{1}{1-x}\sum_{k=0}^{\infty}\left(\frac{x}{1-x}\right)^k \Delta^k a_0$$

Compare with Problem 4.4.

Chapter 5

Difference Equations

DIFFERENTIAL EQUATIONS

In previous chapters we have examined the remarkable analogies which exist between the difference and sum calculus and the differential and integral calculus respectively. It should thus come as no surprise that corresponding to the theory of *differential equations* there is a theory of *difference equations.*

To develop this analogy we recall that a differential equation is a relationship of the form

$$F\left(x, y, \frac{dy}{dx}, \frac{d^2y}{dx^2}, \ldots, \frac{d^ny}{dx^n}\right) = 0 \qquad (1)$$

between x and the derivatives of an unknown function $y = f(x)$. If we can solve for d^ny/dx^n in (*1*) we obtain

$$\frac{d^ny}{dx^n} = G\left(x, y, \frac{dy}{dx}, \ldots, \frac{d^{n-1}y}{dx^{n-1}}\right) \qquad (2)$$

and we call the order n of this highest derivative the *order of the differential equation.*

A *solution* of a differential equation is any function which satisfies the equation, i.e., reduces it to an identity when substituted into the equation. A *general solution* is one which involves exactly n arbitrary independent constants. A *particular solution* is a solution obtained from the general solution by assigning particular values to these constants.

Example 1.

The equation $\dfrac{d^2y}{dx^2} - 3\dfrac{dy}{dx} + 2y = 4x^2$ is a differential equation of order two. A general solution is $y = c_1e^x + c_2e^{2x} + 2x^2 + 6x + 7$ [see Problem 5.1] and a particular solution is $y = 2e^x - 5e^{2x} + 2x^2 + 6x + 7$.

To determine the n arbitrary constants we often prescribe n independent conditions called *boundary conditions* for the function y. The problem of determining solutions of the differential equation subject to these boundary conditions is called a *boundary-value problem* involving differential equations. In many cases differential equations arise from mathematical or physical problems and the boundary conditions arise naturally in the formulation of such problems.

DEFINITION OF A DIFFERENCE EQUATION

By analogy with differential equations we shall define a *difference equation* as a relationship of the form

$$F\left(x, y, \frac{\Delta y}{\Delta x}, \frac{\Delta^2 y}{\Delta x^2}, \ldots, \frac{\Delta^n y}{\Delta x^n}\right) = 0 \qquad (3)$$

Since $\Delta x = h$ [and since by custom or convention $(\Delta x)^p = h^p$ is written Δx^p when appearing in the denominator] we see that the difference equation *(3)* is a relationship connecting $x, y, \Delta y, \Delta^2 y, \ldots, \Delta^n y$ or equivalently $x, f(x), f(x+h), f(x+2h), \ldots, f(x+nh)$ since $y = f(x)$. Thus we can define a difference equation as a relationship of the form

$$G(x, f(x), f(x+h), f(x+2h), \ldots, f(x+nh)) = 0 \tag{4}$$

where $f(x) = y$ is the unknown function.

ORDER OF A DIFFERENCE EQUATION

From the analogy with differential equations it might be suspected that we would define the order of a difference equation as the order of the highest difference present, i.e. n in *(3)* or *(4)*. However this leads to some difficulties [see Problem 5.5] and so instead we define the order to be the difference between the largest and smallest arguments for the function f involved divided by h. Thus in *(4)* if both $f(x)$ and $f(x+nh)$ appear explicitly the order is $[(x+nh)-x]/h = n$. However if $f(x)$ does not appear but $f(x+h)$ and $f(x+nh)$ both appear then the order is $[(x+nh)-(x+h)]/h = n-1$.

SOLUTION, GENERAL SOLUTION AND PARTICULAR SOLUTION OF A DIFFERENCE EQUATION

A *solution* of a difference equation is [as in the case of differential equations] any function which satisfies the equation. A *general solution* of a difference equation of order n is a solution which involves n arbitrary periodic constants [as on page 82]. A *particular solution* is a solution obtained from the general solution by assigning particular periodic constants.

Example 2.

The equation $\dfrac{\Delta^2 y}{\Delta x^2} - 3\dfrac{\Delta y}{\Delta x} + 2y = 4x^{(2)}$, which can be written as

$$f(x+2h) - (3h+2)f(x+h) + (2h^2+3h+1)f(x) = 4h^2x^{(2)}$$

where $y = f(x)$, is a difference equation of order 2. A general solution is given by

$$y = f(x) = c_1(x)(1+2h)^{x/h} + c_2(x)(1+h)^{x/h} + 2x^2 + (6-2h)x + 7$$

as can be verified by substitution [see Problem 5.4]. A particular solution is found for example by letting

$$c_1(x) = 2 + 4h\sin(2\pi x/h), \qquad c_2(x) = 5 + h^2[1 - \cos(2\pi x/h)]$$

both of which have periods h.

To determine the n arbitrary periodic constants of an nth order difference equation we can prescribe n independent *boundary conditions* for the unknown function y. The problem of determining solutions to the difference equation subject to these boundary conditions is called a *boundary-value problem* involving difference equations. As in the case of differential equations these may arise from the nature of a mathematical or physical problem.

DIFFERENTIAL EQUATIONS AS LIMITS OF DIFFERENCE EQUATIONS

It will be noted that the limit of the difference equation of Example 2 as $\Delta x = h$ approaches zero is the differential equation of Example 1. Also the limit of the solution of the difference equation as $\Delta x = h$ approaches zero is the solution of the differential equation [see Problems 5.6 and 5.7].

In general we can express a differential equation as the limit of some corresponding difference equation.

USE OF THE SUBSCRIPT NOTATION

If the subscript notation is used, the difference equation (*4*) with $x = a + kh$, $y_k = f(a+kh)$ can be written as

$$H(k, y_k, y_{k+1}, \ldots, y_{k+n}) = 0 \tag{5}$$

If we are given the n values $y_0, y_1, y_2, \ldots, y_{n-1}$ then we can determine y_k for $k = n, n+1, \ldots$.

Example 3.

The difference equation

$$f(x+3h) - 5f(x+2h) + 6f(x+h) + 3f(x) = 0$$

can be written in subscript notation as

$$y_{k+3} - 5y_{k+2} + 6y_{k+1} + 3y_k = 0$$

Then given $y_0 = 1$, $y_1 = -2$, $y_2 = 0$ we find on putting $k = 0, 1, \ldots$ the values $y_3 = 9$, $y_4 = 51$, ...

The difference equation (*5*) can also be written in terms of the operator E, i.e.

$$H(k, y_k, Ey_k, E^2y_k, \ldots, E^ny_k) = 0 \tag{6}$$

Example 4.

The difference equation of Example 3 can be written as

$$(E^3 - 5E^2 + 6E + 3)y_k = 0$$

LINEAR DIFFERENCE EQUATIONS

An important class of difference equations which arises in practice is a linear difference equation. A *linear difference equation of order n* is a difference equation having the form

$$a_0(k)E^ny_k + a_1(k)E^{n-1}y_k + \cdots + a_n(k)y_k = R(k) \tag{7}$$

where $a_0(k) \neq 0$. This equation can also be written as

$$[a_0(k)E^n + a_1(k)E^{n-1} + \cdots + a_n(k)]y_k = R(k) \tag{8}$$

or simply

$$\phi(E)y_k = R(k) \tag{9}$$

where the linear operator $\phi(E)$ is given by

$$\phi(E) = a_0(k)E^n + a_1(k)E^{n-1} + \cdots + a_n(k) \tag{10}$$

A particularly important case arises when $a_0(k), a_1(k), \ldots, a_n(k)$ are all constants [i.e. independent of k] and in such case we refer to (*7*), (*8*) or (*9*) as a *linear difference equation of order n with constant coefficients.*

A difference equation which does not have the form (*7*), (*8*) or (*9*), i.e. which is not linear, is called a *nonlinear difference equation.*

Example 5.

The equation of Example 4 is a linear difference equation with constant coefficients.

Example 6.

The equation $[(2k+1)E^2 - 3kE + 4]y_k = 4k^2 - 3k$ is a linear difference equation with nonconstant [i.e. *variable*] coefficients.

Example 7.

The equation $y_k y_{k+1} = y_{k-1}^2$ is a nonlinear difference equation.

HOMOGENEOUS LINEAR DIFFERENCE EQUATIONS

If the right side of *(8)* is replaced by zero we obtain

$$(a_0E^n + a_1E^{n-1} + \cdots + a_n)y_k = 0 \tag{11}$$

or

$$\phi(E)y_k = 0 \tag{12}$$

which is called the *homogeneous equation* corresponding to *(9)*. It is also called the *reduced equation* and equation *(9)* is called the *complete equation* or *nonhomogeneous equation.*

Example 8.

The difference equation of Example 4 is a homogeneous linear equation.

Example 9.

The homogeneous equation corresponding to the complete equation of Example 6 is

$$[(2k+1)E^2 - 3kE + 4]y_k = 0$$

It turns out that in order for us to be able to solve the complete equation *(9)* we first need to solve the homogeneous or reduced equation *(12)* [see Theorem 5-3 below].

Since the general theory of linear difference equations is in many ways similar to that of linear difference equations with constant coefficients we shall consider this simplified case first.

HOMOGENEOUS LINEAR DIFFERENCE EQUATIONS WITH CONSTANT COEFFICIENTS

Assume that $y_k = r^k$ is a solution of *(11)* where we suppose that $a_0, a_1, \ldots, a_n$ are constants. Then on substitution we obtain

$$a_0r^{n+k} + a_1r^{n+k-1} + \cdots + a_nr^k = 0$$

or

$$(a_0r^n + a_1r^{n-1} + \cdots + a_n)r^k = 0$$

It follows that r^k is a solution of *(9)* if r is a solution of

$$a_0r^n + a_1r^{n-1} + \cdots + a_n = 0 \tag{13}$$

which is called the *auxiliary equation* and can be written $\phi(r) = 0$. This equation has n roots which we take to be $r_1, r_2, \ldots, r_n$ which may or may not be different. In such case *(11)* can be written in *factored form* as

$$a_0(E-r_1)(E-r_2)\ldots(E-r_n)y_k = 0 \tag{14}$$

Solutions of *(14)* depend on the nature of the roots $r_1, \ldots, r_n$. The following cases can arise.

Case 1. Roots are all real and distinct.

In this case $r_1^k, r_2^k, \ldots, r_n^k$ are all solutions of *(11)*. Since constant multiples of these solutions are also solutions and since sums of solutions are solutions [because $\phi(E)$ is a linear operator] we have the solution

$$y_k = c_1r_1^k + c_2r_2^k + \cdots + c_nr_n^k$$

Case 2. Some of the roots are complex numbers.

If $a_0, a_1, \ldots, a_n$ are assumed to be real it follows that complex roots if they occur must be *conjugate complex numbers,* i.e. if $\alpha + \beta i$ is a root then so also is $\alpha - \beta i$ where α and β are real. In this case $(\alpha+\beta i)^k$ and $(\alpha - \beta i)^k$ are solutions and so we have the solution

$$K_1(\alpha+\beta i)^k + K_2(\alpha-\beta i)^k$$

where K_1, K_2 are constants. To get this in real form we write the complex numbers in polar form, i.e.

$$\alpha + \beta i = \rho(\cos\theta + i\sin\theta), \qquad \alpha - \beta i = \rho(\cos\theta - i\sin\theta)$$

so that by *De Moivre's theorem*

$$(\alpha+\beta i)^k = \rho^k(\cos k\theta + i\sin k\theta), \qquad (\alpha-\beta i)^k = \rho^k(\cos k\theta - i\sin k\theta)$$

Then by choosing appropriate complex constants K_1, K_2 we find the solution in real form given by

$$\rho^k(c_1\cos k\theta + c_2\sin k\theta)$$

Case 3. Some of the roots are equal.

If two roots are equal, say $r_2 = r_1$, then we can show that [see for example Problem 5.17]

$$(c_1 + c_2k)r_1^k$$

is a solution. Similarly if three roots are equal, say $r_3 = r_2 = r_1$, then a solution is

$$(c_1 + c_2k + c_3k^2)r_1^k$$

Generalizations to the case where more than three roots are equal follow a similar pattern.

LINEARLY INDEPENDENT SOLUTIONS

A set of n functions $f_1(k), f_2(k), \ldots, f_n(k)$ is said to be *linearly dependent* if we can find a set of n constants $A_1, A_2, \ldots, A_n$ not all zero such that we have identically

$$A_1f_1(k) + A_2f_2(k) + \cdots + A_nf_n(k) = 0$$

Example 10.

The functions $5k$, k^2, $3k^2+2k$ are linearly dependent since there are constants A_1, A_2, A_3 not all zero such that $A_1(5k) + A_2(k^2) + A_3(3k^2+2k) = 0$ identically, for example $A_1 = -2$, $A_2 = -15$, $A_3 = 5$.

If a set of functions $f_1(k), \ldots, f_n(k)$ is not linearly dependent then the set is called *linearly independent.*

Example 11.

The functions $5k^2$, $4k+2$, $3k-5$ are linearly independent since if $A_1(5k^2) + A_2(4k+2) + A_3(3k-5) = 0$ identically we must have $A_1 = A_2 = A_3 = 0$.

The following theorems are important.

***Theorem 5-1*:** The set of functions $f_1(k), f_2(k), \ldots, f_n(k)$ is linearly independent if and only if the determinant

$$\begin{vmatrix} f_1(0) & f_2(0) & \cdots & f_n(0) \\ f_1(1) & f_2(1) & \cdots & f_n(1) \\ \cdots & \cdots & \cdots & \cdots \\ f_1(n-1) & f_2(n-1) & \cdots & f_n(n-1) \end{vmatrix} \neq 0 \qquad (15)$$

Otherwise the functions are linearly dependent.

The determinant is sometimes called the *Casorati* and is analogous to the *Wronskian* for differential equations. See Problems 5.11 and 5.12.

***Theorem 5-2*:** If $f_1(k), f_2(k), \ldots, f_n(k)$ are n linearly independent solutions of the nth order equation (*11*) or (*12*) the general solution is

$$y_k = c_1f_1(k) + c_2f_2(k) + \cdots + c_nf_n(k) \qquad (16)$$

and all other solutions are special cases of it, i.e. are particular solutions.

We sometimes refer to this theorem as the *superposition principle.*

Example 12.

The equation $(E^2 - 5E + 6)y_k = 0$ has the auxiliary equation $r^2 - 5r + 6 = 0$ with roots $r = 2, 3$ so that $2^k, 3^k$ are solutions which we can show to be linearly independent. Then the general solution is

$$y_k = c_1 2^k + c_2 3^k$$

SOLUTION OF THE NONHOMOGENEOUS OR COMPLETE EQUATION

Now that we know how to solve the homogeneous difference equation we are ready to solve the complete equation. We shall refer to the general solution of the homogeneous equation as the *complementary function* or *complementary solution.* Also we shall refer to any solution which satisfies the complete equation as a *particular solution.* The following theorem is fundamental.

Theorem 5-3: If $Y_c(k)$ is the complementary solution [i.e. the general solution of the homogeneous equation *(12)*] and $Y_p(k)$ is any particular solution of the complete equation *(9)*, then the general solution

$$y_k = Y_c(k) + Y_p(k) \tag{17}$$

and all solutions are special cases of it.

Example 13.

By Example 12 the general solution of $(E^2 - 5E + 6)y_k = 0$ is $c_1 2^k + c_2 3^k$ and a particular solution of $(E^2 - 5E + 6)y_k = 8k$ is $4k + 6$ as can be verified. Then the general solution of $(E^2 - 5E + 6)y_k = 8k$ is $y_k = c_1 2^k + c_2 3^k + 4k + 6$.

METHODS OF FINDING PARTICULAR SOLUTIONS

In order to apply Theorem 5-3 we must know how to find particular solutions of the complete equation. We shall now consider several important methods for obtaining such solutions which will be found useful.

METHOD OF UNDETERMINED COEFFICIENTS

The method of undetermined coefficients is useful in finding particular solutions of the complete equation *(9)* when the right side $R(k)$ consists of terms having certain special forms. Corresponding to each such term which is present in $R(k)$ we consider a *trial solution* containing a number of unknown constant coefficients which are to be determined by substitution into the difference equation. The trial solutions to be used in each case are shown in the following table where the letters $A, B, A_0, A_1, \ldots$ represent the unknown constant coefficients to be determined.

Terms in $R(k)$	Trial Solution
β^k	$A\beta^k$
$\sin \alpha k$ or $\cos \alpha k$	$A \cos \alpha k + B \sin \alpha k$
polynomial $P(k)$ of degree m	$A_0 k^m + A_1 k^{m-1} + \cdots + A_m$
$\beta^k P(k)$	$\beta^k(A_0 k^m + A_1 k^{m-1} + \cdots + A_m)$
$\beta^k \sin \alpha k$ or $\beta^k \cos \alpha k$	$\beta^k(A \cos \alpha k + B \sin \alpha k)$

Fig. 5-1

The only requirement which must be met to guarantee success of the method is that no term of the trial solution can appear in the complementary function. If any term of the trial solution does happen to be in the complementary function then the entire trial solution corresponding to this term must be multiplied by a positive integer power of k which is just large enough so that no term of the new trial solution will appear in the complementary function. The process is illustrated in Problems 5.20-5.22.

SPECIAL OPERATOR METHODS

We have seen that any linear difference equation can be written as

$$\phi(E)y_k = R(k) \tag{18}$$

Let us define the operator $1/\phi(E)$ called the *inverse* of $\phi(E)$ by the relationship

$$\frac{1}{\phi(E)}R(k) = U \quad \text{where} \quad \phi(E)U = R(k) \tag{19}$$

We shall assume that U does not have any arbitrary constants and so is the particular solution of the equation (*18*). We can show that $1/\phi(E)$ is a linear operator [see Problem 5.24].

The introduction of such inverse operators leads to some special operator methods which are very useful and easy to apply in case $R(k)$ takes on special forms as indicated in the following table. In this table $P(k)$ denotes as usual a polynomial of degree m. For illustrations of the methods see Problems 5.23-5.30.

1.	$\dfrac{1}{\phi(E)}\beta^k = \dfrac{\beta^k}{\phi(\beta)}$, $\phi(\beta) \neq 0$, otherwise use method 4.
2.	$\dfrac{1}{\phi(E)}\sin \alpha k$ or $\dfrac{1}{\phi(E)}\cos \alpha k$ Write $\cos \alpha k = \dfrac{e^{ik}+e^{-ik}}{2}$, $\sin \alpha k = \dfrac{e^{i\alpha k}-e^{-i\alpha k}}{2}$ and use 1.
3.	$\dfrac{1}{\phi(E)}P(k) = \dfrac{1}{\phi(1+\Delta)}P(k) = (b_0 + b_1\Delta + \cdots + b_m\Delta^m + \cdots)P(k)$ where the expansion need be carried out only as far as Δ^m since $\Delta^{m+1}P(k) = 0$.
4.	$\dfrac{1}{\phi(E)}\beta^k P(k) = \beta^k \dfrac{1}{\phi(\beta E)}P(k)$. Then use method 3. The result also holds for any function $F(k)$ in place of $P(k)$.

Fig. 5-2

METHOD OF VARIATION OF PARAMETERS

In this method we first find the complementary solution of the given equation

$$\phi(E)y_k = R_k \tag{20}$$

i.e. the general solution of

$$\phi(E)y_k = 0 \tag{21}$$

in the form

$$y_k = c_1u_1 + c_2u_2 + \cdots + c_nu_n \tag{22}$$

We then replace the arbitrary constants $c_1, c_2, \ldots, c_n$ by functions of k denoted by $K_1, K_2, \ldots, K_n$ and seek to determine these so that

$$y_k = K_1u_1 + K_2u_2 + \cdots + K_nu_n \tag{23}$$

satisfies (20). Since n conditions are needed to find $K_1, K_2, \ldots, K_n$ and one of these is that (20) be satisfied there remain $n-1$ arbitrary conditions which can be imposed. These are usually chosen so as to simplify as much as possible the expressions for $\Delta y_k, \Delta^2 y_k, \ldots$. The equations obtained in this way are

$$\left.\begin{array}{rcl} u_1\Delta K_1 + \quad u_2\Delta K_2 + \cdots + \quad u_n\Delta K_n &=& 0 \\ (\Delta u_1)\Delta K_1 + (\Delta u_2)\Delta K_2 + \cdots + (\Delta u_n)\Delta K_n &=& 0 \\ \cdots\cdots\cdots\cdots\cdots\cdots\cdots\cdots\cdots\cdots && \\ (\Delta^{n-2}u_1)\Delta K_1 + (\Delta^{n-2}u_2)\Delta K_2 + \cdots + (\Delta^{n-2}u_n)\Delta K_n &=& 0 \\ (\Delta^{n-1}u_1)\Delta K_1 + (\Delta^{n-2}u_2)\Delta K_2 + \cdots + (\Delta^{n-1}u_n)\Delta K_n &=& R_k \end{array}\right\} \tag{24}$$

which can be solved for $\Delta K_1, \Delta K_2, \ldots, \Delta K_n$ and thus $K_1, K_2, \ldots, K_n$ from which (23) can be found. For an illustration of the method see Problem 5.31. It should be noted that the determinant of the coefficients of (24) is equal to the *Casorati* of the system [see (15)] which is different from zero if $u_1, u_2, \ldots, u_n$ of (23) are linearly independent. See Problem 5.104.

METHOD OF REDUCTION OF ORDER

If an nth order equation can be written as

$$(E-r_1)(E-r_2)\cdots(E-r_n)y_k = R_k \tag{25}$$

then on letting $z_k = (E-r_2)\cdots(E-r_n)y_k$ we are led to the first order equation

$$(E-r_1)z_k = R_k \tag{26}$$

which has the solution

$$z_k = r_1^k\Delta^{-1}\left(\frac{R_k}{r_1^{k+1}}\right) = r_1^k\sum_{p=1}^{k-1}\frac{R_p}{r_1^{p+1}} + c_1r_1^k \tag{27}$$

where c_1 is an arbitrary constant. By continuing in this manner (25) can be solved.

METHOD OF GENERATING FUNCTIONS

The *generating function* for y_k is defined as

$$G(t) = \sum_{k=0}^{\infty} y_k t^k \tag{28}$$

By using this a linear difference equation can be solved. See Problem 5.34.

LINEAR DIFFERENCE EQUATIONS WITH VARIABLE COEFFICIENTS

Up to now we have considered linear difference equations with constant coefficients by using the fundamental Theorem 5-3, page 155. Since this theorem also applies to linear difference equations with variable coefficients it can also be used to solve such equations.

Any linear first order equation with variable coefficients can be written as

$$y_{k+1} - A_ky_k = R_k \quad \text{or} \quad (E-A_k)y_k = R_k \tag{29}$$

and can always be solved. The solution is given by

$$y_k = A_1A_2\cdots A_{k-1}\Delta^{-1}\left(\frac{R_k}{A_1A_2\cdots A_k}\right)$$

$$= A_1A_2\cdots A_{k-1}\sum_{p=1}^{k-1}\frac{R_p}{A_1A_2\cdots A_p} + cA_1A_2\cdots A_{k-1} \tag{30}$$

where c is an arbitrary constant. This solution can be obtained by multiplying both sides of (29) by the *summation factor* $u_k = 1/A_1A_2\cdots A_{k-1}$ analogous to the *integrating factor* for first order differential equations. See Problem 5.35.

Second or higher order linear equations with variable coefficients cannot always be solved exactly. In such cases special methods are used. Among the most common methods are the following.

1. **Factorization of the operator.** In this method we attempt to write the difference equation as
$$(E-A_k)(E-B_k)\cdots(E-U_k)y_k = R_k \tag{31}$$
by finding $A_k, B_k, \ldots, U_k$ and then using the method of reduction of order. See Problems 5.36 and 5.37.

2. **Variation of parameters.** This can be used when the complementary solution can be found. See Problem 5.114.

3. **Generating functions.** See Problem 5.124.

4. **Series solutions.** In this method we assume a solution in the form of a *factorial series*
$$y_k = \sum_{p=-\infty}^{\infty} c_p k^{(p)} \tag{32}$$
where $c_p = 0$ for $p < 0$. See Problem 5.38.

STURM-LIOUVILLE DIFFERENCE EQUATIONS

A *Sturm-Liouville difference equation* is one which has the form

$$\Delta(p_{k-1}\Delta y_{k-1}) + (q_k + \lambda r_k)y_k = 0 \tag{33}$$

where $1 \leqq k \leqq N-1$ and where λ is independent of k.

We shall associate with (33) the boundary conditions

$$\alpha_0 y_0 + \alpha_1 y_1 = 0, \quad \alpha_N y_N + \alpha_{N+1} y_{N+1} = 0 \tag{34}$$

where $\alpha_0, \alpha_1, \alpha_N, \alpha_{N+1}$ are given constants.

We seek nontrivial solutions [i.e. solutions which are not identically zero] of the equation (33) subject to conditions (34). The problem of finding such solutions is called a Sturm-Liouville boundary-value problem. The equations (33) and conditions (34) are called a *Sturm-Liouville system.*

We can show that in general there will be nontrivial solutions only for certain values of the parameter λ. These values are called *characteristic values* or *eigenvalues.* The solutions corresponding to these are then called *characteristic functions* or *eigenfunctions.* In general there will be a discrete set of eigenvalues and eigenfunctions. If λ_m and λ_n are two different eigenvalues we can denote the corresponding eigenfunctions by $\phi_{m,k}$ and $\phi_{n,k}$ respectively.

In case p_k, q_k, r_k are real we can show that the eigenvalues are also real [see Problem 5.40]. If the eigenfunctions are real we can also show that

$$\sum_{k=1}^{N} r_k\phi_{m,k}\phi_{n,k} = 0 \quad \text{if } m \neq n \tag{35}$$

In general we can take $r_k > 0$. Then (35) states that the eigenfunctions $\{\sqrt{r_k}\,\phi_{m,k}\}$ where $m = 1, \ldots, N$ are *mutually orthogonal.* The terminology used is an extension to N dimensions of the idea of orthogonality in 3 dimensions where two vectors $\mathbf{A} = A_1\mathbf{i} + A_2\mathbf{j} + A_3\mathbf{k}$, $\mathbf{B} = B_1\mathbf{i} + B_2\mathbf{j} + B_3\mathbf{k}$ are orthogonal or perpendicular if

$$A_1B_1 + A_2B_2 + A_3B_3 = 0 \quad \text{i.e.} \quad \sum_{k=1}^{3} A_kB_k = 0 \tag{36}$$

The extension to N dimensions is obtained on replacing the upper limit 3 in the sum by N thus yielding (35).

In case (35) holds we also say that the functions $\{\phi_{m,k}\}$ are orthogonal with respect to the *weight* or *density* r_k.

Given a function F_k it is often possible to obtain an expansion in a series of eigenfunctions. We find that

$$F_k = \sum_{m=1}^{N} c_m \phi_{m,k} \tag{37}$$

where the coefficients c_m are given by

$$c_m = \frac{\sum_{k=1}^{N} r_k F_k \phi_{m,k}}{\sum_{k=1}^{N} r_k \phi_{m,k}^2} \qquad m = 1, \ldots, N \tag{38}$$

If we suitably *normalize* the functions $\phi_{m,k}$ so that

$$\sum_{k=1}^{N} r_k \phi_{m,k}^2 = 1 \tag{39}$$

then the coefficients (38) are simplified since the denominator is equal to 1. The results (35) and (39) can be restated as

$$\sum_{k=1}^{N} r_k \phi_{m,k} \phi_{n,k} = \begin{cases} 0 & m \neq n \\ 1 & m = n \end{cases} \tag{40}$$

In such case we say that the set $\{\sqrt{r_k}\,\phi_{m,k}\}$ is an *orthonormal set.*

NONLINEAR DIFFERENCE EQUATIONS

An important class of nonlinear difference equations can be solved by applying suitable transformations which change them into linear difference equations. See Problems 5.43 and 5.44.

SIMULTANEOUS DIFFERENCE EQUATIONS

If two or more difference equations are given with the same number of unknown functions we can solve such equations simultaneously by using a procedure which eliminates all but one of the unknowns. See for example Problem 5.45.

MIXED DIFFERENCE EQUATIONS

In addition to differences which may occur in equations there may also be derivatives or integrals. In such case we refer to the equations as *differential-difference equations, integral-difference equations,* etc. These equations can sometimes be solved by various special techniques. See Problems 5.46 and 5.47 for example.

PARTIAL DIFFERENCE EQUATIONS

Up to now we have been concerned with difference equations involving unknown functions of one variable. These are often called *ordinary difference equations* in contrast to difference equations involving unknown functions of two or more variables which are called *partial difference equations.*

To consider such equations we must generalize the concepts of difference operators to functions of two or more variables. To do this we can consider for example a function $f(x,y)$ and introduce two difference operators Δ_1, Δ_2 defined so that if $h = \Delta x$, $l = \Delta y$ are given,

$$\Delta_1 f(x,y) = f(x+h,y) - f(x,y), \quad \Delta_2 f(x,y) = f(x,y+l) - f(x,y) \tag{41}$$

These are called *partial difference operators.* Similarly we define *partial translation operators* E_1, E_2 so that

$$E_1 f(x,y) = f(x+h,y), \quad E_2 f(x,y) = f(x,y+l) \tag{42}$$

These are related to *partial derivatives* considered in elementary calculus. It is clear that

$$E_1 = 1 + \Delta_1, \quad E_2 = 1 + \Delta_2 \tag{43}$$

We can also define powers of $E_1, E_2, \Delta_1, \Delta_2$.

If we use the subscript notation

$$z_{k,m} = f(a+kh, b+ml) \tag{44}$$

it follows for example that

$$E_1 z_{k,m} = z_{k+1,m}, \quad E_2 z_{k,m} = z_{k,m+1} \tag{45}$$

We can now consider *partial difference equations* such as for example

$$(E_1^2 - 3E_1E_2 + 2E_2^2)z_{k,m} = 0 \tag{46}$$

which can also be written

$$z_{k+2,m} - 3z_{k+1,m+1} + 2z_{k,m+2} = 0 \tag{47}$$

Any function satisfying such an equation is called a *solution* of the equation.

The general linear partial difference equation in two variables can be written as

$$\phi(E_1, E_2)z_{k,m} = R_{k,m} \tag{48}$$

where $\phi(E_1, E_2)$ is a polynomial in E_1 and E_2 of degree n. A solution of (*48*) which contains n arbitrary functions is called a *general solution.* The general solution of (*48*) with the right side replaced by zero is called the *complementary solution.* Any solution which satisfies the complete equation (*48*) is called a *particular solution.* As in the one variable case we have the following fundamental theorem.

***Theorem 5-4*:** The general solution of (*48*) is the sum of its complementary solution and any particular solution.

Various methods are available for finding complementary solutions as in the one variable case. See Problem 5.55.

As might be expected from the one variable case it is possible to express *partial differential equations* as limits of partial difference equations and to obtain their solutions in this manner. See Problem 5.56.

Solved Problems

DIFFERENTIAL EQUATIONS

5.1. Show that the general solution of the differential equation $\frac{d^2y}{dx^2} - 3\frac{dy}{dx} + 2y = 4x^2$ is $y = c_1e^x + c_2e^{2x} + 2x^2 + 6x + 7$.

We have
$$y = c_1e^x + c_2e^{2x} + 2x^2 + 6x + 7 \tag{1}$$
$$\frac{dy}{dx} = c_1e^x + 2c_2e^{2x} + 4x + 6 \tag{2}$$
$$\frac{d^2y}{dx^2} = c_1e^x + 4c_2e^{2x} + 4 \tag{3}$$

Then
$$\frac{d^2y}{dx^2} - 3\frac{dy}{dx} + 2y = (c_1e^x + 4c_2e^{2x} + 4) - 3(c_1e^x + 2c_2e^{2x} + 4x + 6) + 2(c_1e^x + c_2e^{2x} + 2x^2 + 6x + 7) = 4x^2$$

Since the given differential equation is of order 2 and the solution has 2 arbitrary constants it is the general solution.

5.2. Find the particular solution of the differential equation in Problem 5.1 such that $y(0) = 4$, $y'(0) = -3$.

From (*1*) and (*2*) of Problem 5.1 we have
$$y(0) = c_1 + c_2 + 7 = 4 \quad \text{or} \quad c_1 + c_2 = -3$$
$$y'(0) = c_1 + 2c_2 + 6 = -3 \quad \text{or} \quad c_1 + 2c_2 = -9$$
so that $c_1 = 3$, $c_2 = -6$.

Then the required particular solution is
$$y = 3e^x - 6e^{2x} + 2x^2 + 6x + 7$$

DIFFERENCE EQUATIONS

5.3. Show that the difference equation $\frac{\Delta^2 y}{\Delta x^2} - 3\frac{\Delta y}{\Delta x} + 2y = 4x^{(2)}$ can be written as $f(x+2h) - (3h+2)f(x+h) + (2h^2+3h+1)f(x) = 4h^2x^{(2)}$ where $y = f(x)$.

Writing $\Delta x = h$ and $y = f(x)$ the difference equation can be written as
$$\frac{\Delta^2 f(x)}{h^2} - 3\frac{\Delta f(x)}{h} + 2f(x) = 4x^{(2)} \tag{1}$$
or
$$\frac{f(x+2h) - 2f(x+h) + f(x)}{h^2} - 3\left[\frac{f(x+h) - f(x)}{h}\right] + 2f(x) = 4x^{(2)}$$

Multiplying by h^2 and simplifying we obtain
$$f(x+2h) - (3h+2)f(x+h) + (2h^2+3h+1)f(x) = 4h^2x^{(2)} \tag{2}$$

5.4. Show that the general solution of the difference equation in Problem 5.3 is given by
$$y = c_1(x)(1+2h)^{x/h} + c_2(x)(1+h)^{x/h} + 2x^2 + (6-2h)x + 7 \tag{1}$$
where $c_1(x)$ and $c_2(x)$ have periods equal to h.

We have since $c_1(x+h) = c_1(x)$, $c_2(x+h) = c_2(x)$,
$$y = c_1(x)(1+2h)^{x/h} + c_2(x)(1+h)^{x/h} + 2x^2 + (6-2h)x + 7 \tag{2}$$

$$\frac{\Delta y}{\Delta x} = \frac{c_1(x)}{h}[(1+2h)^{(x+h)/h} - (1+2h)^{x/h}] + \frac{c_2(x)}{h}[(1+h)^{(x+h)/h} - (1+h)^{x/h}]$$

$$+ \frac{2[(x+h)^2 - x^2] + (6-2h)[(x+h) - x]}{h}$$

$$= 2c_1(x)(1+2h)^{x/h} + c_2(x)(1+h)^{x/h} + 4x + 6$$

Similarly, $$\frac{\Delta^2 y}{\Delta x^2} = 4c_1(x)(1+2h)^{x/h} + c_2(x)(1+h)^{x/h} + 4$$

Then

$$\begin{aligned}\frac{\Delta^2 y}{\Delta x^2} - 3\frac{\Delta y}{\Delta x} + 2y &= [4c_1(x)(1+2h)^{x/h} + c_2(x)(1+h)^{x/h} + 4] \\ &\quad - 3[2c_1(x)(1+2h)^{x/h} + c_2(x)(1+h)^{x/h} + 4x + 6] \\ &\quad + 2[c_1(x)(1+2h)^{x/h} + c_2(x)(1+h)^{x/h} + 2x^2 + (6-2h)x + 7] \\ &= 4x^2 - 4hx = 4x(x-h) \\ &= 4x^{(2)}\end{aligned}$$

Since the difference equation *(1)* has the difference between the largest and smallest arguments divided by h equal to

$$\frac{(x+2h) - x}{h} = 2$$

it follows that the order is 2. Then since the solution *(1)* has two arbitrary independent periodic constants, it is the general solution.

5.5. Show that $h^2\frac{\Delta^2 y}{\Delta x^2} + 2h\frac{\Delta y}{\Delta x} + y = 2x^{(3)}$ is not a second order difference equation.

We can write the equation if $y = f(x)$ as

$$h^2\left[\frac{f(x+2h) - 2f(x+h) + f(x)}{h^2}\right] + 2h\left[\frac{f(x+h) - f(x)}{h}\right] + f(x) = 2x^{(3)}$$

or $$f(x+2h) = 2x^{(3)}$$

This equation involves only one argument and in fact is equivalent to

$$f(x) = 2(x-2h)^{(3)}$$

which is not really a difference equation.

The result shows that we cannot determine the order of a difference equation by simply looking at the largest value of n in $\Delta^n y/\Delta x^n$. Because of this we must define the order as given on page 151.

DIFFERENTIAL EQUATIONS AS LIMITS OF DIFFERENCE EQUATIONS

5.6. Show that the limit of the difference equation [see Problem 5.3]

$$\frac{\Delta^2 y}{\Delta x^2} - 3\frac{\Delta y}{\Delta x} + 2y = 4x^{(2)}$$

as Δx or h approaches zero is [see Problem 5.1]

$$\frac{d^2 y}{dx^2} - 3\frac{dy}{dx} + 2y = 4x^2$$

This follows at once since

$$\lim_{\Delta x \to 0}\left[\frac{\Delta^2 y}{\Delta x^2} - 3\frac{\Delta y}{\Delta x} + 2y\right] = \frac{d^2 y}{dx^2} - 3\frac{dy}{dx} + 2y$$

and $$\lim_{h\to 0} 4x^{(2)} = \lim_{h \to 0} 4x(x-h) = 4x^2$$

5.7. Show that the limit as Δx or h approaches zero of the solution of the difference equation in Problem 5.6 is the solution of the differential equation in that problem.

By Problem 5.4 the general solution of the difference equation in Problem 5.6 is, if $h = \Delta x$,

$$y = c_1(x)(1+2h)^{x/h} + c_2(x)(1+h)^{x/h} + 2x^2 + (6-2h)x + 7 \qquad (1)$$

From the calculus we have

$$\lim_{n\to\infty}\left(1+\frac{1}{n}\right)^n = e = 2.71828\ldots \qquad (2)$$

or, if $n = a/h$ where a is some given constant,

$$\lim_{h\to 0}\left(1+\frac{h}{a}\right)^{a/h} = e = 2.71828\ldots \qquad (3)$$

Equivalently (*3*) can be written

$$\lim_{h\to 0}\left(1+\frac{h}{a}\right)^{1/h} = e^{1/a} \qquad (4)$$

or

$$\lim_{h\to 0}\left(1+\frac{h}{a}\right)^{x/h} = e^{x/a} \qquad (5)$$

Using (*5*) with $a = 1/2$ we find

$$\lim_{h\to 0}(1+2h)^{x/h} = e^{2x} \qquad (6)$$

Using (*5*) with $a = 1$ we find

$$\lim_{h\to 0}(1+h)^{x/h} = e^{x} \qquad (7)$$

Also since $c_1(x+h) = c_1(x)$, $c_2(x+h) = c_2(x)$ it follows that $\lim_{h\to 0} c_1(x) = c_1$, $\lim_{h\to 0} c_2(x) = c_2$ where c_1 and c_2 are constants, provided that these limits exist.

Then the limit of the solution (*1*) as $h \to 0$ is

$$y = c_1e^{2x} + c_2e^{x} + 2x^2 + 6x + 7$$

which is the general solution of the differential equation of Problem 5.6 [see Problem 5.1].

USE OF THE SUBSCRIPT NOTATION

5.8. Write the difference equation of Problem 5.3 with subscript or k notation.

We have by the definition of the subscript notation

$$y_k = f(x), \quad y_{k+1} = f(x+h), \quad y_{k+2} = f(x+2h)$$

Also using $x = kh$,

$$x^{(2)} = x(x-h) = kh(kh-h) = h^2k(k-1) = h^2k^{(2)}$$

Then the equation

$$f(x+2h) - (3h+2)f(x+h) + (2h^2+3h+1)f(x) = 4x^{(2)}$$

becomes

$$y_{k+2} - (3h+2)y_{k+1} + (2h^2+3h+1)y_k = 4h^2k^{(2)}$$

or using the operator E

$$E^2y_k - (3h+2)Ey_k + (2h^2+3h+1)y_k = 4h^2k^{(2)}$$

which can be written

$$[E^2 - (3h+2)E + (2h^2+3h+1)]y_k = 4h^2k^{(2)}$$

5.9. Write the difference equation

$$f(x+4h) + 2f(x+3h) - 4f(x+2h) + 3f(x+h) - 6f(x) = 0$$

in the subscript or k notation.

We use

$$y_k = f(x), \quad y_{k+1} = f(x+h), \quad y_{k+2} = f(x+2h), \quad y_{k+3} = f(x+3h), \quad y_{k+4} = f(x+4h)$$

Then the difference equation is

$$(E^4 + 2E^3 - 4E^2 + 3E - 6)y_k = 0$$

LINEARLY INDEPENDENT FUNCTIONS

5.10. Determine which of the following sets of functions are linearly dependent and which are linearly independent.

(a) $2^k, 2^{k+3}$ (b) $2^k, 2^{k+3}, 4^k$ (c) $2^k, 4^k$ (d) $2^k, k\cdot 2^k, k^2\cdot 2^k$.

(a) Consider $A_1 2^k + A_2 2^{k+3}$ which can be written as

$$A_1 2^k + A_2 \cdot 2^3 \cdot 2^k = (A_1 + 8A_2)2^k \tag{1}$$

Since we can find A_1 and A_2 not both zero such that (1) is zero [as for example $A_2 = -1, A_1 = 8$] it follows that the functions 2^k, 2^{k+3} are linearly dependent.

(b) Consider $A_1 2^k + A_2 2^{k+3} + A_3 4^k$ which can be written as

$$(A_1 + 8A_2)2^k + A_3 4^k \tag{2}$$

Then since we can find the 3 constants A_1, A_2, A_3 not all zero such that (2) is zero [for example $A_2 = -1$, $A_1 = 8$, $A_3 = 0$] it follows that the functions 2^k, 2^{k+3}, 4^k are linearly dependent.

In general if a set of functions is linearly dependent it remains linearly dependent when one or more functions is added to the set.

(c) Consider $A_1 2^k + A_2 4^k$ where A_1, A_2 are constants. This will be zero if and only if

$$2^k(A_1 + A_2 2^k) = 0 \quad \text{or} \quad A_1 + A_2 2^k = 0$$

However this last equation cannot be true for all k unless $A_1 = A_2 = 0$. Thus it follows that 2^k and 4^k are linearly independent.

(d) Consider $A_1 2^k + A_2 k \cdot 2^k + A_3 k^2 \cdot 2^k$ where A_1, A_2, A_3 are constants. This will be zero if and only if

$$2^k(A_1 + A_2 k + A_3 k^2) = 0 \quad \text{or} \quad A_1 + A_2 k + A_3 k^2 = 0$$

But this last equation cannot be true for all k unless $A_1 = A_2 = A_3 = 0$. Thus the functions are linearly independent.

5.11. Work Problem 5.10 by using Theorem 5-1, page 154.

(a) In this case $f_1(k) = 2^k$, $f_2(k) = 2^{k+3}$ and the Casorati is given by

$$\begin{vmatrix} 2^0 & 2^3 \\ 2^1 & 2^4 \end{vmatrix} = \begin{vmatrix} 1 & 8 \\ 2 & 16 \end{vmatrix} = 0$$

Then by Theorem 5-1 the functions 2^k, 2^{k+3} are linearly dependent.

(b) In this case $f_1(k) = 2^k$, $f_2(k) = 2^{k+3}$, $f_3(k) = 4^k$ and the Casorati is

$$\begin{vmatrix} 2^0 & 2^3 & 4^0 \\ 2^1 & 2^4 & 4^1 \\ 2^2 & 2^5 & 4^2 \end{vmatrix} = \begin{vmatrix} 1 & 8 & 1 \\ 2 & 16 & 4 \\ 4 & 32 & 16 \end{vmatrix} = 8\begin{vmatrix} 1 & 1 & 1 \\ 2 & 2 & 4 \\ 4 & 4 & 16 \end{vmatrix} = 0$$

where we have taken the factor 8 from elements of the second column in the determinant and noted that since the first two columns are identical the determinant is zero.

Thus by Theorem 5-1 the functions 2^k, 2^{k+3}, 4^k are linearly dependent.

(c) In this case $f_1(k) = 2^k$, $f_2(k) = 4^k$ and the Casorati is

$$\begin{vmatrix} 2^0 & 4^0 \\ 2^1 & 4^1 \end{vmatrix} = \begin{vmatrix} 1 & 1 \\ 2 & 4 \end{vmatrix} = 2$$

Then by Theorem 5-1 the functions 2^k, 4^k are linearly independent.

(d) In this case $f_1(k) = 2^k$, $f_2(k) = k \cdot 2^k$, $f_3(k) = k^2 \cdot 2^k$ and the Casorati is

$$\begin{vmatrix} 2^0 & 0 \cdot 2^0 & 0^2 \cdot 2^0 \\ 2^1 & 1 \cdot 2^1 & 1^2 \cdot 2^1 \\ 2^2 & 2 \cdot 2^2 & 2^2 \cdot 2^2 \end{vmatrix} = \begin{vmatrix} 1 & 0 & 0 \\ 2 & 2 & 2 \\ 4 & 8 & 16 \end{vmatrix} = 1\begin{vmatrix} 2 & 2 \\ 8 & 16 \end{vmatrix} = 16$$

Thus the functions 2^k, $k \cdot 2^k$, $k^2 \cdot 2^k$ are linearly independent.

5.12. Prove Theorem 5-1, page 154.

We prove the theorem for the case of 3 functions. The general case can be proved similarly.

By definition the functions $f_1(k)$, $f_2(k)$, $f_3(k)$ are linearly independent if and only if the equation

$$A_1 f_1(k) + A_2 f_2(k) + A_3 f_3(k) = 0 \tag{1}$$

identically holds only when $A_1 = 0$, $A_2 = 0$, $A_3 = 0$. Putting $k = 0, 1, 2$ in (1)

$$\left.\begin{aligned} A_1 f_1(0) + A_2 f_2(0) + A_3 f_3(0) &= 0 \\ A_1 f_1(1) + A_2 f_2(1) + A_3 f_3(1) &= 0 \\ A_1 f_1(2) + A_2 f_2(2) + A_3 f_3(2) &= 0 \end{aligned}\right\} \tag{2}$$

Now (2) will have the sole solution $A_1 = 0$, $A_2 = 0$, $A_3 = 0$ if and only if the determinant

$$\begin{vmatrix} f_1(0) & f_2(0) & f_3(0) \\ f_1(1) & f_2(1) & f_3(1) \\ f_1(2) & f_2(2) & f_3(2) \end{vmatrix} \neq 0 \tag{3}$$

which is the required result.

HOMOGENEOUS LINEAR DIFFERENCE EQUATIONS WITH CONSTANT COEFFICIENTS

5.13. (a) Find linearly independent solutions of the difference equation $y_{k+2} - 6y_{k+1} + 8y_k = 0$ and (b) thus write the general solution.

(a) Let $y_k = r^k$ in the difference equation. Then it becomes

$$r^{k+2} - 6r^{k+1} + 8r^k = 0 \quad \text{or} \quad r^k(r^2 - 6r + 8) = 0$$

Dividing by r^k [assuming $r \neq 0$ which otherwise leads to the trivial solution $y_k = 0$] we obtain the auxiliary equation

$$r^2 - 6r + 8 = 0 \qquad \text{i.e.} \quad (r-2)(r-4) = 0 \quad \text{or} \quad r = 2, 4$$

Thus two solutions are 2^k and 4^k. By Problem 5.11(c) these are linearly independent solutions.

Note that the difference equation can be written as

$$(E^2 - 6E + 8)y_k = 0$$

and that the auxiliary equation can immediately be written down from this on formally replacing E by r in the operator $E^2 - 6E + 8$ and setting the result equal to zero.

(b) Using Theorem 5-2, page 154, i.e. the *superposition principle*, the general solution is

$$y_k = c_1 2^k + c_2 4^k$$

where c_1 and c_2 are arbitrary constants.

5.14. (a) Find that particular solution of $y_{k+2} - 6y_{k+1} + 8y_k = 0$ such that $y_0 = 3$, $y_1 = 2$. (b) Find y_5.

(a) From Problem 5.13 the general solution of the difference equation is

$$y_k = c_1 2^k + c_2 4^k$$

Then from $y_0 = 3$, $y_1 = 2$ we have on putting $k = 0$ and $k = 1$ respectively

$$3 = c_1 + c_2, \quad 2 = 2c_1 + 4c_2$$

from which $c_1 = 5$, $c_2 = -2$.

Thus the required particular solution is

$$y_k = 5 \cdot 2^k - 2 \cdot 4^k$$

(b) Letting $k = 5$ we find

$$y_5 = 5 \cdot 2^5 - 2 \cdot 4^5 = 5(32) - 2(1024) = -1888$$

The value of y_5 can also be found directly from the difference equation [see Problem 5.84].

5.15. Solve the difference equation $y_{k+2} - 2y_{k+1} + 5y_k = 0$.

The difference equation can be written

$$(E^2 - 2E + 5)y_k = 0$$

and the auxiliary equation obtained by letting $y_k = r^k$ in this equation is

$$r^2 - 2r + 5 = 0$$

from which

$$r = \frac{2 \pm \sqrt{4-20}}{2} = \frac{2 \pm 4i}{2} = 1 \pm 2i$$

It follows that solutions are $(1+2i)^k$ and $(1-2i)^k$ which can be shown to be linearly independent [see Problem 5.74]. Since in polar form

$$1 + 2i = \sqrt{5}\left(\frac{1}{\sqrt{5}} + \frac{2}{\sqrt{5}}i\right) = \sqrt{5}(\cos\theta + i\sin\theta) = \sqrt{5}\,e^{i\theta}$$

$$1 - 2i = \sqrt{5}\left(\frac{1}{\sqrt{5}} - \frac{2}{\sqrt{5}}i\right) = \sqrt{5}(\cos\theta - i\sin\theta) = \sqrt{5}\,e^{-i\theta}$$

where

$$\cos\theta = \frac{1}{\sqrt{5}}, \quad \sin\theta = \frac{2}{\sqrt{5}} \qquad (1)$$

it follows that the two linearly independent solutions are

$$(\sqrt{5}\,e^{i\theta})^k = (\sqrt{5})^k e^{ki\theta} = 5^{k/2}e^{ki\theta} = 5^{k/2}(\cos k\theta + i\sin k\theta)$$

$$(\sqrt{5}\,e^{-i\theta})^k = (\sqrt{5})^k e^{-ki\theta} = 5^{k/2}e^{-ki\theta} = 5^{k/2}(\cos k\theta - i\sin k\theta)$$

Then the general solution can be written as

$$y_k = 5^{k/2}(c_1\cos k\theta + c_2\sin k\theta) \qquad (2)$$

5.16. Find the particular solution of the difference equation of Problem 5.15 satisfying the conditions $y_0 = 0,\ y_1 = 1$.

Since $y_0 = 0,\ y_1 = 1$ we have on putting $k = 0,\ k = 1$ respectively in the general solution (*2*) of Problem 5.15

$$0 = c_1, \qquad 1 = 5^{1/2}(c_1 \cos\theta + c_2 \sin\theta)$$

from which

$$c_1 = 0, \qquad c_2 = \frac{1}{\sqrt{5}\sin\theta}$$

Thus

$$y_k = \frac{5^{k/2}}{\sqrt{5}\sin\theta}\sin k\theta = 5^{(k-1)/2}\frac{\sin k\theta}{\sin\theta}$$

Note that θ can be found from (*1*) of Problem 5.15.

5.17. (*a*) Solve the difference equation $y_{k+2} - 4y_{k+1} + 4y_k = 0$ and (*b*) find the solution such that $y_0 = 1,\ y_1 = 3$.

(*a*) The difference equation can be written as $(E^2 - 4E + 4)y_k = 0$ and the auxiliary equation is $r^2 - 4r + 4 = 0$ or $(r-2)^2 = 0$ so that $r = 2, 2$, i.e. 2 is a repeated root.

It follows that $y_k = r^k = 2^k$ is a solution. However we need another solution which is linearly independent. To find it let $y_k = 2^k v_k$ in the given difference equation. Then it becomes

$$v_{k+2} - 2v_{k+1} + v_k = 0 \qquad \text{or} \qquad \Delta^2 v_k = 0$$

Thus

$$v_k = c_1 + c_2 k \qquad \text{and} \qquad y_k = 2^k(c_1 + c_2 k)$$

which gives the required general solution.

(*b*) Putting $k = 0$ and $k = 1$ we find

$$1 = c_1, \qquad 3 = 2(c_1 + c_2)$$

Thus $c_1 = 1,\ c_2 = 1/2$, and the required solution is

$$y_k = 2^k\left(1 + \frac{k}{2}\right) = (k+2)2^{k-1}$$

5.18. (*a*) Solve the difference equation $y_{k+3} + y_{k+2} - y_{k+1} - y_k = 0$ and (*b*) find the solution which satisfies the conditions $y_0 = 2,\ y_1 = -1,\ y_2 = 3$.

(*a*) The difference equation can be written as $(E^3 + E^2 - E - 1)y_k = 0$ and the auxiliary equation is

$$r^3 + r^2 - r - 1 = 0$$

This can be written as

$$r^2(r+1) - (r+1) = 0, \qquad (r+1)(r^2-1) = 0, \qquad \text{or} \qquad (r+1)(r+1)(r-1) = 0$$

so that the roots are $r = -1, -1, 1$.

Corresponding to the two repeated roots $-1, -1$ we have the solution $c_1(-1)^k + c_2 k(-1)^k = (-1)^k(c_1 + c_2 k)$. Corresponding to the root 1 we have the solution $c_3(1)^k = c_3$. Then by the principle of superposition the general solution is

$$y_k = (-1)^k(c_1 + c_2 k) + c_3$$

(*b*) Putting $k = 0, 1, 2$ respectively we find

$$2 = c_1 + c_3, \qquad -1 = -c_1 - c_2 + c_3, \qquad 3 = c_1 + 2c_2 + c_3$$

from which $c_1 = \frac{5}{4},\ c_2 = \frac{1}{2},\ c_3 = \frac{3}{4}$. Thus the required solution is

$$y_k = \frac{3}{4} + \frac{(-1)^k}{4}(2k+5)$$

5.19. Find the general solution corresponding to a linear homogeneous difference equation if the roots of the corresponding auxiliary equation are given by $3, -\frac{1}{2} \pm \frac{\sqrt{3}}{2}i, -3, 3, -2, -3, 3, 4$.

Corresponding to the repeated roots 3, 3, 3 we have the solution $3^k(c_1 + c_2k + c_3k^2)$.

Corresponding to the complex roots $-\frac{1}{2} \pm \frac{\sqrt{3}}{2}i$ [which have polar form $\cos\theta \pm i \sin\theta$ where $\cos\theta = -\frac{1}{2}$, $\sin\theta = \frac{\sqrt{3}}{2}$ or $\theta = 120° = 2\pi/3$ radians] we have the solution

$$c_4 \cos k\theta + c_5 \sin k\theta \quad = \quad c_4 \cos\frac{2\pi k}{3} + c_5 \sin\frac{2\pi k}{3}$$

Corresponding to the repeated roots $-3, -3$ we have the solution $(-3)^k(c_6 + c_7k)$.

Corresponding to the single root -2 we have the solution $c_8(-2)^k$.

Corresponding to the single root 4 we have the solution $c_9 4^k$.

Then the required general solution is by the principle of superposition given by

$$y_k \quad = \quad 3^k(c_1 + c_2k + c_3k^2) + c_4 \cos\frac{2\pi k}{3} + c_5 \sin\frac{2\pi k}{3} + (-3)^k(c_6 + c_7k) + c_8(-2)^k + c_9 4^k$$

METHOD OF UNDETERMINED COEFFICIENTS

5.20. Solve $y_{k+2} - 6y_{k+1} + 8y_k = 3k^2 + 2 - 5 \cdot 3^k$.

The general solution of the homogeneous equation [also called *complementary* or *reduced equation*] is by Problem 5.13

$$y_k \quad = \quad c_1 2^k + c_2 4^k \tag{1}$$

Corresponding to the polynomial $3k^2 + 2$ on the right hand side of the difference equation we assume as trial solution

$$A_1k^2 + A_2k + A_3 \tag{2}$$

since none of these terms occur in the complementary solution (*1*)

Corresponding to the term $-5 \cdot 3^k$ on the right hand side of the difference equation we assume the trial solution

$$A_4 3^k \tag{3}$$

Thus corresponding to the right hand side we assume the trial solution [or particular solution]

$$y_k \quad = \quad A_1k^2 + A_2k + A_3 + A_4 3^k$$

Substituting in the given difference equation we find

$$\begin{aligned} y_{k+2} - 6y_{k+1} + 8y_k \quad &= \quad A_1(k+2)^2 + A_2(k+2) + A_3 + A_4 3^{k+2} \\ &\qquad - 6[A_1(k+1)^2 + A_2(k+1) + A_3 + A_4 3^{k+1}] \\ &\qquad + 8[A_1k^2 + A_2k + A_3 + A_4 3^k] \\ &= \quad 3A_1k^2 + (3A_2 - 8A_1)k + 3A_3 - 4A_2 - 2A_1 - A_4 3^k \end{aligned}$$

Since this must equal the right hand side of the given difference equation we have

$$3A_1k^2 + (3A_2 - 8A_1)k + 3A_3 - 4A_2 - 2A_1 - A_4 3^k \quad = \quad 3k^2 + 2 - 5 \cdot 3^k$$

Equating coefficients of like terms we have

$$3A_1 \;=\; 3, \quad 3A_2 - 8A_1 \;=\; 0, \quad 3A_3 - 4A_2 - 2A_1 \;=\; 2, \quad A_4 \;=\; 5$$

Thus $\qquad A_1 = 1, \quad A_2 = 8/3, \quad A_3 = 44/9, \quad A_4 = 5$

and the particular solution is

$$k^2 + \frac{8}{3}k + \frac{44}{9} + 5 \cdot 3^k$$

Adding this to the complementary solution (*1*) we find

$$y_k \quad = \quad c_1 2^k + c_2 4^k + k^2 + \frac{8}{3}k + \frac{44}{9} + 5 \cdot 3^k$$

5.21. Solve $y_{k+2} - 4y_{k+1} + 4y_k = 3 \cdot 2^k + 5 \cdot 4^k$.

The complementary solution is by Problem 5.17

$$c_1 2^k + c_2 k \cdot 2^k \tag{1}$$

Corresponding to the term $3 \cdot 2^k$ on the right side of the given difference equation we would normally assume the trial solution $A_1 2^k$. This term however is in the complementary solution (1) so that we multiply by k to obtain the trial solution $A_1 k \cdot 2^k$. But this also is in the complementary solution (1). Then we multiply by k again to obtain the trial solution $A_1 k^2 \cdot 2^k$. Since this is not in the complementary solution (1) it is the appropriate trial solution to use.

Corresponding to the term $5 \cdot 4^k$ we can assume the trial solution $A_2 4^k$ since this is not in the complementary solution (1).

The trial solution or particular solution is thus given by

$$y_k = A_1 k^2 \cdot 2^k + A_2 4^k$$

Substituting this into the given difference equation we find

$$\begin{aligned} y_{k+2} - 4y_{k+1} + 4y_k &= A_1(k+2)^2 \cdot 2^{k+2} + A_2 4^{k+2} \\ &\quad - 4[A_1(k+1)^2 \cdot 2^{k+1} + A_2 4^{k+1}] \\ &\quad + 4[A_1 k^2 \cdot 2^k + A_2 4^k] \\ &= 8A_1 2^k + 4A_2 4^k \end{aligned}$$

Thus we must have $$8A_1 2^k + 4A_2 4^k = 3 \cdot 2^k + 5 \cdot 4^k$$

from which $$A_1 = \tfrac{3}{8}, \quad A_2 = \tfrac{5}{4}$$

Then the required solution is

$$y_k = c_1 2^k + c_2 k \cdot 2^k + \tfrac{3}{8} k^2 \cdot 2^k + \tfrac{5}{4} \cdot 4^k \tag{2}$$

or $$y_k = c_1 2^k + c_2 k \cdot 2^k + 3k^2 \cdot 2^{k-3} + 5 \cdot 4^{k-1} \tag{3}$$

5.22. Solve $y_{k+3} - 3y_{k+2} + 3y_{k+1} - y_k = 24(k+2)$.

The complementary equation is

$$(E^3 - 3E^2 + 3E - 1)y_k = 0 \quad \text{or} \quad (E-1)^3 y_k = 0$$

and the complementary solution is thus

$$y_k = c_1 + c_2 k + c_3 k^2$$

Corresponding to the term $24k + 48$ on the right of the given equation we would normally use the trial solution $A_1 k + A_2$. Since these terms occur in the complementary solution we multiply by a power of k which is just sufficient to insure that none of these terms will appear. This is k^3 so that the trial solution is $y_k = A_1 k^4 + A_2 k^3$. Substituting this into the given difference equation we find

$$24A_1 k + 36A_1 + 6A_2 = 24k + 48$$

so that $A_1 = 1$, $A_2 = 2$. Thus the required solution is

$$y_k = c_1 + c_2 k + c_3 k^2 + k^4 + 2k^3$$

SPECIAL OPERATOR METHODS

5.23. Let $\phi(E) = a_0 E^n + a_1 E^{n-1} + \cdots + a_n$. (a) Prove that $\phi(E)\beta^k = \phi(\beta)\beta^k$. (b) Prove that a particular solution of the equation $\phi(E)y_k = \beta^k$ is given by $\dfrac{1}{\phi(E)}\beta^k = \dfrac{\beta^k}{\phi(\beta)}$ if $\phi(\beta) \neq 0$.

(a) We have

$$E\beta^k = \beta^{k+1} = \beta \cdot \beta^k, \quad E^2\beta^k = \beta^{k+2} = \beta^2 \cdot \beta^k, \quad \ldots, \quad E^n\beta^k = \beta^{k+n} = \beta^n \cdot \beta^k$$

Thus

$$\phi(E)\beta^k = (a_0E^n + a_1E^{n-1} + \cdots + a_n)\beta^k$$
$$= (a_0\beta^n + a_1\beta^{n-1} + \cdots + a_n)\beta^k = \phi(\beta)\beta^k$$

(*b*) Since $\phi(E)\beta^k = \phi(\beta)\beta^k$ from part (*a*) it follows that

$$\frac{1}{\phi(E)}\beta^k = \frac{\beta^k}{\phi(\beta)} \quad \text{if } \phi(\beta) \neq 0$$

5.24. Prove that $1/\phi(E)$ is a linear operator.

Let

$$\frac{1}{\phi(E)}R_1(k) = U_1, \quad \frac{1}{\phi(E)}R_2(k) = U_2 \tag{1}$$

Then by definition if $R_1(k)$ and $R_2(k)$ are any functions

$$\phi(E)U_1 = R_1(k), \quad \phi(E)U_2 = R_2(k)$$

and since $\phi(E)$ is a linear operator it follows that

$$\phi(E)\,[U_1 + U_2] = R_1(k) + R_2(k)$$

so that

$$\frac{1}{\phi(E)}[R_1(k) + R_2(k)] = U_1 + U_2$$

or using (*1*)

$$\frac{1}{\phi(E)}[R_1(k) + R_2(k)] = \frac{1}{\phi(E)}R_1(k) + \frac{1}{\phi(E)}R_2(k) \tag{2}$$

Also if α is any constant and $R(k)$ is any function let

$$\frac{1}{\phi(E)}\alpha R(k) = U \tag{3}$$

Then

$$\phi(E)U = \alpha R(k)$$

Dividing by $\alpha \neq 0$ we have since $\phi(E)$ is a linear operator

$$\phi(E)\frac{U}{\alpha} = R(k)$$

or

$$\frac{1}{\phi(E)}R(k) = \frac{U}{\alpha} \tag{4}$$

From (*2*) and (*3*) we have

$$\frac{1}{\phi(E)}\alpha R(k) = \alpha\frac{1}{\phi(E)}R(k) \tag{5}$$

The results (*2*) and (*5*) prove that $1/\phi(E)$ is a linear operator.

5.25. Solve the difference equation $y_{k+2} - 2y_{k+1} + 5y_k = 2\cdot 3^k - 4\cdot 7^k$.

The equation can be written as $(E^2 - 2E + 5)y_k = 2\cdot 3^k - 4\cdot 7^k$. By Problems 5.23 and 5.24 a particular solution is given by

$$\frac{1}{E^2 - 2E + 5}(2\cdot 3^k - 4\cdot 7^k) = 2\cdot\frac{1}{E^2 - 2E + 5}3^k - 4\cdot\frac{1}{E^2 - 2E + 5}7^k$$
$$= 2\cdot\frac{1}{3^2 - 2(3) + 5}3^k - 4\cdot\frac{1}{7^2 - 2(7) + 5}7^k$$
$$= \frac{1}{4}\cdot 3^k - \frac{1}{10}\cdot 7^k$$

Since the complementary solution is as given in equation (*2*) of Problem 5.15 the required general solution is

$$y_k = 5^{k/2}(c_1\cos k\theta + c_2\sin k\theta) + \tfrac{1}{4}\cdot 3^k - \tfrac{1}{10}\cdot 7^k$$

where θ is determined from equation (*1*) of Problem 5.15.

5.26. Let $\phi(E) = a_0E^n + a_1E^{n-1} + \cdots + a_n$. Prove that

$$\phi(E)[\beta^k F(k)] = \beta^k \phi(\beta E) F(k)$$

We have

$$E\beta^k F(k) = \beta^{k+1}F(k+1) = \beta^k[\beta E F(k)]$$
$$E^2\beta^k F(k) = \beta^{k+2}F(k+2) = \beta^k[\beta^2 E^2 F(k)]$$
$$\cdots\cdots\cdots\cdots\cdots\cdots\cdots\cdots$$
$$E^n\beta^k F(k) = \beta^{k+n}F(k+n) = \beta^k[\beta^n E^n F(k)]$$

Then

$$\begin{aligned}\phi(E)[\beta^k F(k)] &= (a_0E^n + a_1E^{n-1} + \cdots + a_n)[\beta^k F(k)] \\ &= \beta^k[a_0(\beta E)^n + a_1(\beta E)^{n-1} + \cdots + a_n]F(k) \\ &= \beta^k \phi(\beta E) F(k)\end{aligned}$$

5.27. Prove that $\dfrac{1}{\phi(E)}\beta^k F(k) = \beta^k \dfrac{1}{\phi(\beta E)} F(k)$.

Let $G(k) = \dfrac{1}{\phi(\beta E)} F(k)$ so that $\phi(\beta E)\, G(k) = F(k)$. Then by Problem 5.26

$$\phi(E)[\beta^k G(k)] = \beta^k \phi(\beta E)\, G(k) = \beta^k F(k)$$

Thus

$$\frac{1}{\phi(E)}\beta^k F(k) = \beta^k G(k) = \beta^k \frac{1}{\phi(\beta E)} F(k)$$

5.28. Find $\dfrac{1}{E^2 - 4E + 4}\, 2^k$.

The method of Problem 5.23 fails since if $\phi(E) = E^2 - 4E + 4$ then $\phi(2) = 0$. However we can use the method of Problem 5.27 with $F(k) = 1$, $\beta = 2$ to obtain

$$\begin{aligned}\frac{1}{\phi(E)} 2^k &= 2^k \frac{1}{\phi(2E)}(1) = 2^k \frac{1}{(2E)^2 - 4(2E) + 4}(1) = 2^k \frac{1}{4E^2 - 8E + 4}(1) \\ &= 2^k \frac{1}{4(1+\Delta)^2 - 8(1+\Delta) + 4}(1) = 2^k \frac{1}{4\Delta^2}(1) = \frac{2^k}{4}\Delta^{-1}[\Delta^{-1}(1)] = \tfrac{1}{4}\Delta^{-1}k^{(1)} \\ &= \frac{2^k}{4}\,\frac{k^{(2)}}{2} = \frac{k(k-1)2^k}{8}\end{aligned}$$

5.29. Find $\dfrac{1}{E^2 - 6E + 8}(3k^2 + 2)$.

We use the method in entry 3 of the table on page 156. Putting $E = 1 + \Delta$ we have

$$\begin{aligned}\frac{1}{E^2 - 6E + 8}(3k^2+2) &= \frac{1}{(1+\Delta)^2 - 6(1+\Delta) + 8}(3k^2+2) \\ &= \frac{1}{3 - 4\Delta + \Delta^2}(3k^2+2) \\ &= \frac{1}{3}\,\frac{1}{1 - (\frac{4}{3}\Delta - \frac{1}{3}\Delta^2)}(3k^2+2) \\ &= \tfrac{1}{3}[1 + (\tfrac{4}{3}\Delta - \tfrac{1}{3}\Delta^2) + (\tfrac{4}{3}\Delta - \tfrac{1}{3}\Delta^2)^2 + \cdots](3k^2+2) \\ &= \tfrac{1}{3}[1 + \tfrac{4}{3}\Delta + \tfrac{13}{9}\Delta^2 + \cdots](3k^2+2) \\ &= \tfrac{1}{3}[3k^2+2] + \tfrac{4}{9}\Delta[3k^{(2)} + 3k^{(1)} + 2] + \tfrac{13}{27}\Delta^2[3k^{(2)} + 3k^{(1)} + 2] \\ &= \tfrac{1}{3}[3k^2+2] + \tfrac{4}{9}[6k^{(1)} + 3] + \tfrac{13}{27}[6] \\ &= k^2 + \tfrac{8}{3}k + \tfrac{44}{9}\end{aligned}$$

5.30. Work Problem 5.21 by operator methods.

The equation can be written $(E^2-4E+4)y_k = 3\cdot 2^k + 5\cdot 4^k$. Then by Problem 5.23 and 5.28 a particular solution is given by

$$\frac{1}{E^2-4E+4}(3\cdot 2^k+5\cdot 4^k) = 3\cdot\frac{1}{E^2-4E+4}(2^k) + 5\cdot\frac{1}{E^2-4E+4}(4^k)$$

$$= \tfrac{3}{8}k(k-1)2^k + \frac{5}{4^2-4(4)+4}4^k$$

$$= \tfrac{3}{8}k(k-1)2^k + \tfrac{5}{4}\cdot 4^k$$

Since the complementary solution is $C_1 2^k + C_2 k\cdot 2^k$ the general solution is

$$y_k = C_1 2^k + C_2 k\cdot 2^k + \tfrac{3}{8}k(k-1)2^k + \tfrac{5}{4}\cdot 4^k$$

This can be written

$$y_k = C_1 2^k + (C_2 - \tfrac{3}{8})k\cdot 2^k + \tfrac{3}{8}k^2 2^k + \tfrac{5}{4}\cdot 4^k$$

and is the same as that obtained in equation (*2*) or (*3*) of Problem 5.21 if we take $c_1 = C_1$, $c_2 = C_2 - \frac{3}{8}$.

METHOD OF VARIATION OF PARAMETERS

5.31. Solve $y_{k+2} - 5y_{k+1} + 6y_k = k^2$ by the method of variation of parameters.

The complementary solution, i.e. the general solution of $y_{k+2}-5y_{k+1}+6y_k = 0$, is $c_1 2^k + c_2 3^k$. Then according to the method of variation of parameters we assume that the general solution of the complete difference equation is

$$y_k = K_1 2^k + K_2 3^k \qquad (1)$$

where K_1 and K_2 are functions of k to be determined.

Now

$$\Delta y_k = K_1 2^k + K_2 2\cdot 3^k + 2^{k+1}\Delta K_1 + 3^{k+1}\Delta K_2 \qquad (2)$$

Since two conditions are needed to determine K_1 and K_2, one of which is that (*1*) satisfy the given difference equation, we are free to impose one arbitrary condition. We shall choose this condition to be that the sum of the last two terms in (*2*) is identically zero, i.e.

$$2^{k+1}\Delta K_1 + 3^{k+1}\Delta K_2 = 0 \qquad (3)$$

so that (*2*) becomes

$$\Delta y_k = K_1 2^k + K_2 2\cdot 3^k \qquad (4)$$

From (*4*) we find

$$\Delta^2 y_k = K_1 2^k + K_2 4\cdot 3^k + 2^{k+1}\Delta K_1 + 2\cdot 3^{k+1}\Delta K_2 \qquad (5)$$

Now the given equation can be written as

$$(E^2-5E+6)y_k = k^2 \qquad (6)$$

or since $E = 1+\Delta$

$$(\Delta^2 - 3\Delta + 2)y_k = k^2 \qquad (7)$$

Using (*1*), (*4*) and (*5*) in (*7*) we find after simplifying

$$2^{k+1}\Delta K_1 + 2\cdot 3^{k+1}\Delta K_2 = k^2 \qquad (8)$$

We must thus determine K_1 and K_2 from the equations (*3*) and (*8*), i.e.

$$\left.\begin{aligned} 2^{k+1}\Delta K_1 + 3^{k+1}\Delta K_2 &= 0 \\ 2^{k+1}\Delta K_1 + 2\cdot 3^{k+1}\Delta K_2 &= k^2 \end{aligned}\right\} \qquad (9)$$

Note that these equations could actually be written down directly by using equations (*24*) on page 157.

From (9) we find

$$\Delta K_1 = \frac{\begin{vmatrix} 0 & 3^{k+1} \\ k^2 & 2\cdot 3^{k+1} \end{vmatrix}}{\begin{vmatrix} 2^{k+1} & 3^{k+1} \\ 2^{k+1} & 2\cdot 3^{k+1} \end{vmatrix}} = \frac{-k^2}{2^{k+1}}, \qquad \Delta K_2 = \frac{\begin{vmatrix} 2^{k+1} & 0 \\ 2^{k+1} & k^2 \end{vmatrix}}{\begin{vmatrix} 2^{k+1} & 3^{k+1} \\ 2^{k+1} & 2\cdot 3^{k+1} \end{vmatrix}} = \frac{k^2}{3^{k+1}}$$

Then

$$K_1 = -\Delta^{-1}\left(\frac{k^2}{2^{k+1}}\right), \qquad K_2 = \Delta^{-1}\left(\frac{k^2}{3^{k+1}}\right)$$

These can be obtained by using the methods of Chapter 4 or the operator method of Problem 5.27. Using the latter we have, omitting the arbitrary additive constant,

$$\begin{aligned} K_1 &= -\frac{1}{\Delta}\left(\frac{k^2}{2^{k+1}}\right) = -\frac{1}{E-1}(\tfrac{1}{2})^k\left(\frac{k^2}{2}\right) \\ &= -(\tfrac{1}{2})^k \frac{1}{\tfrac{1}{2}E-1}\left(\frac{k^2}{2}\right) \\ &= -(\tfrac{1}{2})^k \frac{1}{E-2}(k^2) \\ &= (\tfrac{1}{2})^k \frac{1}{1-\Delta}(k^2) \\ &= (\tfrac{1}{2})^k(1+\Delta+\Delta^2+\cdots)(k^{(2)}+k^{(1)}) \\ &= (\tfrac{1}{2})^k(k^{(2)}+k^{(1)}+2k^{(1)}+1+2) \\ &= (\tfrac{1}{2})^k(k^2+2k+3) \end{aligned}$$

Thus taking into account an arbitrary constant c_1,

$$K_1 = \frac{1}{2^k}(k^2+2k+3) + c_1$$

Similarly we find

$$K_2 = -\frac{1}{2\cdot 3^k}(k^2+k+1) + c_2$$

Then using these in (1) we find the required solution

$$y_k = c_1 2^k + c_2 3^k + \tfrac{1}{2}k^2 + \tfrac{3}{2}k + \tfrac{5}{2}$$

METHOD OF REDUCTION OF ORDER

5.32. Solve $y_{k+1} - ry_k = R_k$ where r is a constant.

Method 1, using variation of parameters.

The complementary or homogeneous equation corresponding to the given equation is

$$y_{k+1} - ry_k = 0 \tag{1}$$

From this we find

$$y_k = ry_{k-1}, \quad y_{k-1} = ry_{k-2}, \quad \ldots, \quad y_2 = ry_1, \quad y_1 = ry_0$$

so that the solution of (1) is

$$y_k = r^k y_0 \quad \text{or} \quad y_k = c_0 r^k$$

where we take $c_0 = y_0$ as the arbitrary constant.

We now replace c_0 by a function of k denoted by $K(k)$, i.e.

$$y_k = K(k)r^k \tag{2}$$

and seek to determine $K(k)$ so that the given equation is satisfied. Substitution of (2) in the given equation yields

$$K(k+1)r^{k+1} - K(k)r^{k+1} = R_k$$

or dividing by r^{k+1},

$$K(k+1) - K(k) = \Delta K = \frac{R_k}{r^{k+1}}$$

Thus

$$K = \Delta^{-1}\left(\frac{R_k}{r^{k+1}}\right)$$

and so from (2)

$$y_k = r^k\Delta^{-1}\left(\frac{R_k}{r^{k+1}}\right) = r^k \sum_{p=1}^{k-1} \frac{R_p}{r^{p+1}} + cr^k$$

Method 2, multiplying by a suitable summation factor.

Multiplying the given equation by $1/r^{k+1}$ it can be written as

$$\frac{y_{k+1}}{r^{k+1}} - \frac{y_k}{r^k} = \frac{R_k}{r^{k+1}} \quad \text{or} \quad \Delta\left(\frac{y_k}{r^k}\right) = \frac{R_k}{r^{k+1}}$$

Thus

$$\frac{y_k}{r^k} = \Delta^{-1}\left(\frac{R_k}{r^{k+1}}\right)$$

or

$$y_k = r^k\Delta^{-1}\left(\frac{R_k}{r^{k+1}}\right) = r^k \sum_{p=1}^{k-1} \frac{R_p}{r^{p+1}} + cr^k$$

where c is an arbitrary constant. The factor $1/r^{k+1}$ used is called a *summation factor.*

5.33. Solve $y_{k+2} - 5y_{k+1} + 6y_k = k^2$ by the method of reduction of order.

Write the equation as $(E-3)(E-2)y_k = k^2$. Letting $z_k = (E-2)y_k$ we have

$$(E-3)z_k = k^2$$

Then by Problem 5.32 the solution is

$$z_k = 3^k\Delta^{-1}\left(\frac{k^2}{3^{k+1}}\right) + c_1 3^k = c_1 3^k - \tfrac{1}{2}(k^2+k+1)$$

Thus

$$(E-2)y_k = c_1 3^k - \tfrac{1}{2}(k^2+k+1)$$

Applying the method of Problem 5.32 again the solution is

$$y_k = 2^k\Delta^{-1}\left(\frac{c_1 3^k - \frac{1}{2}(k^2+k+1)}{3^{k+1}}\right) + c_2 2^k$$

$$= \tfrac{1}{2}k^2 + \tfrac{3}{2}k + \tfrac{5}{2} + c_2 2^k + c_3 3^k$$

We can use the method to find particular solutions by omitting the arbitrary constants.

METHOD OF GENERATING FUNCTIONS

5.34. Solve the difference equation

$$y_{k+2} - 3y_{k+1} + 2y_k = 0 \quad \text{if } y_0 = 2,\ y_1 = 3$$

by the method of generating functions.

Let $G(t) = \sum_{k=0}^{\infty} y_k t^k$. Multiplying the given difference equation by t^k and summing from $k=0$ to ∞ we have

$$\sum_{k=0}^{\infty} y_{k+2}t^k - 3\sum_{k=0}^{\infty} y_{k+1}t^k + 2\sum_{k=0}^{\infty} y_k t^k = 0$$

i.e.

$$(y_2 + y_3t + y_4t^2 + \cdots) - 3(y_1 + y_2t + y_3t^2 + \cdots) + 2(y_0 + y_1t + y_2t^2 + \cdots) = 0$$

This can be written in terms of the generating function $G(t)$ as

$$\frac{G(t) - y_0 - y_1t}{t^2} - 3\left(\frac{G(t) - y_0}{t}\right) + 2G(t) = 0 \qquad (1)$$

Putting $y_0 = 2$, $y_1 = 3$ in (1) and solving for $G(t)$ we obtain

$$G(t) = \frac{2 - 3t}{1 - 3t + 2t^2} = \frac{2 - 3t}{(1-t)(1-2t)}$$

Writing this in terms of partial fractions we find

$$G(t) = \frac{1}{1-t} + \frac{1}{1-2t} = \sum_{k=0}^{\infty} t^k + \sum_{k=0}^{\infty} (2t)^k = \sum_{k=0}^{\infty} (1 + 2^k)t^k$$

Thus

$$\sum_{k=0}^{\infty} y_k t^k = \sum_{k=0}^{\infty} (1 + 2^k)t^k$$

and so

$$y_k = 1 + 2^k \qquad (2)$$

which can be verified as the required solution.

LINEAR DIFFERENCE EQUATIONS WITH VARIABLE COEFFICIENTS

5.35. Solve the first order linear difference equation with variable coefficients given by $y_{k+1} - A_k y_k = R_k$ or $(E - A_k)y_k = R_k$.

Method 1, using variation of parameters.

The complementary or homogeneous equation corresponding to the given equation is

$$y_{k+1} - A_k y_k = 0 \qquad (1)$$

From this we have

$$y_k = A_{k-1}y_{k-1}, \quad y_{k-1} = A_{k-2}y_{k-2}, \quad \ldots, \quad y_2 = A_1y_1$$

so that the solution of (1) is

$$y_k = A_{k-1}A_{k-2}\cdots A_1y_1$$

or

$$y_k = c_1A_1A_2\cdots A_{k-1}$$

where we take $c_1 = y_1$ as the arbitrary constant.

We now replace c_1 by a function of k denoted by $K(k)$, i.e.

$$y_k = K(k)A_1A_2\cdots A_{k-1} \qquad (2)$$

and seek to determine $K(k)$ so that the given equation is satisfied. Substitution of (2) in the given equation yields

$$K(k+1)A_1A_2\cdots A_k - A_kK(k)A_1A_2\cdots A_{k-1} = R_k$$

or dividing by $A_1A_2\cdots A_k$

$$K(k+1) - K(k) = \Delta K = \frac{R_k}{A_1A_2\cdots A_k}$$

Thus

$$K = \Delta^{-1}\left(\frac{R_k}{A_1A_2\cdots A_k}\right)$$

and so from (1) we find the required solution

$$y_k = A_1A_2\cdots A_{k-1}\Delta^{-1}\left(\frac{R_k}{A_1A_2\cdots A_k}\right)$$

$$= A_1A_2\cdots A_{k-1}\sum_{p=1}^{k-1}\frac{R_p}{A_1A_2\cdots A_p} + cA_1A_2\cdots A_{k-1}$$

where c is an arbitrary constant.

Method 2, multiplying by a suitable summation factor.

Multiplying the given equation by $1/A_1A_2\cdots A_k$ it can be written

$$\frac{y_{k+1}}{A_1A_2\cdots A_k} - \frac{y_k}{A_1A_2\cdots A_{k-1}} = \frac{R_k}{A_1A_2\cdots A_k}$$

or

$$\Delta\left(\frac{y_k}{A_1A_2\cdots A_{k-1}}\right) = \frac{R_k}{A_1A_2\cdots A_k}$$

Then

$$y_k = A_1A_2\cdots A_{k-1}\Delta^{-1}\left(\frac{R_k}{A_1A_2\cdots A_k}\right)$$

or

$$y_k = A_1A_2\cdots A_{k-1}\sum_{p=1}^{k-1}\frac{R_p}{A_1A_2\cdots A_p} + cA_1A_2\cdots A_{k-1}$$

We call $1/A_1A_2\cdots A_k$ used to multiply the given equation the *summation factor.*

5.36. Show that the equation $y_{k+2}-(k+2)y_{k+1}+ky_k = k$ can be written in the factored form $(E-A_k)(E-B_k)y_k = k$.

The given equation is in operator form

$$[E^2-(k+2)E+k]y_k = k \tag{1}$$

Let us try to determine A_k and B_k so that (*1*) is the same as

$$(E-A_k)(E-B_k)y_k = k \tag{2}$$

Now the left side of (*2*) can be written as

$$\begin{aligned}(E-A_k)(y_{k+1}-B_ky_k) &= y_{k+2}-B_{k+1}y_{k+1}-A_k(y_{k+1}-B_ky_k)\\ &= y_{k+2}-(A_k+B_{k+1})y_{k+1}+A_kB_ky_k\end{aligned}$$

Comparing with the left side of the given equation (*1*) we have

$$A_k+B_{k+1} = k+2, \quad A_kB_k = k \tag{3}$$

From the second equation of (*3*) we are led to try either $A_k = k$, $B_k = 1$ or $A_k = 1$, $B_k = k$. Since the second set, i.e. $A_k = 1$, $B_k = k$, satisfies the first equation of (*3*) the given equation can be written as

$$(E-1)(E-k)y_k = k$$

5.37. Solve $y_{k+2}-(k+2)y_{k+1}+ky_k = k$, $y_1 = 0$, $y_2 = 1$.

By Problem 5.36 the equation can be written as

$$(E-1)(E-k)y_k = k \tag{1}$$

Let $z_k = (E-k)y_k$. Then (*1*) becomes

$$(E-1)z_k = k \tag{2}$$

which has the solution

$$z_k = \Delta^{-1}k = \Delta^{-1}k^{(1)} = \frac{k^{(2)}}{2}+c_1$$

Then

$$(E-k)y_k = \frac{k^{(2)}}{2}+c_1 = \tfrac{1}{2}k(k-1)+c_1$$

By Problem 5.35 this has the solution

$$y_k = 1\cdot2\cdot\cdots\cdot(k-1)\Delta^{-1}\left(\frac{\frac{1}{2}k(k-1)+c_1}{1\cdot2\cdot\cdots\cdot k}\right) = (k-1)!\,\Delta^{-1}\left(\frac{\frac{1}{2}k(k-1)+c_1}{k!}\right)$$

or

$$\begin{aligned}y_k &= (k-1)!\sum_{p=1}^{k-1}\left(\frac{\frac{1}{2}p(p-1)+c_1}{p!}\right)+c_2(k-1)!\\ &= \frac{(k-1)!}{2}\sum_{p=1}^{k-1}\frac{p(p-1)}{p!}+c_1(k-1)!\sum_{p=1}^{k-1}\frac{1}{p!}+c_2(k-1)!\end{aligned} \tag{3}$$

From $y_1 = 0$, $y_2 = 1$ and the difference equation we find, if $k = 1$, $y_3 - 3y_2 + y_1 = 1$ so that $y_3 = 4$. If we put $k = 2$ and $k = 3$ in (3), we find using $0! = 1$

$$c_1 + c_2 = 1, \quad 1 + 3c_1 + 2c_2 = 4$$

from which $c_1 = 1$, $c_2 = 0$. Thus the required solution is

$$y_k = \frac{(k-1)!}{2}\sum_{p=1}^{k-1}\frac{p(p-1)}{p!} + (k-1)!\sum_{p=1}^{k-1}\frac{1}{p!}$$

This can be written for $k > 3$ as

$$y_k = \frac{(k-1)!}{2}\left[1 + 1 + \frac{1}{2!} + \frac{1}{3!} + \cdots + \frac{1}{(k-3)!}\right] + (k-1)!\left[1 + \frac{1}{2!} + \frac{1}{3!} + \cdots + \frac{1}{(k-1)!}\right]$$

5.38. Solve $(k+1)y_{k+2} - (3k+2)y_{k+1} + (2k-1)y_k = 0.$

The equation can be written as

$$(k+1)\Delta^2 y_k - k\Delta y_k - 2y_k = 0 \tag{1}$$

Assume a solution of the form

$$y_k = \sum_{p=-\infty}^{\infty} c_p k^{(p)} \tag{2}$$

where we take

$$c_p = 0 \quad \text{for } p = -1, -2, -3, \ldots \tag{3}$$

From (2) we obtain

$$\Delta y_k = \sum_{p=-\infty}^{\infty} p c_p k^{(p-1)} \tag{4}$$

$$\Delta^2 y_k = \sum_{p=-\infty}^{\infty} p(p-1) c_p k^{(p-2)} \tag{5}$$

Substituting (2), (4) and (5) into (1), it becomes

$$\sum p(p-1)c_p k k^{(p-2)} + \sum p(p-1)c_p k^{(p-2)} - \sum p c_p k k^{(p-1)} - \sum 2c_p k^{(p)} = 0 \tag{6}$$

where we have omitted the limits of summation. On using the result

$$kk^{(m)} = k^{(m+1)} + mk^{(m)}$$

(6) can be written

$$\sum p(p-1)c_p[k^{(p-1)} + (p-2)k^{(p-2)}] + \sum p(p-1)c_p k^{(p-2)}$$
$$- \sum p c_p[k^{(p)} + (p-1)k^{(p-1)}] - \sum 2c_p k^{(p)} = 0$$

This in turn becomes on collecting coefficients of $k^{(p)}$

$$\sum \{(p+2)(p+1)^2 c_{p+2} - (p+2)c_p\} k^{(p)} = 0 \tag{7}$$

Since (7) is an identity each coefficient must be zero, i.e.

$$(p+2)(p+1)^2 c_{p+2} - (p+2)c_p = 0 \tag{8}$$

Using (3) it follows that we must have $p = 0, 1, 2, \ldots$ so that (8) becomes on dividing by $p + 2 \neq 0$

$$(p+1)^2 c_{p+2} - c_p = 0 \tag{9}$$

Putting $p = 0, 1, 2, \ldots$ in (9) we find

$$1^2 c_2 - c_0 = 0, \quad 2^2 c_3 - c_1 = 0, \quad 3^2 c_4 - c_2 = 0, \quad 4^2 c_5 - c_3 = 0, \quad \ldots$$

Then

$$c_2 = \frac{c_0}{1^2}, \quad c_3 = \frac{c_1}{2^2}, \quad c_4 = \frac{c_2}{3^2} = \frac{c_0}{1^2 \cdot 3^2}, \quad c_5 = \frac{c_3}{4^2} = \frac{c_1}{2^2 \cdot 4^2}, \quad \ldots$$

and so the required solution (*2*) becomes

$$y_k = c_0\left[1 + \frac{k^{(2)}}{1^2} + \frac{k^{(4)}}{1^2 \cdot 3^2} + \frac{k^{(6)}}{1^2 \cdot 3^2 \cdot 5^2} + \cdots\right]$$

$$+ c_1\left[k^{(1)} + \frac{k^{(3)}}{2^2 \cdot 4^2} + \frac{k^{(5)}}{2^2 \cdot 4^2 \cdot 6^2} + \cdots\right]$$

STURM-LIOUVILLE DIFFERENCE EQUATIONS

5.39. Prove that the eigenfunctions $\phi_{m,k}$, $\phi_{n,k}$ belonging to two different eigenvalues λ_m, λ_n satisfy the orthogonality condition

$$\sum_{k=1}^{N} r_k \phi_{m,k} \phi_{n,k} = 0$$

Since λ_m, $\phi_{m,k}$ and λ_n, $\phi_{n,k}$ are the corresponding eigenvalues and eigenfunctions they must by their definition satisfy the equations

$$\Delta(p_{k-1}\Delta\phi_{m,k-1}) + (q_k + \lambda_m r_k)\phi_{m,k} = 0 \tag{1}$$

$$\Delta(p_{k-1}\Delta\phi_{n,k-1}) + (q_k + \lambda_n r_k)\phi_{n,k} = 0 \tag{2}$$

for $k = 1, \ldots, N$.

If we multiply these equations by $\phi_{n,k}$ and $\phi_{m,k}$ respectively and then subtract we find

$$(\lambda_m - \lambda_n) r_k \phi_{m,k}\phi_{n,k} = \phi_{m,k}\Delta[p_{k-1}\Delta\phi_{n,k-1}] - \phi_{n,k}\Delta[p_{k-1}\Delta\phi_{m,k-1}] \tag{3}$$

Summing from $k = 1$ to N then yields

$$(\lambda_m - \lambda_n)\sum_{k=1}^{N} r_k\phi_{m,k}\phi_{n,k} = \sum_{k=1}^{N} \phi_{m,k}\Delta[p_{k-1}\Delta\phi_{n,k-1}] - \sum_{k=1}^{N} \phi_{n,k}\Delta[p_{k-1}\Delta\phi_{m,k-1}]$$

Using summation by parts this can be written

$$\begin{aligned}(\lambda_m - \lambda_n)\sum_{k=1}^{N} r_k\phi_{m,k}\phi_{n,k} &= p_{k-1}[\phi_{m,k}\Delta\phi_{n,k-1} - \phi_{n,k}\Delta\phi_{m,k-1}]\Big|_{k=1}^{N+1} \\ &= p_N[\phi_{m,N+1}\Delta\phi_{n,N} - \phi_{n,N+1}\Delta\phi_{m,N}] \\ &\qquad - p_0[\phi_{m,1}\Delta\phi_{n,0} - \phi_{n,1}\Delta\phi_{m,0}] \\ &= p_N[\phi_{m,N}\phi_{n,N+1} - \phi_{n,N}\phi_{m,N+1}] \\ &\qquad - p_0[\phi_{m,0}\phi_{n,1} - \phi_{n,0,}\phi_{m,1}]\end{aligned} \tag{4}$$

From the boundary conditions (*34*), page 158, which must be satisfied by $\phi_{m,k}$ and $\phi_{n,k}$ we have

$$\alpha_0\phi_{m,0} + \alpha_1\phi_{m,1} = 0, \quad \alpha_N\phi_{m,N} + \alpha_{N+1}\phi_{m,N+1} = 0$$

$$\alpha_0\phi_{n,0} + \alpha_1\phi_{n,1} = 0, \quad \alpha_N\phi_{n,N} + \alpha_{N+1}\phi_{n,N+1} = 0$$

By eliminating α_0, α_1, α_N, α_{N+1} from these equations we see that they are equivalent to

$$\phi_{m,0}\phi_{n,1} - \phi_{n,0}\phi_{m,1} = 0, \quad \phi_{m,N}\phi_{n,N+1} - \phi_{n,N}\phi_{m,N+1} = 0$$

Using these in (*4*) we thus have

$$(\lambda_m - \lambda_n)\sum_{k=1}^{N} r_k\phi_{m,k}\phi_{n,k} = 0$$

and if $\lambda_m \neq \lambda_n$ we must have as required

$$\sum_{k=1}^{N} r_k\phi_{m,k}\phi_{n,k} = 0 \tag{5}$$

5.40. Prove that if p_k, q_k, r_k in equation (*33*), page 158, are real then any eigenvalues of the Sturm-Liouville system must also be real.

Suppose that λ_m is an eigenvalue which may be complex and that $\phi_{m,k}$ is the corresponding eigenfunction which may be complex. Then by definition,

$$\Delta(p_{k-1}\Delta\phi_{m,k-1}) + (q_k + \lambda_m r_k)\phi_{m,k} = 0 \tag{1}$$

Taking the complex conjugate of this equation [denoted by a bar over the letter] we have

$$\Delta(p_{k-1}\Delta\bar{\phi}_{m,k-1}) + (q_k + \bar{\lambda}_m r_k)\bar{\phi}_{m,k} = 0 \tag{2}$$

using the fact that p_{k-1}, q_k, r_k are real.

By proceeding as in Problem 5.39 multiplying the first equation $\bar{\phi}_{m,k}$, the second equation by $\phi_{m,k}$ and then subtracting, we find [compare equation (*3*) of Problem 5.39]

$$(\lambda_m - \bar{\lambda}_m)\sum_{k=1}^{N} r_k\phi_{m,k}\bar{\phi}_{m,k} = 0$$

or

$$(\lambda_m - \bar{\lambda}_m)\sum_{k=1}^{N} r_k|\phi_{m,k}|^2 = 0 \tag{3}$$

Since $r_k > 0$, the sum in (*3*) cannot be zero and so we must have $\lambda_m - \bar{\lambda}_m = 0$ or $\lambda_m = \bar{\lambda}_m$, i.e. λ_m is real.

5.41. Let $F_k = \sum_{m=1}^{N} c_m\phi_{m,k}$ where $\phi_{m,k}$ are the eigenfunctions of the Sturm-Liouville system on page 158 assumed to be real.

(*a*) Show that formally

$$c_m = \frac{\sum_{k=1}^{N} r_k F_k \phi_{m,k}}{\sum_{k=1}^{N} r_k \phi_{m,k}^2} \qquad m = 1, \ldots, N$$

(*b*) What limitations if any are there in the expansion obtained in part (*a*)?

(*a*) If

$$F_k = \sum_{m=1}^{N} c_m\phi_{m,k}$$

then multiplying both sides by $r_k\phi_{n,k}$ and summing from $k = 1$ to N we have

$$\begin{aligned}\sum_{k=1}^{N} r_k F_k \phi_{n,k} &= \sum_{k=1}^{N}\sum_{m=1}^{N} c_m r_k \phi_{m,k}\phi_{n,k} \\ &= \sum_{m=1}^{N} c_m \left\{\sum_{k=1}^{N} r_k\phi_{m,k}\phi_{n,k}\right\} \\ &= c_n \sum_{k=1}^{N} r_k\phi_{n,k}^2\end{aligned}$$

where we have used the fact that

$$\sum_{k=1}^{N} r_k\phi_{m,k}\phi_{n,k} = 0 \qquad m \neq n$$

by Problem 5.39. We thus have

$$c_n = \frac{\sum_{k=1}^{N} r_k F_k \phi_{n,k}}{\sum_{k=1}^{N} r_k \phi_{n,k}^2}$$

which is the same as the required result on replacing n by m.

For the case where the eigenfunctions may be complex see Problem 5.129.

(b) In order to understand the limitations for the expansion in (a) it is useful to use the analogy of vectors in 3 dimensions as indicated on page 159. In 3 dimensions **i**, **j**, **k** represent unit vectors. If we wish to expand an arbitrary 3-dimensional vector in terms of **i**, **j**, **k** we seek constants A_1, A_2, A_3 such that

$$\mathbf{A} = A_1\mathbf{i} + A_2\mathbf{j} + A_3\mathbf{k}$$

We find
$$A_1 = \mathbf{A}\cdot\mathbf{i}, \quad A_2 = \mathbf{A}\cdot\mathbf{j}, \quad A_3 = \mathbf{A}\cdot\mathbf{k}$$
as the solution of this problem.

Now if the only vectors given to us were **i** and **j** [with **k** not present] it would not be possible to expand the general 3-dimensional vector **A** in terms of **i** and **j**, i.e. we could not have

$$\mathbf{A} = A_1\mathbf{i} + A_2\mathbf{j}$$

We say in such case that the set of unit vectors **i** and **j** are not a *complete set* since we have in fact omitted **k**.

In an analogous manner it is necessary for us to have a *complete set* of *eigenfunctions* in order to be able to write the expansion in (a).

5.42. Illustrate the results of Problems 5.39-5.41 by considering the Sturm-Liouville system

$$\Delta^2 y_{k-1} + \lambda y_k = 0, \qquad y_0 = 0, \quad y_{N+1} = 0$$

The system is a special case of equations (33) and (34) on page 158 with

$$p_{k-1} = 1, \quad q_k = 0, \quad r_k = 1, \quad \alpha_0 = 1, \quad \alpha_1 = 0, \quad \alpha_N = 0, \quad \alpha_{N+1} = 1$$

The equation can be written

$$y_{k+1} - (2-\lambda)y_k + y_{k-1} = 0 \tag{1}$$

Letting $y_k = r^k$ we find that

$$r^2 - (2-\lambda)r + 1 = 0$$

so that

$$r = \frac{2-\lambda \pm \sqrt{(2-\lambda)^2 - 4}}{2} \tag{2}$$

It is convenient to write

$$2 - \lambda = 2\cos\theta \tag{3}$$

so that (2) becomes

$$r = \cos\theta \pm i\sin\theta \tag{4}$$

Then the general solution of the equation is

$$y_k = c_1 \cos k\theta + c_2 \sin k\theta \tag{5}$$

From the first boundary condition $y_0 = 0$ we have $c_1 = 0$ so that

$$y_k = c_2 \sin k\theta \tag{6}$$

From the second boundary condition $y_{N+1} = 0$ we have since $c_2 \neq 0$

$$\sin(N+1)\theta = 0 \tag{7}$$

From this we have

$$(N+1)\theta = n\pi \quad \text{or} \quad \theta = \frac{n\pi}{N+1} \qquad n = 1, 2, 3, \ldots \tag{8}$$

where we have omitted $n = 0$ since this leads to the trivial solution $y_k = 0$ and we have omitted $n = -1, -2, -3, \ldots$ since these lead to essentially the same solutions as $n = 1, 2, 3, \ldots$. From (3) and (8) we see that

$$\lambda = 2\left(1 - \cos\frac{n\pi}{N+1}\right) = 4\sin^2\frac{n\pi}{2(N+1)} \qquad n = 1, 2, 3, \ldots$$

Since these values start to repeat after $n = N$, there will be only N different values given by

$$\lambda_n = 4\sin^2\frac{n\pi}{2(N+1)} \qquad n = 1, 2, \ldots, N \tag{9}$$

These represent the eigenvalues of the Sturm-Liouville system. The corresponding eigenfunctions are obtained from (6) as

$$y_k = c_2 \sin\frac{kn\pi}{N+1}$$

From the orthogonality property of Problem 5.39 we would have

$$\sum_{k=1}^{N} \sin\frac{km\pi}{N+1}\sin\frac{kn\pi}{N+1} = 0 \qquad m \neq n \tag{10}$$

which can in fact be shown directly [see Problem 5.131(*a*)].

Similarly if we are given a function F_k we can obtain the expansion

$$F_k = \sum_{m=1}^{N} c_m \sin\frac{km\pi}{N+1} \tag{11}$$

where

$$c_m = \frac{2}{N+1}\sum_{k=1}^{N} F_k \sin\frac{km\pi}{N+1} \tag{12}$$

These can also be shown directly [see Problem 5.131(*b*),(*c*)].

NONLINEAR DIFFERENCE EQUATIONS

5.43. Solve $y_{k+1} - y_k + ky_{k+1}y_k = 0, \quad y_1 = 2.$

Dividing by $y_k y_{k+1}$, the equation can be written as

$$\frac{1}{y_{k+1}} - \frac{1}{y_k} = k \tag{1}$$

Then letting $1/y_k = v_k$ the difference equation to be solved is

$$v_{k+1} - v_k = k, \qquad v_1 = \tfrac{1}{2} \tag{2}$$

The solution to the linear difference equation in (2) is

$$v_k = \Delta^{-1}k = \frac{k^{(2)}}{2} + c_1 = \frac{k(k-1)}{2} + c_1 \tag{3}$$

and using the fact that $v_1 = \frac{1}{2}$ we find $c_1 = \frac{1}{2}$. Thus

$$v_k = \frac{k(k-1)}{2} + \frac{1}{2} = \frac{k^2-k+1}{2} \tag{4}$$

and the required solution is

$$y_k = \frac{2}{k^2-k+1} \tag{5}$$

5.44. Solve $y_{k+2}y_k^2 = y_{k+1}^3$ if $y_1 = 1, \; y_2 = 2.$

Taking logarithms of the given difference equation it becomes

$$\ln y_{k+2} - 3\ln y_{k+1} + 2\ln y_k = 0 \tag{1}$$

Letting

$$v_k = \ln y_k \tag{2}$$

the equation (1) can be written as a linear difference equation

$$v_{k+2} - 3v_{k+1} + 2v_k = 0, \qquad v_1 = 0, \; v_2 = \ln 2 \tag{3}$$

The general solution to the difference equation in (3) is

$$v_k = c_1 + c_2 2^k \tag{4}$$

Using the conditions in (3) we find $c_1 = -\ln 2,\ c_2 = \frac{1}{2}\ln 2$ so that

$$v_k = (\ln 2)(2^{k-1} - 1) \tag{5}$$

The required solution of the given difference equation is thus

$$\ln y_k = (\ln 2)(2^{k-1} - 1)$$

or

$$y_k = e^{(\ln 2)(2^{k-1}-1)}$$

which can be written

$$y_k = 2^{2^{k-1}-1} \tag{6}$$

SIMULTANEOUS DIFFERENCE EQUATIONS

5.45. Solve the system $\begin{cases} y_{k+1} + z_k - 3y_k = k \\ 3y_k + z_{k+1} - 5z_k = 4^k \end{cases}$ subject to the conditions $y_1 = 2, z_1 = 0$.

Write the system of equations as

$$\left.\begin{aligned} (E-3)y_k + z_k &= k \\ 3y_k + (E-5)z_k &= 4^k \end{aligned}\right\} \tag{1}$$

Operating on the first equation in (1) with $E - 5$ and leaving the second equation alone the system (1) can be written

$$\left.\begin{aligned} (E-5)(E-3)y_k + (E-5)z_k &= 1 - 4k \\ 3y_k + (E-5)z_k &= 4^k \end{aligned}\right\} \tag{2}$$

Subtracting the equations in (2) yields

$$(E^2 - 8E + 12)y_k = 1 - 4k - 4^k \tag{3}$$

The general solution of (3) found by any of the usual methods is

$$y_k = c_1 2^k + c_2 6^k + \frac{1}{4}\cdot 4^k - \frac{4}{5}k - \frac{19}{25} \tag{4}$$

and so from the first equation in (1)

$$z_k = c_1 2^k - 3c_2 6^k - \frac{1}{4}\cdot 4^k - \frac{3}{5}k - \frac{34}{25} \tag{5}$$

Putting $k = 1$ in (4) and (5) we have since $y_1 = 2,\ z_1 = 0$

$$2c_1 + 6c_2 = \frac{64}{25}, \qquad 2c_1 - 18c_2 = \frac{74}{25} \tag{6}$$

from which $c_1 = 133/100,\ c_2 = -1/60$.

Thus the required solution is

$$y_k = \frac{133}{100}2^k - \frac{1}{60}6^k + 4^{k-1} - \frac{4}{5}k - \frac{19}{25} \tag{7}$$

$$z_k = \frac{133}{100}2^k + \frac{1}{20}6^k - 4^{k-1} - \frac{3}{5}k - \frac{34}{25} \tag{8}$$

MIXED DIFFERENCE EQUATIONS

5.46. Solve the differential-difference equation

$$y'_{k+1}(t) = y_k(t), \qquad y_0(t) = t, \qquad y_k(0) = k$$

Putting $k = 0$ we find

$$y'_1(t) = y_0(t) = t \quad \text{or} \quad y_1(t) = \frac{t^2}{2} + c_1$$

Then since $y_1(0) = 1$, we have $c_1 = 1$ so that

$$y_1(t) = 1 + \frac{t^2}{2}$$

Putting $k = 1$ we find

$$y_2'(t) = y_1(t) = 1 + \frac{t^2}{2} \qquad \text{or} \qquad y_2(t) = t + \frac{t^3}{3!} + c_2$$

Then since $y_2(0) = 2$, $c_2 = 2$ so that

$$y_2(t) = 2 + t + \frac{t^3}{3!}$$

Continuing in this manner we find

$$y_3(t) = 3 + 2t + \frac{t^2}{2!} + \frac{t^4}{4!}$$

$$y_4(t) = 4 + 3t + \frac{2t^2}{2!} + \frac{t^3}{3!} + \frac{t^5}{5!}$$

$$y_5(t) = 5 + 4t + \frac{3t^2}{2!} + \frac{2t^3}{3!} + \frac{t^4}{4!} + \frac{t^6}{6!}$$

and in general

$$y_n(t) = n + (n-1)t + \frac{(n-2)t^2}{2!} + \frac{(n-3)t^3}{3!} + \cdots + \frac{t^{n-1}}{(n-1)!} + \frac{t^{n+1}}{(n+1)!}$$

which is the required solution.

5.47. Solve Problem 5.46 by using the method of generating functions.

Let the generating function of $y_k(t)$ be

$$Y(s,t) = \sum_{k=0}^{\infty} y_k(t)s^k \tag{1}$$

Then from the difference equation,

$$\sum_{k=0}^{\infty} y_{k+1}'(t)s^k = \sum_{k=0}^{\infty} y_k(t)s^k \tag{2}$$

which can be written in terms of (1) as

$$\frac{1}{s}\frac{\partial}{\partial t}\{Y(s,t) - y_0(t)\} = Y(s,t) \tag{3}$$

or since $y_0(t) = t$

$$\frac{\partial Y}{\partial t} - sY = 1 \tag{4}$$

Putting $t = 0$ in (1) and using $y_k(0) = k$ yields

$$Y(s,0) = \sum_{k=0}^{\infty} ks^k \tag{5}$$

Equation (4) can be written as

$$\frac{1}{s}\frac{\partial}{\partial t}\ln(sY+1) = 1 \qquad \text{or} \qquad \ln(sY+1) = st + c_1$$

so that

$$Y(s,t) = \frac{ce^{st}-1}{s} \tag{6}$$

Using (5) and (6) we find

$$\begin{aligned} Y(s,t) &= \frac{1}{s}\left\{e^{st}\left(1 + \sum_{k=0}^{\infty} ks^{k+1}\right) - 1\right\} \\ &= \frac{1}{s}\left\{\sum_{p=0}^{\infty}\frac{(st)^p}{p!}\left(1 + \sum_{k=0}^{\infty} ks^{k+1}\right) - 1\right\} \\ &= \frac{1}{s}\left\{\sum_{p=1}^{\infty}\frac{(st)^p}{p!} + \sum_{p=0}^{\infty}\sum_{k=0}^{\infty}\frac{ks^{p+k+1}t^p}{p!}\right\} \end{aligned} \tag{7}$$

Since the coefficient of s^n in this expansion is $y_n(t)$ we find it to be

$$y_n(t) = \frac{t^{n+1}}{(n+1)!} + \sum_{p=0}^{n} \frac{(n-p)t^p}{p!} \tag{8}$$

This is obtained by taking the term corresponding to $p = n+1$ in the first summation of (7) and $p+k+1 = n+1$ in the second summation of (7).

The required solution (8) with n replaced by k agrees with that of Problem 5.46.

PARTIAL DIFFERENCE EQUATIONS

5.48. Write the partial difference equation $u(x+h, y) - 4u(x, y+l) = 0$ (a) in terms of the operators E_1, E_2 and (b) in terms of subscript notation.

(a) Since $E_1u(x,y) = u(x+h, y)$, $E_2u(x,y) = u(x, y+l)$, the given equation can be written as

$$E_1u - 4E_2u = 0 \quad \text{or} \quad (E_1 - 4E_2)u = 0 \tag{1}$$

where we have denoted $u(x,y)$ briefly by u.

(b) In subscript notation

$$u(x+h, y) = u_{k+1, m}, \quad u(x, y+l) = u_{k, m+1}, \quad u(x, y) = u_{k,m}$$

and the given equation can be written as

$$u_{k+1, m} - 4u_{k, m+1} = 0 \quad \text{or} \quad (E_1 - 4E_2)u_{k,m} = 0 \tag{2}$$

5.49. Solve the difference equation of Problem 5.48.

We can write equation (2) of Problem 5.48 as

$$E_1u_{k,m} = 4E_2u_{k,m} \tag{1}$$

Considering m as fixed (1) can be thought of as a first order linear difference equation with constant coefficients in the single variable k. Calling $u_{k,m} = U_k$, $4E_2 = a$ the equation becomes

$$EU_k = aU_k \tag{2}$$

A solution of (2) is

$$U_k = a^kC \tag{3}$$

where C is an arbitrary constant as far as k is concerned but may depend on m, i.e. it is an arbitrary function of m. We can thus write (3) as

$$U_k = a^kC_m \tag{4}$$

Restoring the former notation, i.e. $U_k = u_{k,m}$, $a = 4E_2$, (4) becomes

$$u_{k,m} = (4E_2)^kC_m$$

But this can be written as

$$u_{k,m} = 4^kE_2^kC_m = 4^kC_{m+k} \tag{5}$$

To obtain this in terms of x and y we use the fact that $x = a + kh$, $y = b + ml$. Then (5) becomes

$$u(x,y) = 4^{(x-a)/h}C\left(\frac{x-a}{h} + \frac{y-b}{l}\right)$$

which can be written

$$u(x,y) = 4^{x/h}H\left(\frac{x}{h} + \frac{y}{l}\right) \tag{6}$$

where H is an arbitrary function.

We can show that (6) satisfies the equation of Problem 5.48. Then since it involves one arbitrary function it is the general solution.

5.50. Work Problem 5.49 by finding solutions of the form $\lambda^k \mu^m$ where λ, μ are arbitrary constants.

Substituting the assumed solution

$$u_{k,m} = \lambda^k \mu^m$$

in the given difference equation (*2*) of Problem 5.48 we find

$$\lambda^{k+1}\mu^m - 4\lambda^k\mu^{m+1} = 0 \quad \text{or} \quad \lambda - 4\mu = 0$$

on dividing by $\lambda^k\mu^m \neq 0$.

From $\lambda = 4\mu$ it follows that a solution is given by

$$\lambda^k\mu^m = (4\mu)^k\mu^m = 4^k\mu^{k+m}$$

Since sums of these solutions over any values of μ are also solutions, we are led to the solutions

$$\sum_\mu 4^k\mu^{k+m} = 4^k\sum_\mu \mu^{k+m} = 4^kF(k+m)$$

where F is an arbitrary function. Thus we are led to the same result as given in Problem 5.49 where the arbitrary function $F(k+m) = C_{k+m}$.

5.51. Solve $u(x+1, y) - 4u(x, y+1) = 0, \quad u(0, y) = y^2$.

The difference equation is the same as that of Problem 5.48 with $h = 1$, $l = 1$. Thus the general solution is from equation (*6*) of Problem 5.49,

$$u(x, y) = 4^xH(x+y) \tag{1}$$

Using the boundary condition $u(0, y) = y^2$ we have from (*1*)

$$u(0, y) = H(y) = y^2$$

Thus

$$H(x+y) = (x+y)^2$$

and the required solution is

$$u(x, y) = 4^x(x+y)^2 \tag{2}$$

5.52. Solve $u(x+1, y) - 2u(x, y+1) = 3u(x, y)$.

We can write the equation in subscript notation as

$$E_1u_{k,m} - 2E_2u_{k,m} = 3u_{k,m} \quad \text{or} \quad (E_1 - 2E_2 - 3)u_{k,m} = 0$$

Rewriting it as

$$E_1u_{k,m} = (2E_2 + 3)u_{k,m}$$

we can consider it as a first order linear difference equation with constant coefficients as in Problem 5.49 having solution

$$u_{k,m} = (2E_2+3)^kC_m$$

This can be written as

$$u_{k,m} = 3^k(1 + \tfrac{2}{3}E_2)^kC_m$$

or on expanding by the binomial theorem,

$$\begin{aligned} u_{k,m} &= 3^k\left[1 + k(\tfrac{2}{3}E_2) + \frac{k(k-1)}{2!}(\tfrac{2}{3}E_2)^2 + \frac{k(k-1)(k-2)}{3!}(\tfrac{2}{3}E_2)^3 + \cdots\right]C_m \\ &= 3^k[C_m + \tfrac{2}{3}kC_{m+1} + \tfrac{2}{9}k(k-1)C_{m+2} + \tfrac{4}{81}k(k-1)(k-2)C_{m+3} + \cdots] \end{aligned}$$

To obtain the result in terms of x and y we can use $x = a + kh$, $y = b + ml$, choosing $h = 1$, $l = 1$, $a = 0$, $b = 0$ so that $k = x$, $m = y$. Then writing $C_m = H(y)$ we obtain

$$u(x, y) = 3^x[H(y) + \tfrac{2}{3}xH(y+1) + \tfrac{2}{9}x(x-1)H(y+2) + \tfrac{4}{81}x(x-1)(x-2)H(y+3) + \cdots]$$

This contains one arbitrary function $H(y)$ and is the general solution.

5.53. Solve $u_{k,m} = pu_{k-1,m} + qu_{k,m-1}$ if $u_{k,0} = 0$ for $k > 0$, $u_{0,m} = 1$ for $m > 0$ and p, q are given positive constants such that $p + q = 1$.

In operator notation the given equation can be written

$$(pE_1^{-1} + qE_2^{-1} - 1)u_{k,m} = 0$$

This can also be written as

$$pE_1^{-1}u_{k,m} = (1 - qE_2^{-1})u_{k,m} \quad \text{or} \quad E_1 u_{k,m} = \frac{p}{1 - qE_2^{-1}}\, u_{k,m}$$

Considering this last equation as one in which m is fixed the solution is

$$\begin{aligned} u_{k,m} &= \left(\frac{p}{1 - qE_2^{-1}}\right)^k C_m \\ &= p^k(1 - qE_2^{-1})^{-k}C_m \\ &= p^k\left[1 + k(qE_2^{-1}) + \frac{k(k+1)}{2!}(qE_2^{-1})^2 + \frac{k(k+1)(k+2)}{3!}(qE_2^{-1})^3 + \cdots\right]C_m \end{aligned}$$

or

$$u_{k,m} = p^k\left[C_m + kqC_{m-1} + \frac{k(k+1)}{2!}q^2C_{m-2} + \frac{k(k+1)(k+2)}{3!}q^3C_{m-3} + \cdots\right]$$

Now if $m = 0$, $k > 0$, this becomes

$$0 = p^k\left[C_0 + kqC_{-1} + \frac{k(k+1)}{2!}q^2C_{-2} + \cdots\right]$$

from which

$$C_0 = 0, \quad C_{-1} = 0, \quad C_{-2} = 0, \quad \ldots$$

If $k = 0$, $m > 0$, it becomes

$$C_m = 1, \quad m > 0$$

Then for all $k > 0$, $m > 0$,

$$u_{k,m} = p^k\left[1 + kq + \frac{k(k+1)}{2!}q^2 + \cdots + \frac{k(k+1)\cdots(k+m-2)}{(m-1)!}q^{m-1}\right]$$

5.54. Solve Problem 5.53 by the method of generating functions.

Let

$$G_k(t) = u_{k,0} + u_{k,1}t + u_{k,2}t^2 + \cdots = \sum_{m=0}^{\infty} u_{k,m}t^m \tag{1}$$

Then multiplying the given difference equation by t^m and summing from $m = 1$ to ∞ we have

$$\sum_{m=1}^{\infty} u_{k,m}t^m = p\sum_{m=1}^{\infty} u_{k-1,m}t^m + q\sum_{m=1}^{\infty} u_{k,m-1}t^m$$

This can be written in terms of the generating function (*1*) as

$$G_k(t) - u_{k,0} = p[G_{k-1}(t) - u_{k-1,0}] + qtG_k(t) \tag{2}$$

Since $u_{k,0} = 0$ and $u_{k-1,0} = 0$ for $k \geqq 1$ and $u_{0,0} = 0$, (*2*) becomes

$$G_k(t) = pG_{k-1}(t) + qtG_k(t) \tag{3}$$

or

$$G_k(t) = \frac{p}{1 - qt}G_{k-1}(t) \tag{4}$$

This equation has solution

$$G_k(t) = \left(\frac{p}{1 - qt}\right)^k G_0(t) \tag{5}$$

and since

$$G_0(t) = u_{0,0} + u_{0,1}t + u_{0,2}t^2 + \cdots = t + t^2 + t^3 + \cdots = \frac{t}{1 - t}$$

(*5*) becomes

$$G_k(t) = \frac{p^k t}{(1 - qt)^k(1 - t)} \tag{6}$$

We can write this as

$$G_k(t) \quad = \quad p^k(1-qt)^{-k}\frac{t}{1-t}$$

and expand into a power series in t to obtain

$$G_k(t) \quad = \quad p^k\left[1 + kqt + \frac{k(k+1)}{2!}q^2t^2 + \frac{k(k+1)(k+2)}{3!}q^3t^3\right][t + t^2 + t^3 + \cdots]$$

The coefficient of t^m in this product is then given by

$$u_{k,m} \quad = \quad p^k\left[1 + kq + \frac{k(k+1)}{2!}q^2 + \cdots + \frac{k(k+1)\cdots(k+m-2)}{(m-1)!}q^{m-1}\right]$$

which is the required solution in agreement with that of Problem 5.53.

5.55. Solve $u(x+1, y) - 4u(x, y+1) \;=\; 6x^2y + 4$.

By Problem 5.49 the equation with the right hand side replaced by zero, i.e. the homogeneous equation

$$u(x+1, y) - 4u(x, y+1) \;=\; 0 \qquad \text{or} \qquad (E_1 - 4E_2)u(x, y) \;=\; 0$$

has the solution

$$4^xH(x+y)$$

where H is an arbitrary function. Then to solve the complete equation we need only find a particular solution.

Method 1, using undetermined coefficients.

Assume a particular solution having the form

$$u(x, y) \;=\; A_1x^2y + A_2xy + A_3x^2 + A_4x + A_5y + A_6$$

where $A_1, \ldots, A_6$ are constants. Substitution in the complete equation yields

$$-3A_1x^2y + (2A_1 - 3A_2)xy - (4A_1 + 3A_3)x^2 + (2A_3 - 3A_4 - 4A_2)x$$
$$+ (A_1 + A_2 - 3A_5)y + A_3 + A_4 - 4A_5 - 3A_6 \;=\; 6x^2y + 4$$

Then equating coefficients of corresponding terms

$$-3A_1 = 6,\;\; 2A_1 - 3A_2 = 0,\;\; 4A_1 + 3A_3 = 0,\;\; 2A_3 - 3A_4 - 4A_2 = 0,\;\; A_1 + A_2 - 3A_5 = 0,$$
$$A_3 + A_4 - 4A_5 - 3A_6 = 4$$

so that

$$A_1 = -2,\;\; A_2 = -\frac{4}{3},\;\; A_3 = \frac{8}{3},\;\; A_4 = \frac{32}{9},\;\; A_5 = -\frac{10}{9},\;\; A_6 = \frac{20}{9}$$

Then the required solution is

$$u(x, y) \;=\; 4^xH(x+y) - 2x^2y - \frac{4}{3}xy + \frac{8}{3}x^2 + \frac{32}{9}x - \frac{10}{9}y + \frac{20}{9} \tag{1}$$

Method 2, using operators.

Since the complete equation is $(E_1 - 4E_2)u(x, y) \;=\; 6x^2y + 4$ a particuar solution is, using $E_1 = 1 + \Delta_1$, $E_2 = 1 + \Delta_2$,

$$\begin{aligned}
\frac{1}{E_1 - 4E_2}(6x^2y+4) &= \frac{1}{-3 + \Delta_1 - 4\Delta_2}(6x^2y+4) \\
&= -\frac{1}{3}\frac{1}{1 - (\Delta_1 - 4\Delta_2)/3}(6x^2y+4) \\
&= -\frac{1}{3}\left[1 + \frac{\Delta_1 - 4\Delta_2}{3} + \frac{(\Delta_1 - 4\Delta_2)^2}{9} + \cdots\right](6x^2y+4) \\
&= -2x^2y - \frac{4}{3}xy + \frac{8}{3}x^2 + \frac{32}{9}x - \frac{10}{9}y + \frac{20}{9}
\end{aligned}$$

and so the required solution is the same as that obtained in equation (*1*) of Method 1.

MISCELLANEOUS PROBLEMS

5.56. Solve the partial differential equation

$$\frac{\partial u}{\partial x} + 2\frac{\partial u}{\partial y} = 0, \qquad u(x,0) = x^3 - 3\sin x$$

The given partial differential equation is the limit as $h \to 0,\ l \to 0$ of

$$\frac{u(x+h,y)-u(x,y)}{h} + 2\left[\frac{u(x,y+l)-u(x,y)}{l}\right] = 0$$

which can be written in operator and subscript notation as

$$(lE_1 + 2hE_2 - l - 2h)u_{k,m} = 0 \quad \text{or} \quad E_1 u_{k,m} = \left(1 + \frac{2h}{l} - \frac{2h}{l}E_2\right)u_{k,m}$$

This can be solved as in Problem 5.49 to yield

$$u_{k,m} = \left(1 + \frac{2h}{l} - \frac{2h}{l}E_2\right)^k C_m$$

or in terms of x and y as

$$u(x,y) = \left(1 + \frac{2h}{l} - \frac{2h}{l}E_2\right)^{x/h} H\left(\frac{y}{l}\right) \tag{1}$$

In order to obtain meaningful results as $h \to 0,\ l \to 0$ choose $1 + 2h/l = 0$ so that (1) becomes

$$u(x,y) = H\left(\frac{y-2x}{l}\right) = J(2x-y)$$

where J is an arbitrary function. The required general solution is thus

$$u(x,y) = J(2x-y) \tag{2}$$

Since
$$u(x,0) = J(2x) = x^3 - 3\sin x$$

we have
$$J(x) = \tfrac{1}{8}x^3 - 3\sin(x/2)$$

Then from (2) we have
$$u(x,y) = \tfrac{1}{8}(2x-y)^3 - 3\sin\left(\frac{2x-y}{2}\right)$$

5.57. Solve the mixed difference equation

$$u(x+1,y) - u(x,y) = 2\frac{\partial u}{\partial y}$$

The given equation can be written as

$$E_1 u - u = 2D_2 u \tag{1}$$

or
$$E_1 u = (1+2D_2)u \tag{2}$$

Then treating $1 + 2D_2$ as a constant we find as in Problem 5.49

$$u(x,y) = (1+2D_2)^x H(y) \tag{3}$$

where $H(y)$ is an arbitrary function of y. Expanding by the binomial theorem we then find

$$\begin{aligned} u(x,y) &= \left[1 + \binom{x}{1}2D_2 + \binom{x}{2}(2D_2)^2 + \cdots + \binom{x}{m}(2D_2)^m + \cdots\right]H(y) \\ &= H(y) + 2\binom{x}{1}H'(y) + 2^2\binom{x}{2}H''(y) + \cdots + 2^m\binom{x}{m}H^{(m)}(y) + \cdots \end{aligned}$$

Then if x is a positive integer the series terminates and we have

$$u(x,y) = H(y) + 2\binom{x}{1}H'(y) + 2^2\binom{x}{2}H''(y) + \cdots + 2^x\binom{x}{x}H^{(x)}(y) \tag{4}$$

5.58. Solve the preceding problem subject to $u(0, y) = y^3$.

From (4) of Problem 5.57 we see that

$$u(0, y) = H(y) = y^3$$

Thus

$$H'(y) = 3y^2, \quad H''(y) = 6y, \quad H'''(y) = 6, \quad H^{(IV)}(y) = 0, \quad \ldots$$

and so

$$u(x, y) = y^3 + 6\binom{x}{1}y^2 + 24\binom{x}{2}y + 48\binom{x}{3}$$

$$= y^3 + 6xy^2 + 12x(x-1)y + 8x(x-1)(x-2)$$

5.59. Find the general solution of

$$u_{k,m} = pu_{k+1,m-1} + qu_{k-1,m+1}$$

where p and q are given constants such that $p + q = 1$.

Method 1.

Assume a solution of the form

$$u_{k,m} = \lambda^k \mu^m$$

Then by substitution in the given difference equation we find after dividing by $\lambda^k \mu^m$

$$p\lambda^2 - \lambda\mu + q\mu^2 = 0$$

from which

$$\lambda = \frac{\mu \pm \sqrt{\mu^2 - 4pq\mu^2}}{2p} = \frac{\mu \pm \mu\sqrt{1-4pq}}{2} = \mu, \frac{\mu q}{p}$$

Thus solutions are given by $\mu^k \cdot \mu^m = \mu^{k+m}$ and $(\mu q/p)^k \mu^m = \mu^{k+m}(q/p)^k$. Since sums of such solutions are also solutions, it follows that $F(k+m)$ and $(q/p)^k G(k+m)$ are solutions and the general solution is

$$u_{k,m} = F(k+m) + (q/p)^k G(k+m) \tag{1}$$

where F and G are arbitrary functions.

Method 2.

It is noted that the sum of the subscripts in the given difference equation is a constant. If we denote this by c we have $k + m = c$. The equation is thus given by

$$u_{k,\,c-k} = pu_{k+1,\,c-(k+1)} + qu_{k-1,\,c-(k-1)} \tag{2}$$

If we let

$$u_{k,\,c-k} = v_k \tag{3}$$

(2) becomes

$$v_k = pv_{k+1} + qv_{k-1} \tag{4}$$

which involves only one subscript and can be solved by the usual methods. We find

$$v_k = c_1 + c_2(q/p)^k \tag{5}$$

Now the constants c_1 and c_2 can be considered as arbitrary functions of the constant $c = k + m$: $c_1 = F(k+m)$, $c_2 = G(k+m)$. Thus we obtain from (3) and (5)

$$u_{k,m} = F(k+m) + G(k+m)(q/p)^k$$

Supplementary Problems

DIFFERENTIAL EQUATIONS AND RELATIONSHIPS TO DIFFERENCE EQUATIONS

5.60. (*a*) Show that the general solution of the differential equation

$$\frac{dy}{dx} - y = x^2$$

is $y = ce^x - x^2 - 2x - 2$.

(*b*) Find that particular solution to the equation in (*a*) which satisfies the condition $y(0) = 3$.

5.61. (*a*) Show that the equation

$$\frac{\Delta y}{\Delta x} - y = x^{(2)}$$

[compare Problem 5.60] where $y = f(x)$, $\Delta x = h$ can be written as

$$f(x+h) - (1+h)f(x) = hx^2 - h^2x$$

(*b*) Show that the general solution of the equation in (*a*) is

$$y(x) = f(x) = C(x)(1+h)^{x/h} - x^2 + (h-2)x - 2$$

where $C(x+h) = C(x)$.

(*c*) Find that particular solution for which $y(0) = 3$.

5.62. Explain clearly the relationship between the difference equation and the differential equation of Problems 5.60 and 5.61 discussing in particular the limiting case as $h \to 0$.

5.63. (*a*) Show that the general solution of

$$\frac{d^2y}{dx^2} + y = 3x^2 - 5x + 4$$

is $y = c_1 \cos x + c_2 \sin x + 3x^2 - 5x - 2$.

(*b*) Find that particular solution to the equation in (*a*) which satisfies the conditions $y(0) = 2$, $y'(0) = 0$.

5.64. (*a*) Show that the equation

$$\frac{\Delta^2 y}{\Delta x^2} + y = 3x^{(2)} - 5x^{(1)} + 4$$

can be written as

$$f(x+2h) - 2f(x+h) + (1+h^2)f(x) = 3h^2x^2 - (3h^3 + 5h^2)x + 4h^2$$

where $y = f(x)$.

(*b*) Show that the general solution of the equation in (*a*) is

$$y = f(x) = C_1(x)(1+ih)^{x/h} + C_2(x)(1-ih)^{x/h} + 3x^2 - (3h+5)x - 2$$

where $C_1(x)$ and $C_2(x)$ are periodic constants with period h.

5.65. (*a*) Explain clearly the relationships between Problems 5.63 and 5.64, in particular the limiting case as $h \to 0$. (*b*) What are the boundary conditions for the difference equation in Problem 5.64 which correspond to those for the differential equation in Problem 5.63? (*c*) Obtain the particular solution of Problem 5.64 corresponding to the boundary conditions just found in (*b*), and explain the relationship between this and the particular solution of Problem 5.63.

THE SUBSCRIPT NOTATION FOR DIFFERENCE EQUATIONS

5.66. Write the difference equation of (*a*) Problem 5.61 and (*b*) Problem 5.64 with subscript notation.

5.67. Write each of the following in subscript notation.

(*a*) $f(x+3h) - 3f(x+2h) + 3f(x+h) - f(x) = x^{(3)} - 2x^{(2)}$

(*b*) $f(x+4h) + f(x) = \cos x$

5.68. Change each of the following equations to one involving the "x notation".

(a) $2y_{k+2} - 3y_{k+1} + y_k = 0$

(b) $y_{k+3} + 4y_{k+2} - 5y_{k+1} + 2y_k = k^2 - 4k + 1$

LINEARLY INDEPENDENT FUNCTIONS

5.69. Determine which of the following sets of functions are linearly dependent and which are linearly independent.

(a) $2k,\ k-3$

(b) $k+4,\ k-2,\ 2k+1$

(c) $k^2-2k,\ 3k,\ 4$

(d) $2^k,\ 2^{2k},\ 2^{3k}$

(e) $1,\ k,\ k^2,\ k^3,\ k^4$

(f) $k^2-3k+2,\ k^2+5k-4,\ 2k^2,\ 3^k$

(g) $2^k \cos k\theta,\ 2^k \sin k\theta$

(h) $k^2,\ (k+1)^2,\ (k+2)^2,\ (k+3)^2$

5.70. Work Problem 5.69 using Theorem 5-1, page 154.

5.71. Show that if 0 is added to any linearly independent set of functions then the new set is linearly dependent.

5.72. Prove Theorem 5-1, page 154, for any number of functions.

5.73. Prove Theorem 5-2, page 155.

5.74. If $a \neq b$ show that a^k and b^k are linearly independent.

HOMOGENEOUS LINEAR DIFFERENCE EQUATIONS WITH CONSTANT COEFFICIENTS

5.75. Find the solution of each of the following subject to given conditions (if any).

(a) $y_{k+2} + 5y_{k+1} + 6y_k = 0$

(b) $y_{k+2} - 3y_{k+1} + 2y_k = 0,\ y_0 = 1,\ y_1 = 2$

(c) $3y_{k+2} + 8y_{k+1} + 4y_k = 0$

(d) $y_{k+2} - 9y_k = 0,\ y_0 = 2,\ y_1 = -1$

(e) $2y_{k+1} + 3y_k = 0,\ y_0 = 4$

(f) $y_{k+2} + 2y_{k+1} + 2y_k = 0,\ y_0 = 0,\ y_1 = -1$

(g) $y_{k+2} + 16y_k = 0,\ y_0 = 0,\ y_1 = 1$

(h) $4y_{k+2} + 25y_k = 0$

(i) $y_{k+2} - 6y_{k+1} + 9y_k = 0$

(j) $4y_{k+2} + 4y_{k+1} + y_k = 0$

5.76. Solve $y_{k+3} - 6y_{k+2} + 11y_{k+1} - 6y_k = 0$ subject to the conditions $y_0 = 0,\ y_1 = 1,\ y_2 = 1$.

5.77. Solve $y_{k+4} + y_k = 0$.

5.78. Solve $4y_{k+4} - 25y_k = 0$.

5.79. Solve $y_{k+3} - 8y_k = 0$.

5.80. Solve $y_{k+4} + 12y_{k+2} - 64y_k = 0$.

5.81. Solve $y_{k+3} - 3y_{k+2} + 4y_{k+1} - 2y_k = 0$.

5.82. Find the general solution corresponding to a linear homogeneous difference equation if the roots of the corresponding auxiliary equation are given by $2, -2, -3 \pm 4i, -2, -2, 4, -2, 2$.

5.83. What is the difference equation of Problem 5.82?

5.84. Work Problem 5.14(b) directly by using the difference equation.

METHOD OF UNDETERMINED COEFFICIENTS

5.85. Solve each of the following subject to given conditions if any.

(a) $y_{k+2} - 3y_{k+1} + 2y_k = 4^k$

(b) $y_{k+2} - 4y_{k+1} + 4y_k = k^2$

(c) $4y_{k+2} + y_k = 2k^2 - 5k + 3$

(d) $y_{k+1} - y_k = k,\ y_0 = 1$

(e) $y_{k+2} + 2y_{k+1} + y_k = k + 2^k$

(f) $y_{k+2} - 8y_{k+1} + 16y_k = 3 \cdot 4^k,\ y_0 = 0,\ y_1 = 0$

(g) $2y_{k+2} - 3y_{k+1} + y_k = k^2 - 4k + 3$

(h) $y_{k+2} + 4y_k = 5(-3)^k + 10k$

(i) $y_{k+2} - y_k = 1/3^k$

5.86. Solve $y_{k+1} - \alpha y_k = \beta$ for all values of the constants α and β.

5.87. Solve $y_{k+2} + y_k = 4 \cos 2k$.

5.88. Solve $y_{k+3} - 6y_{k+2} + 11y_{k+1} - 6y_k = 4k + 3 \cdot 2^k - 5^k$.

5.89. Solve $y_{k+4} - 16y_k = k^2 - 5k + 2 - 4 \cdot 3^k$.

5.90. Solve $y_{k+3} + y_k = 2^k \cos 3k$.

5.91. Solve $3y_{k+4} - 4y_{k+2} + y_k = 3^k(k-1)$.

SPECIAL OPERATOR METHODS

5.92. Solve the following difference equations by operator methods.

(a) $y_{k+2} - 3y_{k+1} + 2y_k = 4 \cdot 3^k - 2 \cdot 5^k$
(b) $y_{k+2} + 4y_{k+1} + 4y_k = k^2 - 3k + 5$
(c) $3y_{k+2} - 8y_{k+1} - 3y_k = 3^k - 2k + 1$
(d) $y_{k+2} - 8y_{k+1} + 16y_k = 3 \cdot 4^k - 5(-2)^k$
(e) $2y_{k+2} - 2y_{k+1} + y_k = k^3 - 4k + 5^k$
(f) $y_{k+2} - 4y_{k+1} + 4y_k = k^2 \cdot 2^k$
(g) $y_{k+3} - 3y_{k+2} + 3y_{k+1} - y_k = 2k + 4 - 3^{k-1}$
(h) $y_{k+4} - 16y_k = 3k + 2^k + 3^k$

5.93. Show that (a) $\dfrac{1}{\phi(E)} \cos k\theta = \text{Re}\left\{\dfrac{e^{ik\theta}}{\phi(e^{i\theta})}\right\}$ (b) $\dfrac{1}{\phi(E)} \sin k\theta = \text{Im}\left\{\dfrac{e^{ik\theta}}{\phi(e^{i\theta})}\right\}$

where Re denotes "real part of" and Im denotes "imaginary part of".

5.94. Use Problem 5.93 to solve $y_{k+2} + 4y_{k+1} - 12y_k = 5 \cos(\pi k/3)$.

5.95. (a) Determine $\dfrac{1}{E - a} R(k)$.

(b) Show that $\dfrac{1}{E^2 - 6E + 8} = \dfrac{1}{2}\left(\dfrac{1}{E-4} - \dfrac{1}{E-2}\right)$ i.e. the method of partial fractions can be used for operators.

(c) Show how the results of (a) and (b) can be used to solve the equation

$$y_{k+2} - 6y_{k+1} + 8y_k = k^2 + 3^k$$

5.96. Show how operator methods can be used to solve

$$(E-1)^2(E+2)(E-3)y_k = k^2 - 3k + 5 + 3^k$$

5.97. Solve $(E^2 + 1)^2 y_k = 2k - 3$.

5.98. Solve Problem 5.85(a)-(i) by operator methods.

5.99. Solve (a) Problem 5.87, (b) Problem 5.88, (c) Problem 5.89, (d) Problem 5.90 and (e) Problem 5.91 by operator methods.

METHOD OF VARIATION OF PARAMETERS

5.100. Solve each of the following difference equations by using the method of variation of parameters.

(a) $y_{k+2} - 5y_{k+1} + 6y_k = k^2$
(b) $y_{k+2} - 2y_{k+1} + y_k = 3 + k + 4^k$
(c) $4y_{k+2} - 4y_{k+1} + y_k = 3/2^k$
(d) $y_{k+2} + y_k = 1/k$
(e) $5y_{k+2} - 3y_{k+1} - 2y_k = 3k + (-2)^k$
(f) $y_{k+2} - y_{k+1} + y_k = 1/k!$

5.101. Solve Problem 5.85(*a*)-(*i*) by using variation of parameters.

5.102. Solve Problem 5.92(*a*)-(*h*) by using variation of parameters.

5.103. Solve $y_{k+3} - 6y_{k+2} + 11y_{k+1} - 6y_k = 5 \cdot 2^k + k^2$ by using variation of parameters.

5.104. Verify that the determinant of the coefficients (*24*) on page 157 is equal to the *Casorati* of the system [see (*15*), page 154]. Discuss the significance of the case where the determinant is or is not zero.

METHOD OF GENERATING FUNCTIONS

5.105. Verify each of the following generating functions.

(*a*) $\dfrac{1}{1-t} = \sum_{k=0}^{\infty} t^k$

(*b*) $(1+t)^n = \sum_{k=0}^{n} \binom{n}{k} t^k$

(*c*) $e^t = \sum_{k=0}^{\infty} \dfrac{t^k}{k!}$

(*d*) $-\ln(1-t) = \sum_{k=0}^{\infty} \dfrac{t^{k+1}}{k+1}$

(*e*) $\sin t = \sum_{k=0}^{\infty} \dfrac{(-1)^k t^{2k+1}}{(2k+1)!}$

(*f*) $\cos t = \sum_{k=0}^{\infty} \dfrac{(-1)^k t^{2k}}{(2k)!}$

(*g*) $\dfrac{1}{(1-t)^2} = \sum_{k=0}^{\infty} k t^{k-1}$

(*h*) $\tan^{-1} t = \sum_{k=0}^{\infty} \dfrac{(-1)^k t^{2k+1}}{2k+1}$

5.106. Solve each of the following by the method of generating functions.

(*a*) $y_{k+2} - 6y_{k+1} + 8y_k = 0, \quad y_0 = 0, \ y_1 = 2$

(*b*) $2y_{k+1} - y_k = 1, \quad y_0 = 0,$

(*c*) $y_{k+2} + 4y_{k+1} + 4y_k = 0, \quad y_0 = 1, \ y_1 = 0$

(*d*) $6y_{k+2} - 5y_{k+1} + y_k = 0$

(*e*) $y_{k+2} + 2y_{k+1} + y_k = 0, \quad y_0 = 2, \ y_1 = -1$

(*f*) $y_{k+3} - 3y_{k+2} + 3y_{k+1} - y_k = 0$

(*g*) $y_{k+2} + 2y_{k+1} + y_k = 1 + k$

(*h*) $4y_{k+2} - 4y_{k+1} + y_k = 2^{-k}, \quad y_0 = 0, \ y_1 = 1$

5.107. Solve $y_{k+2} + 3y_{k+1} - 4y_k = \dfrac{1}{k!}, \quad y_0 = 0, \ y_1 = 0.$

5.108. Show how to obtain the generating function of the sequence (*a*) $u_k = k$, (*b*) $u_k = k^2$, (*c*) $u_k = k^3$. [*Hint.* Use Problem 5.105(*a*) and differentiate both sides with respect to *t*.]

5.109. If $G(t)$ and $H(t)$ are the generating functions of u_k and v_k respectively show that $G(t)H(t)$ is the generating function of

$$u_k^* v_k = \sum_{p=0}^{k} u_{k-p} v_p$$

which is called the *convolution* of u_k and v_k.

5.110. Show that

(*a*) $u_k^* v_k = v_k^* u_k$

(*b*) $u_k^* (v_k + w_k) = u_k^* v_k + u_k^* w_k$

(*c*) $u_k^* (v_k^* w_k) = (u_k^* v_k)^* w_k$

What laws of algebra does the convolution satisfy?

LINEAR DIFFERENCE EQUATIONS WITH VARIABLE COEFFICIENTS

5.111. Solve each of the following.

(*a*) $y_{k+1} - k y_k = k, \quad y_1 = 0$

(*b*) $k y_{k+1} + 2y_k = 1$

(*c*) $y_{k+1} + k y_k = k^{(2)}$

(*d*) $y_{k+1} + 4k y_k = k!, \quad y_1 = 1$

(*e*) $k y_{k+1} - y_k = 2^k, \quad y_1 = 0$

5.112. Solve $y_{k+2} - (k+1)y_{k+1} + ky_k = k, \quad y_1 = 0, \; y_2 = 1.$

5.113. Solve $y_{k+2} - (2k+1)y_{k+1} + k^2 y_k = 0.$

5.114. Work Problem 5.113 if the right side is replaced by 2^k, using variation of parameters.

5.115. Solve $(k+3)y_{k+2} - 3(k+2)y_{k+1} + (2k+1)y_k = 0.$

5.116. Solve $(k-1)y_{k+2} - 2ky_{k+1} + \frac{3}{4}(k+1)y_k = 0.$

5.117. Let $Y_k \neq 0$ be any solution of the difference equation

$$y_{k+2} + a_k y_{k+1} + b_k y_k = 0$$

Show that this second order linear difference equation can be reduced to a first order linear difference equation by means of the substitution $y_k = Y_k u_k$ and thus obtain the solution.

5.118. Show that if Y_k is a solution of the equation

$$y_{k+2} + a_k y_{k+1} + b_k y_k = 0$$

then the substitution $y_k = Y_k u_k$ can be used to solve the equation

$$y_{k+2} + a_k y_{k+1} + b_k y_k = R_k$$

5.119. Use Problems 5.117 and 5.118 to solve the equations

$$(a) \quad ky_{k+2} + (1-k)y_{k+1} - 2y_k = 0 \qquad (b) \quad ky_{k+2} + (1-k)y_{k+1} - 2y_k = 1$$

[*Hint.* A solution of (*a*) is $y_k = k$.]

5.120. Discuss the relationship of the methods of Problems 5.117 and 5.118 with the method of variation of parameters.

5.121. Solve

$$y_{k+2} + \left(a + \frac{b}{k+2}\right) y_{k+1} + \frac{ab}{k+1} y_k = 0$$

5.122. Solve

$$a_{k+2} y_{k+2} - (\alpha a_{k+1} + b_{k+1}) y_{k+1} + \alpha b_k y_k = 0$$

where α is a given constant.

5.123. Solve $y_{k+2} + ky_k = k, \; y_0 = 0, \; y_1 = 1$ by the method of generating functions.

5.124. Solve Problem 5.116 by the method of generating functions.

5.125. (*a*) Solve Problem 5.37 by making the substitution $y_k = (k-1)!\, v_k$. (*b*) Show that $\lim_{k \to \infty} y_k = \frac{3}{2} e$.

STURM-LIOUVILLE DIFFERENCE EQUATIONS

5.126. Find eigenvalues and eigenfunctions for the Sturm-Liouville system

$$\Delta^2 y_{k-1} + \lambda y_k = 0, \qquad y_1 = y_0, \; y_{N+1} = y_N$$

5.127. Work Problem 5.126 if the boundary conditions are $y_1 = y_0, \; y_{N+1} = 0$.

5.128. Write the orthogonality conditions for the eigenfunctions of (*a*) Problem 5.126 and (*b*) Problem 5.127.

5.129. (*a*) Show that if the eigenfunctions of a Sturm-Liouville system are not necessarily real then the orthogonality of page 159 must be rewritten as

$$\sum_{k=1}^{N} r_k \bar{\phi}_{m,k} \phi_{n,k} = 0 \qquad m \neq n$$

(*b*) If $F_k = \sum c_m \phi_{m,k}$ determine the coefficients c_m for this case.

5.130. Can any second order linear difference equation

$$a_k \Delta^2 y_k + b_k \Delta y_k + c_k y_k = 0$$

be written as a Sturm-Liouville difference equation? Justify your answer.

5.131. Prove the results (*a*) (*10*), (*b*) (*11*) and (*c*) (*12*) on page 181 directly without using Sturm-Liouville theory.

5.132. Explain how you could solve the equation $\Delta^2 y_{k-1} + \lambda y_k = k$ if $y_0 = 0$, $y_{N+1} = 0$. Discuss the relationship with Sturm-Liouville theory.

NONLINEAR DIFFERENCE EQUATIONS

5.133. Solve $y_k + y_{k+1} = y_k y_{k+1}$. [*Hint.* Let $1/y_k = v_k$.]

5.134. Solve $y_k y_{k+1} = 2y_k + 1$ by letting $y_k = c + (1/v_k)$ and choosing c.

5.135. Solve $y_k y_{k+1} y_{k+2} = y_k + y_{k+1} + y_{k+2}$. [*Hint.* Let $y_k = \tan u_k$.]

5.136. Solve $y_k y_{k+1} y_{k+2} = K(y_k + y_{k+1} + y_{k+2})$ where K is a given constant.

5.137. Solve $y_{k+1}^3 = 3y_k$, $y_0 = 1$. [*Hint.* Take logarithms.]

5.138. Solve (*a*) $y_{k+2} = y_k y_{k+1}$ (*b*) $y_{k+1} = 2y_k^2 - 1$.

5.139. Solve $y_k^2 = 1/y_{k-1} y_{k+1}$.

5.140. Solve $y_{k+1}^2 - 5y_{k+1} y_k + 6y_k^2 = 0$.

5.141. Solve $y_{k+1} = \sqrt{1 - y_k^2}$.

5.142. (*a*) Solve $y_{k+1} = \sqrt{2 + y_k}$, $y_0 = 0$ and (*b*) find $\lim_{k \to \infty} y_k$.

5.143. (*a*) Solve $y_{k+1} = \dfrac{1}{2}\left(y_k + \dfrac{1}{y_k}\right)$, $y_0 = a$ and (*b*) find $\lim_{k \to \infty} y_k$.

SIMULTANEOUS DIFFERENCE EQUATIONS

5.144. Solve $\begin{cases} 2u_{k+1} + v_{k+1} = u_k + 3v_k \\ u_{k+1} + v_{k+1} = u_k + v_k \end{cases}$.

5.145. Solve $\begin{cases} y_{k+1} - kz_k = 0 \\ z_{k+1} - 4ky_k = 0 \end{cases}$.

5.146. Solve the system of equations $\begin{cases} u_{k+1} = pu_k + qv_k \\ v_{k+1} = (1-p)u_k + (1-q)v_k \end{cases}$ if $1 - p + q \neq 0$.

5.147. Solve Problem 5.146 if $1 - p + q = 0$.

5.148. Solve $\begin{cases} 2u_{k+1} + v_{k+2} = 16v_k \\ u_{k+2} - 2v_{k+1} = 4u_k \end{cases}$.

MIXED DIFFERENCE EQUATIONS

5.149. Solve $y'_{k+1}(t) = y_k(t)$ subject to the conditions $y_0(t) = 1$, $y_k(0) = 1$.

5.150. (*a*) Determine $\lim_{k \to \infty} y_k(t)$ for Problem 5.149. (*b*) Could this limit have been found without solving the equation? Explain.

5.151. (*a*) Solve $y_k'(t) = y_{k+1}(t)$ subject to the conditions $y_0(t) = 1$, $y_k(0) = 1$. (*b*) Find $\lim_{k\to\infty} y_k(t)$. (*c*) Discuss the relationship of this problem with Problems 5.149 and 5.150.

5.152. (*a*) Solve $y_{k+1}''(t) + y_k(t) = 0$ subject to the conditions $y_0(t) = 1$, $y_0'(t) = 0$, $y_k(0) = 1$. (*b*) Discuss the limiting case of (*a*) as $k \to \infty$.

PARTIAL DIFFERENCE EQUATIONS

5.153. Solve $u(x+1, y) + 2u(x, y+1) = 0$.

5.154. Solve $u(x+1, y) - 2u(x, y+1) = 0$, $u(0, y) = 3y + 2$.

5.155. Solve $u(x, y+1) + 2u(y, x+1) = 3xy - 4x + 2$.

5.156. Solve $u(x+h, y) - 3u(x, y+k) = x + y$.

5.157. Solve $(E_1E_2 - 2)(E_1 - E_2)u = 0$.

5.158. Solve $u_{k+1,m+2} - u_{k,m} = 2^{k-m}$.

5.159. Solve $u_{k+1,m+1} - u_{k+1,m} = u_{k,m}$ if $u_{k,0} = \begin{cases} 0 & k > 0 \\ 1 & k = 0 \end{cases}$.

5.160. Solve $u_{k+2,m+2} - 3u_{k+1,m+1} + 2u_{k,m} = 0$.

5.161. Solve $u_{k+1,m} + u_{k,m+1} = 0$.

5.162. Solve $u_{k,m} = u_{k-1,m-1} + u_{k+1,m+1}$.

5.163. Solve $u_{k+1}(y) - u_k(y) + 3\dfrac{\partial u_k(y)}{\partial y} = 0$, $u_0(y) = 3e^{-2y/3}$.

MISCELLANEOUS PROBLEMS

5.164. Show that the general solution of

$$y_{k+2} - 2(\cos\alpha)y_{k+1} + y_k = R_k$$

is given by

$$y_k = C_1 \cos k\alpha + C_2 \sin k\alpha + \sum_{p=1}^{k} \frac{R_p \sin(k-p)\alpha}{\sin\alpha}$$

where it is assumed that $\alpha \neq 0, \pm\pi, \pm 2\pi, \ldots$.

5.165. Obtain the general solution in Problem 5.164 in the case $\alpha = 0, \pm\pi, \pm 2\pi, \ldots$.

5.166. Let $y_{k+2} + 3y_{k+1} + 2y_k = 6$. Prove that if $\lim_{k\to\infty} y_k$ exists then it must be equal to 1.

5.167. (*a*) Express the differential equation $y'' - xy' - 2y = 0$

as a limiting case of a difference equation having differencing interval equal to h.

(*b*) Solve the difference equation of (*a*) by assuming a factorial series expansion.

(*c*) Use the limiting form of (*b*) as $h \to 0$ to obtain the solution of the differential equation in (*a*).

5.168. (*a*) Solve the difference equation

$$\frac{\Delta y}{\Delta x} + P(x)y = Q(x)$$

(*b*) By finding the limit of the results in (*a*) as $\Delta x = h \to 0$ show how to solve the differential equation

$$\frac{dy}{dx} + P(x)y = Q(x)$$

5.169. (*a*) Solve the difference equation

$$\frac{\Delta^2 y}{\Delta x^2} - 4\frac{\Delta y}{\Delta x} + 3y = x + 2$$

(*b*) Use an appropriate limiting procedure in (*a*) to solve the differential equation

$$\frac{d^2 y}{dx^2} - 4\frac{dy}{dx} + 3y = x + 2$$

5.170. Obtain the solution of the differential equation

$$\frac{d^2 y}{dx^2} + 2\frac{dy}{dx} + y = e^{-x}$$

by using difference equation methods.

5.171. Let $f(x, y, c) = 0$ denote a one-parameter family of curves. The differential equation of the family is obtained by finding dy/dx from $f(x, y, c) = 0$ and then eliminating c.

(*a*) Find the differential equation of the family $y = cx^2$ and discuss the geometric significance.

(*b*) Explain how you might find the difference equation of a family and illustrate by using $y = cx^2$.

(*c*) Give a geometric interpretation of the results in (*b*).

5.172. Work Problem 5.171 for $y = cx + c^2$.

5.173. Generalize Problem 5.171 to two-parameter families of curves $f(x, y, c_1, c_2) = 0$ and illustrate by using $y = c_1 a^x + c_2 b^x$ where a and b are given constants.

5.174. (*a*) Show that two solutions of $\quad y = x\dfrac{\Delta y}{\Delta x} + \left(\dfrac{\Delta y}{\Delta x}\right)^2$

are $y = cx + c^2$ and $y = c^2 - \frac{1}{4}x^2 + \frac{1}{16}h^2 - \frac{1}{2}ch(-1)^{x/h}$.

(*b*) Discuss the relationship of your results in (*a*) with those of Problem 5.172.

5.175. (*a*) Show that two solutions of the differential equation

$$y = x\frac{dy}{dx} + \left(\frac{dy}{dx}\right)^2$$

are $y = cx + c^2$ and $y = -x^2/4$.

(*b*) By obtaining the graph of the solutions in (*a*) discuss the relationship of these solutions.

(*c*) Explain the relationship between the solutions in part (*a*) and Problem 5.172.

5.176. The difference equation $u_k = k\Delta u_k + F(\Delta u_k)$ or in "x notation"

$$y = x\frac{\Delta y}{\Delta x} + F\left(\frac{\Delta y}{\Delta x}\right)$$

is called *Clairaut's difference equation* by analogy with *Clairaut's differential equation*

$$y = x\frac{dy}{dx} + F\left(\frac{dy}{dx}\right)$$

(*a*) Solve Clairaut's differential equation by differentiating both sides of it with respect to x. Illustrate by using the equation

$$y = x\frac{dy}{dx} + \left(\frac{dy}{dx}\right)^2$$

and explain the relationship with Problem 5.175.

(*b*) Is there an analogous method of obtaining the solution of Clairaut's difference equation? Discuss with reference to Problem 5.174.

5.177. Solve $y_{k+1} + (-2)^k y_k = 0$.

5.178. Solve $y_k y_{k+1} + 1 = 2^k(y_{k+1} - y_k)$.

5.179. Show that the nonlinear equation

$$y_{k+1}y_k + A_k y_k + B_k y_{k+1} = C_k$$

can be reduced to a linear equation by the substitution

$$y_k = \frac{v_{k+1}}{v_k} - A_k$$

Illustrate by solving Problem 5.134.

5.180. (a) Show that if $\alpha_1, \alpha_2, \alpha_3$ are constants then the equation

$$y_{k+3} + \alpha_1 \beta_k y_{k+2} + \alpha_2 \beta_k \beta_{k-1} y_{k+1} + \alpha_3 \beta_k \beta_{k-1} \beta_{k-2} y_k = R_k$$

can be reduced to one with constant coefficients by means of the substitution

$$y_k = \beta_1 \beta_2 \cdots \beta_{k-3} u_k$$

(b) What is the corresponding result for first and second order equations? What is the result for nth order equations?

5.181. Solve $y_{k+1} = \dfrac{a}{b + y_k}$ by letting $b + y_k = \dfrac{u_{k+1}}{u_k}$.

5.182. Solve $y_k y_{k+1} = a/(k+1)$.

5.183. Solve $u_{k+3,m} - 3u_{k+2,m+1} + 3u_{k+1,m+2} - u_{k,m+3} = 0$.

5.184. Solve $u_{k,m+1} = u_{k-1,m} - m u_{k,m}$ where $u_{0,m} = 0$ for $m = 1, 2, 3, \ldots$ and $u_{k,m} = 0$ for $k > m$.

5.185. Solve $u_{k,m+1} = u_{k-1,m} + k u_{k,m}$ where $u_{0,m} = 0$ for $m = 1, 2, 3, \ldots$ and $u_{k,m} = 0$ for $k > m$.

5.186. Solve (a) $\dfrac{\partial u}{\partial x} + 2\dfrac{\partial u}{\partial y} = 4$, (b) $2\dfrac{\partial u}{\partial x} = \dfrac{\partial u}{\partial y} + x - y$ by using difference equation methods.

5.187. Solve $y_k y_{k+1} + \sqrt{(1-y_k^2)(1-y_{k+1}^2)} = \cos(4\pi/k)$ subject to the condition $y_1 = 0$.

5.188. Solve $Y(x) + \dfrac{Y(x+1)}{1!} + \dfrac{Y(x+2)}{2!} + \cdots = x$.

5.189. Solve $Y(x) - \dfrac{Y(x+2)}{2!} + \dfrac{Y(x+4)}{4!} - \cdots = x^2 - 4x + 1$.

5.190. Solve $k u_k = 1 + u_1 + u_2 + \cdots + u_{k-2}$, $u_1 = 1$, $u_2 = 1/2$.

5.191. Solve the partial differential equations

(a) $2\dfrac{\partial u}{\partial x} - 3\dfrac{\partial u}{\partial y} = 0$, $u(0, y) = y^2 + e^{-y}$

(b) $\dfrac{\partial u}{\partial x} + 3\dfrac{\partial u}{\partial y} = x - 2y$, $u(x, 0) = x^2 + x + 1$

5.192. Solve the partial differential equations

(a) $\dfrac{\partial^2 u}{\partial x^2} - 3\dfrac{\partial^2 u}{\partial x\, \partial y} + 2\dfrac{\partial^2 u}{\partial y^2} = 0$, $u(x, 0) = x^2$, $u(0, y) = y$

(b) $\dfrac{\partial^2 u}{\partial x^2} - 4\dfrac{\partial^2 u}{\partial y^2} = x + y$, $u(x, 0) = 0$, $u(0, y) = y$

Chapter 6

Applications of Difference Equations

FORMULATION OF PROBLEMS INVOLVING DIFFERENCE EQUATIONS

Various problems arising in mathematics, physics, engineering and other sciences can be formulated by the use of difference equations. In this chapter we shall consider applications to such fields as mechanics, electricity, probability and others.

APPLICATIONS TO VIBRATING SYSTEMS

There are many applications of difference equations to the mechanics of vibrating systems. As one example suppose that a string of negligible mass is stretched between two fixed points P and Q and is loaded at equal intervals h by N particles of equal mass m as indicated in Fig. 6-1. Suppose further that the particles are set into motion so that they vibrate transversely in a plane, i.e. in a direction perpendicular to PQ. Then the equation of motion of the kth particle is, assuming that the vibrations are small and that no external forces are present, given by

$$m\frac{d^2y_k}{dt^2} = \frac{\tau}{h}(y_{k-1} - 2y_k + y_{k+1}) \tag{1}$$

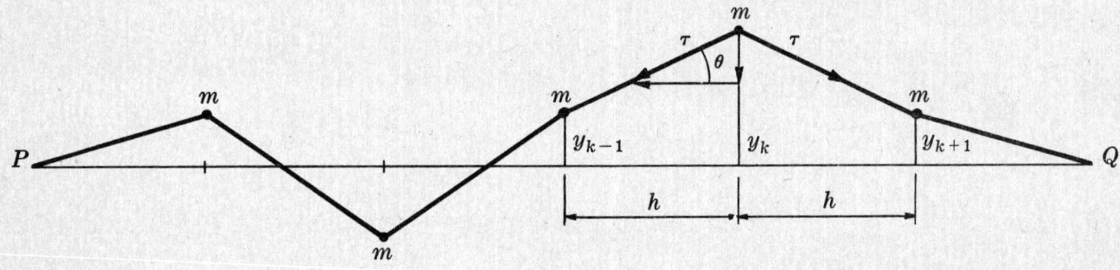

Fig. 6-1

where y_k denotes the displacement of the kth particle from PQ and τ is the tension in the string assumed constant.

Assuming that $y_k = A_k \cos(\omega t + \alpha)$ so that the particles all vibrate with frequency $2\pi/\omega$ we obtain the difference equation

$$A_{k+1} + \left(\frac{m\omega^2 h}{\tau} - 2\right) A_k + A_{k-1} = 0 \tag{2}$$

This leads to a set of *natural frequencies* for the string given by

$$f_n = \frac{1}{2\pi}\sqrt{\frac{4\tau}{mh}} \sin\frac{n\pi}{2(N+1)} \qquad n = 1, 2, \ldots, N \tag{3}$$

Each of these corresponds to a particular *mode of vibration* sometimes called *natural modes* or *principal modes*. Any complex vibration is a combination of these modes and involves more than one of the natural frequencies.

In a similar manner we can formulate and solve other problems involving vibrating systems. See Problems 6.36 and 6.39.

APPLICATIONS TO ELECTRICAL NETWORKS

In Fig. 6-2, a set of N capacitors or condensers having capacitance or capacity C and N resistors having resistance R are connected by electrical wires as shown to a generator supplying voltage V. The problem is to find the current and voltage through any of the resistors.

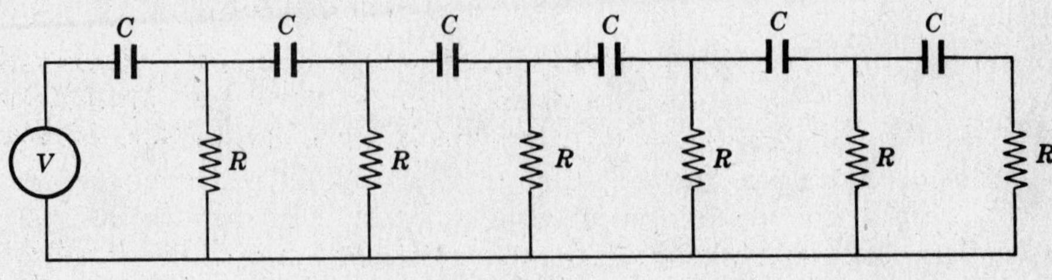

Fig. 6-2

The formulation of this problem and other problems of a similar nature involving various combinations of capacitors, resistors or inductors is based on Kirchhoff's two laws. These laws can be stated as follows.

Kirchhoff's laws.

1. The algebraic sum of the currents at any junction point of a network is zero.
2. The algebraic sum of the voltage drops around any closed loop is zero.

If I denotes the current, Q denotes the charge on a capacitor and t denotes time then we have:

Voltage drop across a resistor of resistance $R = IR$

Voltage drop across a capacitor of capacitance $C = \dfrac{Q}{C}$

Voltage drop across an inductor of inductance $L = L\dfrac{dI}{dt}$

APPLICATIONS TO BEAMS

Suppose that we have a continuous uniform beam which is simply supported at points which are equidistant as indicated in Fig. 6-3. If there is no load acting between the supports we can show that the bending moments at these successive supports satisfy the equation

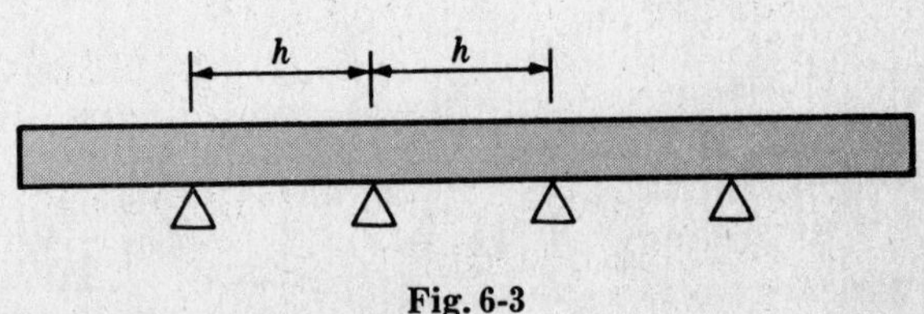

Fig. 6-3

$$M_{k-1} + 4M_k + M_{k+1} = 0 \tag{4}$$

sometimes called the *three moment equation.* The boundary conditions to be used together with this equation depend on the particular physical situation. See Problems 6.5 and 6.6.

For cases where there is a load acting between supports equation (4) is modified by having a suitable term on the right side.

APPLICATIONS TO COLLISIONS

Suppose that two objects, such as spheres, have a collision. If we assume that the spheres are smooth the forces due to the impact are exerted along the common normal to the spheres, i.e. along the line passing through their centers.

In solving problems involving collisions we make use of Newton's collision rule which can be stated as follows.

Newton's collision rule. If v_{12} and v_{12}' are the relative velocities of the first sphere with respect to the second before and after impact respectively then

$$v_{12} = -\epsilon v_{12}' \tag{5}$$

where ϵ, called the *coefficient of restitution,* is generally taken as a constant between 0 and 1. If $\epsilon = 0$ the collision is called *perfectly inelastic* while if $\epsilon = 1$ the collision is called *perfectly elastic.*

Another important principle which is used is the *principle of conservation of momentum.* The momentum of an object is defined as the mass multiplied by the velocity and the principle can be stated as follows.

Conservation of momentum. The total momentum of a system before a collision is equal to the total momentum after the collision.

Some applications which lead to difference equations are given in Problems 6.7, 6.8 and 6.53.

APPLICATIONS TO PROBABILITY

If we denote two events by A_1 and A_2 then the probabilities of the events are indicated by $P(A_1)$ and $P(A_2)$ respectively. It is convenient to denote the event that *either* A_1 *or* A_2 *occurs or both occur* by $A_1 + A_2$. Thus $P(A_1 + A_2)$ is the probability that at least one of the events A_1, A_2 occurs.

We denote the event that *both* A_1 *and* A_2 *occur* by A_1A_2 so that $P(A_1A_2)$ is the probability that both A_1 and A_2 occur.

We denote by $P(A_2 | A_1)$ the probability that A_2 *occurs given that* A_1 *has occurred.* This is often called the *conditional probability.*

An important result is that

$$P(A_1A_2) = P(A_1)P(A_2 | A_1) \tag{6}$$

In words this states that the probability of both A_1 and A_2 occurring is the same as the probability that A_1 occurs multiplied by the probability that A_2 occurs given that A_1 is known to have occurred.

Now it may happen that $P(A_2 | A_1) = P(A_2)$. In such case the fact that A_1 has occurred is irrelevant and we say that the events are independent. Then (*6*) becomes

$$P(A_1A_2) = P(A_1)P(A_2) \qquad A_1, A_2 \text{ independent} \tag{7}$$

We also have

$$P(A_1 + A_2) = P(A_1) + P(A_2) - P(A_1A_2) \tag{8}$$

i.e. the probability that at least one of the events occurs is equal to the probability that A_1 occurs plus the probability that A_2 occurs minus the probability that both occur.

It may happen that *both* A_1 and A_2 cannot occur simultaneously. In such case we say that A_1 and A_2 are *mutually exclusive events* and $P(A_1A_2) = 0$. Then (*8*) becomes

$$P(A_1 + A_2) = P(A_1) + P(A_2) \qquad A_1, A_2 \text{ mutually exclusive} \tag{9}$$

The ideas presented above can be generalized to any number of events.

Many problems in probability lead to difference equations. See Problems 6.9-6.12 for example.

THE FIBONACCI NUMBERS

The *Fibonacci numbers* F_k, $k = 0, 1, 2, \ldots$, are defined by the equation

$$F_k = F_{k-1} + F_{k-2} \tag{10}$$

where

$$F_0 = 0, \quad F_1 = 1 \tag{11}$$

It follows that the numbers are members of the sequence

$$0, 1, 1, 2, 3, 5, 8, 13, 21, \ldots \tag{12}$$

where each number after the second is the sum of the two preceding ones. For a table of Fibonacci numbers see Appendix G, page 238.

We can show [see Problem 6.13] that the kth Fibonacci number in the sequence (*12*) is given by

$$F_k = \frac{1}{\sqrt{5}}\left[\left(\frac{1+\sqrt{5}}{2}\right)^k - \left(\frac{1-\sqrt{5}}{2}\right)^k\right] \qquad k = 0, 1, 2, \ldots \tag{13}$$

These numbers have many interesting and remarkable properties which are considered in the problems.

MISCELLANEOUS APPLICATIONS

Difference equations can be used in solving various special problems. Among these are

1. **Evaluation of integrals.** See Problem 6.15.
2. **Evaluation of determinants.** See Problem 6.16.
3. **Problems involving principal and interest.** See Problems 6.17 and 6.18.
4. **Numerical solution of differential equations.** See Problems 6.19-6.24.

Solved Problems

APPLICATIONS TO VIBRATING SYSTEMS

6.1. A string of negligible mass is stretched between two points P and Q and is loaded at equal intervals h by N particles of equal mass m as indicated in Fig. 6-1, page 199. The particles are then set into motion so that they vibrate transversely. Assuming that the displacements are small compared with h find (*a*) the equations of motion and (*b*) natural frequencies of the system.

(*a*) Let y_k denote the displacement of the kth particle from PQ. Let us call τ the tension in the string which we shall take as constant.

The vertical force acting on the kth particle due to the $(k-1)$th particle is

$$-\tau \sin\theta = -\frac{\tau(y_k - y_{k-1})}{\sqrt{(y_k - y_{k-1})^2 + h^2}} \tag{1}$$

and since we assume that the displacements are small compared with h, we can replace the result in (1) by

$$-\frac{\tau(y_k - y_{k-1})}{h} \tag{2}$$

to a high degree of approximation.

Similarly the vertical force on the kth particle due to the $(k+1)$th particle is given to a high degree of approximation by

$$-\frac{\tau(y_k - y_{k+1})}{h} \tag{3}$$

The total vertical force on the kth particle is thus

$$-\frac{\tau(y_k - y_{k-1})}{h} - \frac{\tau(y_k - y_{k+1})}{h} = \frac{\tau}{h}(y_{k-1} - 2y_k + y_{k+1}) \tag{4}$$

Assuming that there are no other forces acting on the kth particle we see that by Newton's law the equation of motion is

$$m\frac{d^2y_k}{dt^2} = \frac{\tau}{h}(y_{k-1} - 2y_k + y_{k+1}) \tag{5}$$

If there are other forces acting, such as gravity for example, we must take them into account.

(b) To find the natural frequencies assume that

$$y_k = A_k \cos(\omega t + \alpha) \tag{6}$$

i.e. assume that all the particles vibrate with the same frequency $\omega/2\pi$. Then (5) becomes

$$-m\omega^2 A_k = \frac{\tau}{h}(A_{k-1} - 2A_k + A_{k+1}) \tag{7}$$

which can be written as

$$A_{k+1} + \left(\frac{m\omega^2 h}{\tau} - 2\right)A_k + A_{k-1} = 0 \tag{8}$$

Now we can consider the points P and Q as fixed particles whose displacements are zero. Thus we have

$$A_0 = 0, \quad A_{N+1} = 0 \tag{9}$$

Letting $A_k = r^k$ in (8) we find that

$$r^2 + \left(\frac{m\omega^2 h}{\tau} - 2\right)r + 1 = 0 \tag{10}$$

or

$$r = \frac{1}{2}\left\{\left(2 - \frac{m\omega^2 h}{\tau}\right) \pm \sqrt{\left(2 - \frac{m\omega^2 h}{\tau}\right)^2 - 4}\right\} \tag{11}$$

We can simplify (11) if we let

$$2 - \frac{m\omega^2 h}{\tau} = 2\cos\theta \tag{12}$$

so that

$$r = \cos\theta \pm i \sin\theta \tag{13}$$

In this case we find that the general solution is

$$A_k = c_1 \cos k\theta + c_2 \sin k\theta \tag{14}$$

Using the first of conditions (9) we find $c_1 = 0$ so that

$$A_k = c_2 \sin k\theta \tag{15}$$

From the second of conditions (9) we must have

$$c_2 \sin(N+1)\theta = 0 \tag{16}$$

or since $c_2 \neq 0$,

$$\theta = \frac{n\pi}{N+1} \qquad n = 1, 2, \ldots \tag{17}$$

Thus (12) yields

$$2 - \frac{m\omega^2 h}{\tau} = 2\cos\frac{n\pi}{N+1} \tag{18}$$

from which

$$\omega^2 = \frac{2\tau}{mh}\left(1 - \cos\frac{n\pi}{N+1}\right) = \frac{4\tau}{mh}\sin^2\frac{n\pi}{2(N+1)} \tag{19}$$

Denoting the N different values of ω^2 corresponding to $n = 1, 2, 3, \ldots, N$ by $\omega_1, \omega_2, \omega_3, \ldots, \omega_N$ it then follows that

$$\omega_n = \sqrt{\frac{4\tau}{mh}}\sin\frac{n\pi}{2(N+1)} \qquad n = 1, 2, \ldots, N \tag{20}$$

Thus the natural frequencies are

$$f_n = \frac{1}{2\pi}\sqrt{\frac{4\tau}{mh}}\sin\frac{n\pi}{2(N+1)} \qquad n = 1, 2, \ldots, N \tag{21}$$

6.2. Find the displacement of the kth particle of the string in Problem 6.1.

Let us consider the particular mode of vibration in which the frequency corresponds to ω_n, i.e. the nth mode of vibration. In this case we see from Problem 6.1 that the displacement of the kth particle is given by

$$y_{k,n} = A_{k,n}\cos(\omega_n t + \alpha_n) \tag{1}$$

where

$$A_{k,n} = c_n \sin\frac{n\pi}{N+1} \tag{2}$$

the additional subscript n being used for ω, α and c to emphasize that the result is true for the nth mode of vibration.

Now since the difference equation is linear, sums of solutions are also solutions so that we have for the general displacement of the kth particle

$$y_k = \sum_{n=1}^{N} c_n \sin\frac{nk\pi}{N+1}\cos(\omega_n t + \alpha_n) \tag{3}$$

Since this satisfies the boundary conditions at the ends of the string, i.e. $y_0 = 0$, $y_{N+1} = 0$, it represents the required displacement.

The $2N$ constants c_n, α_n, $n = 1, \ldots, N$, can be found if we specify the positions and velocities of the N particles at the time $t = 0$.

6.3. Find the natural frequencies of the string of Problems 6.1 and 6.2 if the number of particles N becomes infinite while the total mass M and length L of the string remain constant.

The total mass M and length L of the string are given by

$$M = Nm, \qquad L = (N+1)h \tag{1}$$

Then the natural frequencies are given by

$$f_n = \frac{1}{2\pi}\sqrt{\frac{4\tau N(N+1)}{ML}}\sin\frac{n\pi}{2(N+1)} \tag{2}$$

or

$$f_n = \frac{1}{2\pi}\sqrt{\frac{4\tau}{ML}}\sqrt{N(N+1)}\sin\frac{n\pi}{2(N+1)} \tag{3}$$

Now for any particular value of n we have

$$\lim_{N\to\infty} \sqrt{N(N+1)}\, \sin\frac{n\pi}{2(N+1)} = \frac{n\pi}{2}$$

[by using for example the fact that for small angles θ, $\sin\theta$ is approximately equal to θ]. Thus the required natural frequencies are given by

$$f_n = \frac{n}{2}\sqrt{\frac{\tau}{ML}} \qquad n = 1, 2, \ldots \tag{4}$$

If we let ρ denote the density in this case, then

$$\frac{M}{L} = \rho \quad \text{or} \quad M = \rho L \tag{5}$$

and (4) becomes

$$f_n = \frac{n}{2L}\sqrt{\frac{\tau}{\rho}} \qquad n = 1, 2, \ldots \tag{6}$$

APPLICATIONS TO ELECTRICAL NETWORKS

6.4. An electrical network has the form shown in Fig. 6-4 where R_1, R_2 and R are given constant resistances and V is the voltage supplied by a generator. (*a*) Use Kirchhoff's laws to express the relationship among the various currents. (*b*) By solving the equations obtained in (*a*) find an expression for the current in any of the loops.

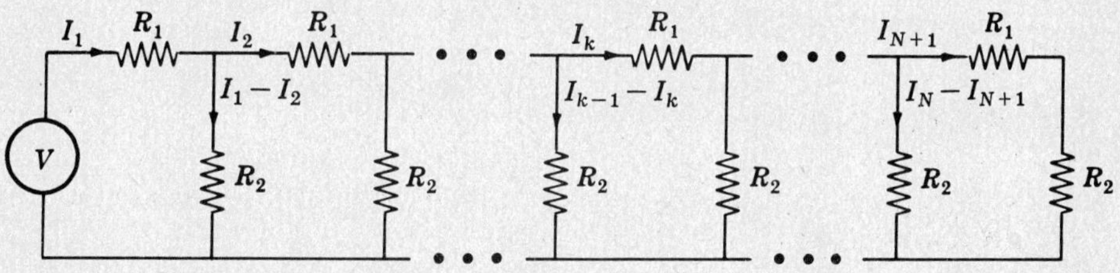

Fig. 6-4

(*a*) For the first loop,

$$I_1R_1 + (I_1 - I_2)R_2 - V = 0 \tag{1}$$

For the kth loop,

$$I_kR_1 + (I_k - I_{k+1})R_2 - (I_{k-1} - I_k)R_2 = 0 \tag{2}$$

For the $(N+1)$th loop,

$$I_{N+1}R_1 - (I_N - I_{N+1})R_2 = 0 \tag{3}$$

(*b*) Equation (2) can be written as

$$I_{k+1} - \left(2 + \frac{R_1}{R_2}\right) I_k + I_{k-1} = 0 \tag{4}$$

Letting $I_k = r^k$ in (4) it can be written as

$$r^2 - \left(2 + \frac{R_1}{R_2}\right) r + 1 = 0 \tag{5}$$

so that

$$r = \frac{2 + R_1/R_2 \pm \sqrt{(2 + R_1/R_2)^2 - 4}}{2} = 1 + \frac{R_1}{2R_2} \pm \sqrt{\frac{R_1}{R_2} + \frac{R_1^2}{4R_2^2}}$$

Then the general solution of (4) is

$$I_k = c_1 r_1^k + c_2 r_2^k \tag{6}$$

where $$r_1 = 1 + \frac{R_1}{2R_2} + \sqrt{\frac{R_1}{R_2} + \frac{R_1^2}{4R_2^2}}, \qquad r_2 = 1 + \frac{R_1}{2R_2} - \sqrt{\frac{R_1}{R_2} + \frac{R_1^2}{4R_2^2}} \tag{7}$$

From (1),

$$I_2 = \frac{I_1(R_1 + R_2) - V}{R_2} \tag{8}$$

From (3),

$$I_N = \frac{(R + R_2)I_{N+1}}{R_2} \tag{9}$$

Now if we put $k = 2$ and $k = N$ in (6) we have

$$I_2 = c_1 r_1^2 + c_2 r_2^2 \tag{10}$$

$$I_N = c_1 r_1^N + c_2 r_2^N \tag{11}$$

from which

$$c_1 = \frac{I_2 r_2^N - I_N r_2^2}{r_1^2 r_2^N - r_2^2 r_1^N}, \qquad c_2 = \frac{I_N r_1^2 - I_2 r_1^N}{r_1^2 r_2^N - r_2^2 r_1^N}$$

so that

$$I_n = \left(\frac{I_2 r_2^N - I_N r_2^2}{r_1^2 r_2^N - r_2^2 r_1^N}\right) r_1^n + \left(\frac{I_N r_1^2 - I_2 r_1^N}{r_1^2 r_2^N - r_2^2 r_1^N}\right) r_2^n$$

where r_1, r_2 are given in (7) and I_2, I_N are given by (10) and (11).

APPLICATIONS TO BEAMS

6.5. Derive the equation of three moments (4) on page 200.

Assume that the spacing between supports is h and consider three successive supports as shown in Fig. 6-5. Choosing origin O at the center support and letting x be the distance from this origin we find that the bending moment at x is given by

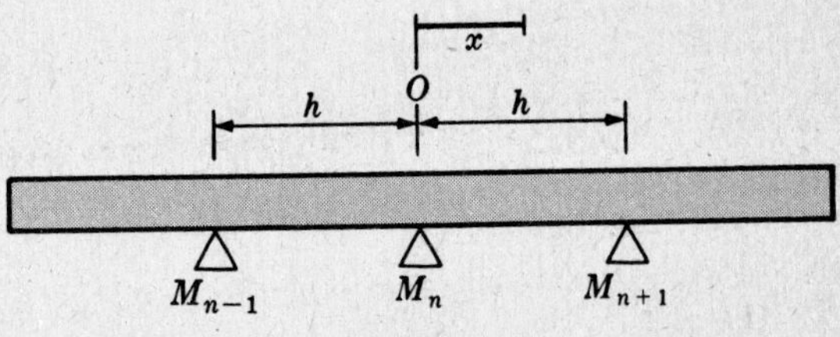

Fig. 6-5

$$M(x) = \begin{cases} M_n + (M_n - M_{n-1})\dfrac{x}{h} & -h \leqq x \leqq 0 \\[2ex] M_n + (M_{n+1} - M_n)\dfrac{x}{h} & 0 \leqq x \leqq h \end{cases} \tag{1}$$

where M_{n-1}, M_n, M_{n+1} are the bending moments at the $(n-1)$th, nth and $(n+1)$th supports.

Now we know from the theory of beams that the curve of deflection of a beam is $y(x)$ where

$$YIy'' = M(x) \tag{2}$$

and YI, called the *flexural rigidity* of the beam, is assumed to be constant. Here Y is *Young's modulus of elasticity* and I is the moment of inertia about an axis through the centroid.

Integrating (2) using (1) we find

$$YIy' = \begin{cases} M_n x + (M_n - M_{n-1})\dfrac{x^2}{2h} + c_1 & -h \leqq x \leqq 0 \\[2ex] M_n x + (M_{n+1} - M_n)\dfrac{x^2}{2h} + c_2 & 0 \leqq x \leqq h \end{cases} \tag{3}$$

Now since y' must be continuous at $x = 0$, it follows that $c_1 = c_2 = c$

Integrating (3) we have similarly

$$YIy = \begin{cases} \dfrac{M_n x^2}{2} + (M_n - M_{n-1})\dfrac{x^3}{6h} + cx & -h \leqq x \leqq 0 \\ \dfrac{M_n x^2}{2} + (M_{n+1} - M_n)\dfrac{x^3}{6h} + cx & 0 \leqq x \leqq h \end{cases} \tag{4}$$

Since the deflection y is zero at $x = -h, 0, h$ we find from (4)

$$\frac{M_n h^2}{2} - (M_n - M_{n-1})\frac{h^2}{6} - ch = 0 \tag{5}$$

$$\frac{M_n h^2}{2} + (M_{n+1} - M_n)\frac{h^2}{6} + ch = 0 \tag{6}$$

Addition of these equations leads to the result

$$M_{n-1} + 4M_n + M_{n+1} = 0 \tag{7}$$

as required.

6.6. Suppose that the beam of page 200 is of infinite extent on the right and that at the distance h to the left of the first support there is a load W which acts on the beam [see Fig. 6-6]. Prove that the bending moment at the nth support is given by

$$M_n = (-1)^{n+1} Wh(2-\sqrt{3})^n$$

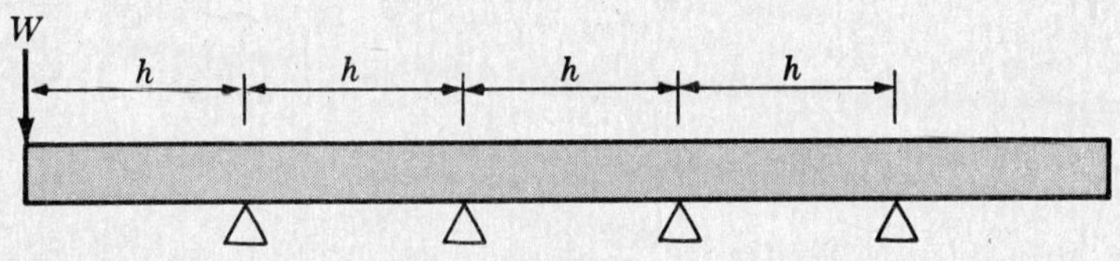

Fig. 6-6

The bending moment at the distance h to the left of the first support is given by

$$M_0 = -Wh \tag{1}$$

Also at infinity the bending moment should be zero so that

$$\lim_{n\to\infty} M_n = 0 \tag{2}$$

Now from the difference equation

$$M_{n-1} + 4M_n + M_{n+1} = 0 \tag{3}$$

we have on making the substitution $M_n = r^n$

$$r^{n-1} + 4r^n + r^{n+1} = 0 \quad \text{or} \quad r^2 + 4r + 1 = 0 \tag{4}$$

i.e. $$r = \frac{-4 \pm \sqrt{16-4}}{2} = -2 \pm \sqrt{3}$$

Then the general solution of (3) is

$$M_n = c_1(-2+\sqrt{3})^n + c_2(-2-\sqrt{3})^n \tag{5}$$

Now since $-2+\sqrt{3}$ is a number between -1 and 0 while $-2-\sqrt{3}$ is approximately -3.732, it follows that condition (2) will be satisfied if and only if $c_2 = 0$. Then the solution (5) becomes

$$M_n = c_1(-2+\sqrt{3})^n \tag{6}$$

Also from (1) we have $c_1 = -Wh$ so that

$$M_n = -Wh(-2+\sqrt{3})^n = (-1)^{n+1} Wh(2-\sqrt{3})^n \tag{7}$$

APPLICATIONS TO COLLISIONS

6.7. **Two billiard balls of equal mass m lie on a billiards table with their line of centers perpendicular to two sides of the table. One of the balls is hit so that it travels with constant speed S to the other ball. Set up equations for the speeds of the balls before the kth impact occurs assuming that the table is smooth and the sides are perfectly elastic.**

Denote by U_k and V_k the speeds of the balls before the kth impact occurs. Then assuming that U_k refers to the ball which is hit initially we have

$$U_1 = S, \quad V_1 = 0 \tag{1}$$

Since the sides are perfectly elastic, the speeds of the balls after the kth impact are U_{k+1} and V_{k+1} respectively. Then the total momentum before and after the kth impact are given by $mU_k + mV_k$ and $mU_{k+1} - mV_{k+1}$ respectively. Thus by the conservation of momentum we have

$$mU_k + mV_k = mU_{k+1} - mV_{k+1}$$

or

$$U_k + V_k = U_{k+1} - V_{k+1} \tag{2}$$

In addition we have by Newton's collision rule

$$U_{k+1} + V_{k+1} = -\epsilon(U_k - V_k) \tag{3}$$

where ϵ is the coefficient of restitution between the two billiard balls.

The required equations are given by (*2*) and (*3*) which are to be solved subject to the conditions (*1*).

6.8. **Determine the speeds of the balls of Problem 6.7 before the kth impact.**

Writing the equations (*2*) and (*3*) of Problem 6.7 in terms of the shifting operator E we obtain

$$(E-1)U_k - (E+1)V_k = 0 \tag{1}$$

$$(E+\epsilon)U_k + (E-\epsilon)V_k = 0 \tag{2}$$

Then operating on (*1*) with $E - \epsilon$, on (*2*) with $E+1$ and adding we obtain

$$(E^2 + \epsilon)U_k = 0 \tag{3}$$

Letting $U_k = r^k$, (*3*) yields

$$r^2 + \epsilon = 0, \quad r = \pm i\sqrt{\epsilon} = \sqrt{\epsilon}\, e^{\pm \pi i/2}$$

so that the general solution of (*3*) is

$$U_k = \epsilon^{k/2}(Ae^{\pi k i/2} + Be^{-\pi k i/2}) = \epsilon^{k/2}\left(c_1 \cos\frac{\pi k}{2} + c_2 \sin\frac{\pi k}{2}\right) \tag{4}$$

By eliminating V_{k+1} between equations (*2*) and (*3*) of Problem 6.7 we find

$$(\epsilon + 1)V_k = 2U_{k+1} + (\epsilon - 1)U_k \tag{5}$$

Then using (*4*)

$$V_k = \frac{\epsilon^{k/2}}{\epsilon+1}\left[\{(\epsilon-1)c_2 - 2\sqrt{\epsilon}\,c_1\} \sin\frac{\pi k}{2} + \{(\epsilon-1)c_1 + 2\sqrt{\epsilon}\,c_2\} \cos\frac{\pi k}{2}\right] \tag{6}$$

From conditions (*1*) of Problem 6.7 we have

$$\sqrt{\epsilon}\left(c_1 \cos\frac{\pi}{2} + c_2 \sin\frac{\pi}{2}\right) = S \tag{7}$$

$$(\epsilon - 1)c_2 - 2\sqrt{\epsilon}\,c_1 = 0 \tag{8}$$

so that

$$c_1 = \frac{(\epsilon-1)S}{2\epsilon} \qquad c_2 = \frac{S}{\sqrt{\epsilon}} \tag{9}$$

Using these in (4) and (6) we obtain the required results

$$U_k = \tfrac{1}{2}S\epsilon^{(k-2)/2}\left[(\epsilon-1)\cos\frac{\pi k}{2} + 2\sqrt{\epsilon}\sin\frac{\pi k}{2}\right] \tag{10}$$

$$V_k = \tfrac{1}{2}S\epsilon^{(k-2)/2}(\epsilon+1)\cos\frac{\pi k}{2} \tag{11}$$

APPLICATIONS TO PROBABILITY

6.9. A and B play a game. In each step of the game A can win a penny from B with probability p while B can win a penny from A with probability q, where p and q are positive constants such that $p+q=1$. We assume that no tie can occur. Suppose that the game is started with A and B having a and b pennies respectively, the total being $a+b=N$, and that the game ends when one or the other has M pennies. Formulate (a) a difference equation and (b) boundary conditions for the probability that A wins the game.

(a) Let u_k be the probability that A wins the game when he has k pennies. Now A will win if one of the following two mutually exclusive events occurs:

1. On the next step A wins a penny and then wins the game.
2. On the next step A loses a penny and then wins the game.

The probability of the first event occurring is given by

$$pu_{k+1} \tag{1}$$

since if A wins on the next step [for which the probability is p] he has $k+1$ pennies and can then win the game with probability u_{k+1}.

Similarly the probability of the second event occurring is given by

$$qu_{k-1} \tag{2}$$

since if A loses on the next step [for which the probability is q] he has $k-1$ pennies and can then win the game with probability u_{k-1}.

It follows that the probability of A winning the game is the sum of the probabilities (1) and (2), i.e.

$$u_k = pu_{k+1} + qu_{k-1} \tag{3}$$

(b) To determine boundary conditions for u_k we note that A wins the game when he has M pennies, i.e.

$$u_M = 1 \tag{4}$$

Similarly A loses the game when B has M pennies and thus A has $N-M$ pennies, i.e.

$$u_{N-M} = 0 \tag{5}$$

6.10. Obtain the probability that A wins the game of Problem 6.9 if (a) $p \neq q$, (b) $p=q=\tfrac{1}{2}$.

We must solve the difference equation (3) of Problem 6.9 subject to the conditions (4) and (5). To do this we proceed according to the usual methods of Chapter 5 by letting $u_k = r^k$ in equation (3) of Problem 6.9. Then we obtain

$$pr^2 - r + q = 0 \tag{1}$$

from which

$$r = \frac{1 \pm \sqrt{1-4pq}}{2p} = \frac{1 \pm \sqrt{1-4p+4p^2}}{2p} = \frac{1 \pm (1-2p)}{2p} = 1,\ q/p \tag{2}$$

(a) If $p \neq q$, the roots of (1) are 1 and q/p so that

$$u_k = c_1 + c_2(q/p)^k \tag{3}$$

Using the condition (4) of Problem 6.9, i.e. $u_M = 1$, we have

$$1 = c_1 + c_2(q/p)^M \tag{4}$$

Similarly using condition (5) of Problem 6.9, i.e. $u_{N-M} = 0$, we have

$$0 = c_1 + c_2(q/p)^{N-M} \tag{5}$$

Then solving (4) and (5) simultaneously we obtain

$$c_1 = \frac{-(q/p)^{N-M}}{(q/p)^M - (q/p)^{N-M}}, \qquad c_2 = \frac{1}{(q/p)^M - (q/p)^{N-M}} \tag{6}$$

and so from (3) we have

$$u_k = \frac{(q/p)^k - (q/p)^{N-M}}{(q/p)^M - (q/p)^{N-M}} \tag{7}$$

Since A starts with $k = a$ pennies, the probability that he will win the game is

$$u_a = \frac{(q/p)^a - (q/p)^{N-M}}{(q/p)^M - (q/p)^{N-M}} \tag{8}$$

while the probability that he loses the game is

$$1 - u_a = \frac{(q/p)^M - (q/p)^a}{(q/p)^M - (q/p)^{N-M}} \tag{9}$$

(b) If $p = q = \frac{1}{2}$, the roots of (1) are 1, 1, i.e. they are repeated. In this case the solution of equation (3) of Problem 6.9 is

$$u_k = c_3 + c_4 k \tag{10}$$

Then using $u_M = 1$, $u_{N-M} = 0$, we obtain

$$c_3 + c_4 M = 1, \qquad c_3 + c_4(N-M) = 0 \tag{11}$$

i.e.

$$c_3 = \frac{M-N}{2M-N}, \qquad c_4 = \frac{1}{2M-N} \tag{12}$$

Substituting in (10)

$$u_k = \frac{M-N+k}{2M-N} \tag{13}$$

Thus the probability that A wins the game is

$$u_a = \frac{M-N+a}{2M-N} \tag{14}$$

while the probability that he loses the game is

$$1 - u_a = \frac{M-a}{2M-N} \tag{15}$$

6.11. Suppose that in Problem 6.9 the game is won when either A or B wins all the money, i.e. $M = N$. Find the probability of A being "ruined", i.e. losing all his money if (a) $p \neq q$, (b) $p = q = 1/2$.

(a) Putting $M = N$ in equation (9) of Problem 6.10 we find for the probability of A being ruined

$$1 - u_a = \frac{(q/p)^N - (q/p)^a}{(q/p)^N - 1} = \frac{(q/p)^{a+b} - (q/p)^a}{(q/p)^{a+b} - 1} \tag{1}$$

(b) Putting $M = N$ in equation (14) of Problem 6.10 we find for the probability of A being ruined

$$1 - u_a = \frac{N-a}{N} = \frac{b}{a+b} \tag{2}$$

The case where $p = q = 1/2$ is sometimes called a "fair game" since A and B then have equal chances of getting a point at each step. From (2) we note that even in a fair game the chance of A being ruined is proportional to the initial capital of B. It follows as a consequence that if B has many more pennies than A initially then A has a very great chance of being ruined.

This is illustrated by events which take place at many gambling casinos where a person is competing in a game [which often is not even fair] against an "infinitely rich" adversary.

6.12. A and B play a game. A needs k points in order to win the game while B needs m points to win. If the probability of A getting one point is p while the probability of B getting one point is q, where p and q are positive constants such that $p+q=1$, find the probability that A wins the game.

Let $u_{k,m}$ denote the probability that A wins the game. The event that A wins the game can be realized if one of the following two possible mutually exclusive events occurs:

1. A wins the next point and then wins the game.
2. A loses the next point and then wins the game.

The probability of the first event occurring is given by

$$pu_{k-1,m} \tag{1}$$

since if A wins the next point [for which the probability is p] he needs only $k-1$ points to win the game [for which the probability is $u_{k-1,m}$].

Similarly the probability of the second event occurring is

$$qu_{k,m-1} \tag{2}$$

since if A loses the next point [for which the probability is q] he still needs k points to win the game while B needs only $m-1$ points to win [for which the probability is $u_{k,m-1}$].

It follows that the probability of A winning the game is the sum of the probabilities (1) and (2), i.e.

$$u_{k,m} = pu_{k-1,m} + qu_{k,m-1} \tag{3}$$

To determine boundary conditions for $u_{k,m}$ we note that if $k=0$ while $m>0$, i.e. if A needs no more points to win, then he has already won. Thus

$$u_{0,m} = 1 \qquad m>0 \tag{4}$$

Also if $k>0$ while $m=0$, then B has already won so that

$$u_{k,0} = 0 \qquad k>0 \tag{5}$$

The solution of the difference equation (3) subject to conditions (4) and (5) has already been found [see Problems 5.53 and 5.54, page 186] and the result is

$$u_{k,m} = p^k\left[1 + kq + \frac{k(k+1)}{2!}q^2 + \cdots + \frac{k(k+1)\cdots(k+m-2)}{(m-1)!}q^{m-1}\right]$$

FIBONACCI NUMBERS

6.13. Prove that the kth Fibonacci number is given by

$$F_k = \frac{1}{\sqrt{5}}\left[\left(\frac{1+\sqrt{5}}{2}\right)^k - \left(\frac{1-\sqrt{5}}{2}\right)^k\right]$$

We must solve the difference equation

$$F_k = F_{k-1} + F_{k-2} \tag{1}$$

subject to the conditions

$$F_0 = 0, \qquad F_1 = 1 \tag{2}$$

Letting $F_k = r^k$ in (1) we obtain the auxiliary equation

$$r^2 - r - 1 = 0 \quad \text{or} \quad r = \frac{1 \pm \sqrt{5}}{2} \tag{3}$$

Then the general solution is

$$F_k = c_1\left(\frac{1+\sqrt{5}}{2}\right)^k + c_2\left(\frac{1-\sqrt{5}}{2}\right)^k \tag{4}$$

From the conditions (2) we find

$$c_1 + c_2 = 0, \qquad (1+\sqrt{5})c_1 + (1-\sqrt{5})c_2 = 2 \tag{5}$$

Solving (5) we find

$$c_1 = \frac{1}{\sqrt{5}}, \qquad c_2 = -\frac{1}{\sqrt{5}} \tag{6}$$

and so from (4) we have

$$F_k = \frac{1}{\sqrt{5}}\left[\left(\frac{1+\sqrt{5}}{2}\right)^k - \left(\frac{1-\sqrt{5}}{2}\right)^k\right] \tag{7}$$

6.14. Prove that $\lim_{k\to\infty} \dfrac{F_{k+1}}{F_k} = \dfrac{1+\sqrt{5}}{2}$.

We have using Problem 6.13

$$\begin{aligned}\frac{F_{k+1}}{F_k} &= \frac{[(1+\sqrt{5})/2]^{k+1} - [(1-\sqrt{5})/2]^{k+1}}{[(1+\sqrt{5})/2]^k - [(1-\sqrt{5})/2]^k} \\ &= \frac{[(1+\sqrt{5})/2]^{k+1}\{1 - [(1-\sqrt{5})/(1+\sqrt{5})]^{k+1}\}}{[(1+\sqrt{5})/2]^k\{1 - [(1-\sqrt{5})/(1+\sqrt{5})]^k\}} \\ &= [(1+\sqrt{5})/2]\frac{\{1 - [(1-\sqrt{5})/(1+\sqrt{5})]^{k+1}\}}{\{1 - [(1-\sqrt{5})/(1+\sqrt{5})]^k\}}\end{aligned}$$

Then since $|(1-\sqrt{5})/(1+\sqrt{5})| < 1$ we have

$$\lim_{k\to\infty} \frac{F_{k+1}}{F_k} = \frac{1+\sqrt{5}}{2}$$

This limit, which is equal to 1.618033988..., is often called the *golden mean*. It is supposed to represent the ratio of the sides of a rectangle which is "most pleasing" to the eye.

MISCELLANEOUS APPLICATIONS

6.15. Evaluate the integral

$$S_k = \int_0^\pi \frac{\cos k\theta - \cos k\alpha}{\cos\theta - \cos\alpha}\,d\theta$$

Let a, b, c be arbitrary constants and consider

$$\begin{aligned}aS_{k+1} + bS_k + cS_{k-1} &= \int_0^\pi \frac{a[\cos(k+1)\theta - \cos(k+1)\alpha]}{\cos\theta - \cos\alpha}\,d\theta \\ &\quad + \int_0^\pi \frac{b[\cos k\theta - \cos k\alpha]}{\cos\theta - \cos\alpha}\,d\theta \\ &\quad + \int_0^\pi \frac{c[\cos(k-1)\theta - \cos(k-1)\alpha]}{\cos\theta - \cos\alpha}\,d\theta \\ &= \int_0^\pi \frac{[(a+c)\cos\theta + b]\cos k\theta + [(c-a)\sin\theta]\sin k\theta}{\cos\theta - \cos\alpha}\,d\theta\end{aligned}$$

If we now choose a, b and c so that

$$c = a, \qquad b = -2a \cos \alpha$$

the last integral becomes

$$2a \int_0^{\pi} \cos k\theta \, d\theta = 0 \quad \text{if } k = 1, 2, 3, \ldots$$

It thus follows that

$$S_{k+1} - 2(\cos \alpha) S_k + S_{k-1} = 0 \qquad k = 1, 2, 3, \ldots \tag{1}$$

From the definition of S_k we have

$$S_0 = 0, \quad S_1 = \pi \tag{2}$$

We must thus solve the difference equation (*1*) subject to the conditions (*2*).

According to the usual methods for solving difference equations we let $S_k = r^k$ in (*1*) and obtain

$$r^{k+1} - 2(\cos \alpha) r^k + r^{k-1} = 0 \quad \text{or} \quad r^2 - 2(\cos \alpha) r + 1 = 0$$

so that

$$r = \frac{2 \cos \alpha \pm \sqrt{4 \cos^2 \alpha - 4}}{2} = \cos \alpha \pm i \sin \alpha$$

Then by de Moivre's theorem,

$$r^k = \cos k\alpha \pm i \sin k\alpha$$

so that the general solution is

$$S_k = c_1 \cos k\alpha + c_2 \sin k\alpha \tag{3}$$

Using the conditions (*2*) we have

$$c_1 = 0, \quad c_1 \cos \alpha + c_2 \sin \alpha = \pi \tag{4}$$

so that $c_2 = \pi/\sin \alpha$ and thus

$$S_k = \frac{\pi \sin k\alpha}{\sin \alpha} \tag{5}$$

which is the required value of the integral.

6.16. Find the value of the determinant

$$\begin{vmatrix} b & a & 0 & 0 & 0 & \ldots & 0 & 0 & 0 \\ a & b & a & 0 & 0 & \ldots & 0 & 0 & 0 \\ 0 & a & b & a & 0 & \ldots & 0 & 0 & 0 \\ \cdots & & & & & & & & \cdots \\ 0 & 0 & 0 & 0 & 0 & \ldots & a & b & a \\ 0 & 0 & 0 & 0 & 0 & \ldots & 0 & a & b \end{vmatrix}$$

Denote the value of the kth order determinant by A_k. Then by the elementary rules for evaluating determinants using the elements of the first row we have

$$A_k = bA_{k-1} - a^2 A_{k-2} \quad \text{or} \quad A_k - bA_{k-1} + a^2 A_{k-2} = 0 \tag{1}$$

Letting $A_k = r^k$ we obtain

$$r^k - br^{k-1} + a^2 r^{k-2} = 0 \quad \text{or} \quad r^2 - br + a^2 = 0$$

from which

$$r = \frac{b \pm \sqrt{b^2 - 4a^2}}{2} \tag{2}$$

It is convenient to let

$$b = 2a\cos\theta \tag{3}$$

so that (2) becomes

$$r = a\cos\theta \pm ai\sin\theta \tag{4}$$

Thus since $r^k = a^k(\cos k\theta \pm i\sin k\theta)$ we can write the general solution as

$$A_k = a^k[c_1\cos k\theta + c_2\sin k\theta] \tag{5}$$

Now if $k=1$, the determinant of order 1 has the value $A_1 = b$, while if $k=2$, the determinant of order 2 has the value

$$A_2 = \begin{vmatrix} b & a \\ a & b \end{vmatrix} = b^2 - a^2$$

Putting $k=2$ in (5) we find that

$$A_2 - bA_1 + a^2A_0 = 0$$

from which we see on putting $A_1 = b$, $A_2 = b^2 - a^2$ that $A_0 = 1$. It follows that we can solve (1) subject to the conditions

$$A_0 = 1, \quad A_1 = b \tag{6}$$

These values together with (5) lead to

$$c_1 = 1, \qquad c_2 = \frac{b - a\cos\theta}{a\sin\theta} = \frac{\cos\theta}{\sin\theta} \tag{7}$$

using (3) so that (5) becomes

$$A_k = a^k\left[\cos k\theta + \frac{\sin k\theta\cos\theta}{\sin\theta}\right] = \frac{a^k\sin(k+1)\theta}{\sin\theta} \tag{8}$$

Using the fact that

$$\theta = \cos^{-1}(b/2a) \tag{9}$$

we can write the value of the determinant in the form

$$A_k = \frac{a^k\sin[(k+1)\cos^{-1}(b/2a)]}{\sin[\cos^{-1}(b/2a)]} \tag{10}$$

or

$$A_k = \frac{2a^{k+1}\sin[(k+1)\cos^{-1}(b/2a)]}{\sqrt{4a^2-b^2}} \tag{11}$$

The above results hold for $|b| < 2|a|$. In case $|b| > 2|a|$ we must replace $\cos^{-1}$ by $\cosh^{-1}$.

For the case $|b| = 2|a|$ see Problem 6.97.

6.17. A man borrows an amount of money A. He is to pay back the loan in equal sums S to be paid periodically [such as every month, every 3 months, etc.]. This payment includes one part which is interest on the loan and the other part is used to reduce the principal which is outstanding. If the interest rate is i per payment period, (*a*) set up a difference equation for the principal P_k which is outstanding after the kth payment and (*b*) solve for P_k.

(*a*) Since the principal outstanding after the kth payment is P_k, it follows that the principal outstanding after the $(k+1)$st payment is P_{k+1}. Now this last principal to be paid is the principal outstanding from the kth payment plus the interest due on this payment for the period minus the sum S paid for this period. Thus

$$P_{k+1} = P_k + iP_k - S \tag{1}$$

or

$$P_{k+1} - (1+i)P_k = -S$$

The outstanding principal when $k=0$ is the total debt A so that we have

$$P_0 = A \tag{2}$$

(*b*) Solving the difference equation (*1*) we find

$$P_k = c(1+i)^k + S/i$$

and using condition (*2*) we find $c = A - S/i$ so that

$$P_k = \left(A - \frac{S}{i}\right)(1+i)^k + \frac{S}{i} = A(1+i)^k - S\left[\frac{(1+i)^k - 1}{i}\right] \tag{3}$$

6.18. In Problem 6.17 determine what sum S is to be paid back each period so as to pay back the debt in exactly n payments.

We must determine S so that $P_n = 0$. From Problem 6.17 this is seen to be equivalent to

$$A(1+i)^n - S\left[\frac{(1+i)^n - 1}{i}\right] = 0 \tag{1}$$

or

$$S = A\left[\frac{i}{1-(1+i)^{-n}}\right] \tag{2}$$

The process of paying back a debt according to (*2*) and Problem 6.17 is often called *amortization.* The factor

$$a_{\overline{n}|i} = \frac{1-(1+i)^{-n}}{i} \tag{3}$$

is often called the *amortization factor* and is tabulated for various values of n and i. In terms of this we can write (*2*) as

$$S = \frac{A}{a_{\overline{n}|i}} \tag{4}$$

6.19. Given the differential equation

$$\frac{dy}{dx} = f(x,y), \qquad y(x_0) = y_0$$

Show that if $y(x_k) = y_k$, $y'(x_k) = y'_k = f(x_k, y_k)$ then the differential equation is approximated by the difference equation

$$y_{k+1} = y_k + hy'_k = y_k + hf(x_k, y_k)$$

where $h = \Delta x = x_{k+1} - x_k$.

From the differential equation we have on integrating from $x = x_k$ to x_{k+1} for which the corresponding values of y are y_k and y_{k+1} respectively

$$\int_{y_k}^{y_{k+1}} dy = \int_{x_k}^{x_{k+1}} f(x,y)\,dx \tag{1}$$

or

$$y_{k+1} = y_k + \int_{x_k}^{x_{k+1}} f(x,y)\,dx \tag{2}$$

If we consider $f(x,y)$ to be approximately constant between x_k and x_{k+1}, (*2*) becomes

$$\begin{aligned} y_{k+1} &= y_k + \int_{x_k}^{x_{k+1}} f(x_k, y_k)\,dx \\ &= y_k + f(x_k, y_k)\int_{x_k}^{x_{k+1}} dx \qquad (3) \\ &= y_k + hf(x_k, y_k) \end{aligned}$$

as required.

6.20. (*a*) Use the method of Problem 6.19, sometimes called the *Euler method,* to find the value of y corresponding to $x = 0.5$ for the differential equation

$$\frac{dy}{dx} = 2x + y, \quad y(0) = 1$$

(*b*) Compare the result obtained in (*a*) with the exact value.

(*a*) We have for initial conditions $x_0 = 0$, $y_0 = 1$. Then by Problem 6.19 with $k = 0, 1, 2, \ldots$ we find

$$\begin{aligned} y_1 &= y_0 + hf(x_0, y_0) \\ y_2 &= y_1 + hf(x_1, y_1) \\ y_3 &= y_2 + hf(x_2, y_2) \\ &\text{etc.} \end{aligned} \tag{1}$$

We can use a convenient value for h, for example $h = 0.1$, and construct the table of Fig. 6-7 using equations (*1*) with $f(x_k, y_k) = 2x_k + y_k$.

k	x_k	y_k	$y'_k = f(x_k, y_k)$	$y_{k+1} = y_k + hy'_k$
0	0	1.0000	1.0000	1.1000
1	0.1	1.1000	1.3000	1.2300
2	0.2	1.2300	1.6300	1.3930
3	0.3	1.3930	1.9930	1.5923
4	0.4	1.5923	2.3923	1.8315
5	0.5	1.8315		

Fig. 6-7

From the table we can see that the value of y corresponding to $x = 0.5$ is 1.8315.

(*b*) The differential equation can be written as

$$\frac{dy}{dx} - y = 2x \tag{2}$$

a first order linear differential equation having integrating factor $e^{\int(-1)\,dx} = e^{-x}$. Multiplying by e^{-x} we have

$$e^{-x}\frac{dy}{dx} - ye^{-x} = 2xe^{-x} \quad \text{or} \quad \frac{d}{dx}(e^{-x}y) = 2xe^{-x}$$

Thus on integrating,

$$e^{-x}y = \int 2xe^{-x}\,dx = -2xe^{-x} - 2e^{-x} + c$$

or

$$y = -2x - 2 + ce^{x} \tag{3}$$

Since $y = 1$ when $x = 0$, we find $c = 3$. Thus the required solution is

$$y = 3e^{x} - 2x - 2 \tag{4}$$

Putting $x = 0.5$ we have for the required value of y

$$y = 3e^{0.5} - 2(0.5) - 2 = 3(1.6487) - 3 = 1.9461 \tag{5}$$

It is seen that the result in (*a*) is not too accurate since the percent error is about $5\frac{1}{2}\%$. Better results can be obtained by using smaller values of h for example $h = 0.05$ [see Problem 6.101].

This method, also called the *step by step method,* is important because of its simplicity and also theoretical value. There are of course many methods which provide greater accuracy. One of these which is a modification of the Euler method presented here is given in Problems 6.21 and 6.22. Other methods are given in the supplementary problems.

6.21. Given the differential equation of Problem 6.19 show that a better approximation is given by the difference equation

$$y_{k+1} = y_k + \tfrac{1}{2}h(y'_k + y'_{k+1})$$

Using equation (*2*) of Problem 6.19 we have

$$y_{k+1} = y_k + \int_{x_k}^{x_{k+1}} f(x,y)\,dx \qquad (1)$$

Instead of assuming that $f(x,y)$ is practically constant over the range from x_k to x_{k+1}, we use the trapezoidal rule to approximate the integral in (*1*). We thus obtain

$$y_{k+1} = y_k + \frac{h}{2}[f(x_k, y_k) + f(x_{k+1}, y_{k+1})] = y_k + \tfrac{1}{2}h(y'_k + y'_{k+1}) \qquad (2)$$

as required.

6.22. Use the method of Problem 6.21, sometimes called the *modified Euler method*, to work Problem 6.20.

The initial values of x and y are $x_0 = 0$, $y_0 = 1$ and are entered in the table of Fig. 6-7. In order to start the procedure using Problem 6.21 we must first find an approximate value of y corresponding to $x = 0.1$, assuming that we choose $h = 0.1$ as in Problem 6.20. For this purpose we use the ordinary Euler method of Problem 6.20 and find

$$\begin{aligned} y_1 = y(0.1) &= y_0 + hf(x_0, y_0) = y_0 + hy'_0 \\ &= 1.0000 + 0.1(1.0000) \\ &= 1.1000 \end{aligned}$$

This is entered in the table of Fig. 6-8 under Approximation I.

		Approximation I		Approximation II		Approximation III		
k	x_k	y_k	y'_k	y_k	y'_k	y_k	y'_k	y_k
0	0	1.0000	1.0000					
1	0.1	1.1000	1.3000	1.1150	1.3150	1.1158	1.3158	1.1158
2	0.2	1.2474	1.6474	1.2640	1.6640	1.2648	1.6648	1.2648
3	0.3	1.4313	2.0313	1.4496	2.0496	1.4505	2.0505	1.4506
4	0.4	1.6557	2.4557	1.6759	2.4759	1.6769	2.4769	1.6770
5	0.5	1.9247	2.9247	1.9471	2.9471	1.9482	2.9482	1.9483

Fig. 6-8

From the differential equation we now find

$$\begin{aligned} y'_1 = f(x_1, y_1) &= 2x_1 + y_1 \\ &= 2(0.1) + 1.1000 \\ &= 1.3000 \end{aligned}$$

which is also entered under Approximation I in Fig. 6-8.

From the modified Euler formula of Problem 6.21 we have on putting $k = 0$

$$\begin{aligned} y_1 &= y_0 + \tfrac{1}{2}h(y'_0 + y'_1) \\ &= 1.0000 + \tfrac{1}{2}(0.1)(1.0000 + 1.3000) \\ &= 1.1150 \end{aligned}$$

This better approximation to y_1 is entered in the table of Fig. 6-8 under Approximation II.

Repeating the process we then find

$$y_1' = f(x_1, y_1) = 2x_1 + y_1 = 2(0.1) + 1.1150 = 1.3150$$

and so Approximation III for y_1 is given by

$$y_1 = y_0 + \tfrac{1}{2}h(y_0' + y_1') = 1.0000 + \tfrac{1}{2}(0.1)(1.0000 + 1.3150) = 1.1158$$

which yields

$$y_1' = f(x_1, y_1) = 2x_1 + y_1 = 2(0.1) + 1.1158 = 1.3158$$

We then find

$$y_1 = y_0 + \tfrac{1}{2}h(y_0' + y_1') = 1.0000 + \tfrac{1}{2}(0.1)(1.0000 + 1.3158) = 1.1158$$

Since this gives the same value as that obtained above for y_1 the desired accuracy to four decimal places is achieved.

We now proceed in the same manner to find $y(0.2) = y_2$. Using the value $y_1 = 1.1158$ corresponding to $x_1 = 0.1$ we have for Approximation I

$$y_2 = y_1 + hf(x_1, y_1) = y_1 + hy_1' = 1.1158 + 0.1(1.3158) = 1.2474$$

and

$$y_2' = f(x_2, y_2) = 2x_2 + y_2 = 2(0.2) + 1.2474 = 1.6474$$

By continuing the process over and over again we finally obtain the required value $y_5 = y(0.5) = 1.9483$ as indicated in the table. This compares favorably with the exact value 1.9461 and is accurate to two decimal places.

It will be noticed that in the modified Euler method the equation

$$y_{k+1} = y_k + \tfrac{1}{2}h(y_k' + y_{k+1}')$$

is used to "predict" the value of y_{k+1}. The differential equation

$$y_k' = f(x_k, y_k)$$

is used to "correct" the value of y_{k+1}. Because of this we often refer to the method as a "predictor-corrector" method. Many such methods are available which yield better accuracy. See for example Problem 6.105.

6.23. Show how the boundary-value problem

$$\frac{\partial^2 U}{\partial x^2} + \frac{\partial^2 U}{\partial y^2} = 0 \qquad 0 \leqq x \leqq 1,\ 0 \leqq y \leqq 1$$

$$U(x,0) = 1, \quad U(x,1) = 0, \quad U(0,y) = 0, \quad U(1,y) = 0$$

can be approximated by a difference equation.

It is convenient to approximate the derivatives by using central differences. If we let δ_x and E_x denote the central difference operator and shifting operator with respect to x, then the partial derivative operator with respect to x is given by

$$D_x = \frac{\partial}{\partial x} \sim \frac{\delta_x}{h} = \frac{E_x^{1/2} - E_x^{-1/2}}{h}$$

where the symbol $\sim$ denotes "corresponds to". Thus by squaring the operators and operating on U we have

$$\frac{\partial^2 U}{\partial x^2} \sim \frac{\delta_x^2}{h^2}U = \frac{1}{h^2}(E_x^{1/2} - E_x^{-1/2})^2 U$$
$$= \frac{1}{h^2}(E_x - 2 + E_x^{-1})U(x,y)$$
$$= \frac{1}{h^2}[U(x+h,\,y) - 2U(x,\,y) + U(x-h,\,y)]$$

Similarly using operators with respect to y we have

$$\frac{\partial^2 U}{\partial y^2} \sim \frac{\delta_y^2}{h^2}U = \frac{1}{h^2}(E_y^{1/2} - E_y^{-1/2})^2 U$$
$$= \frac{1}{h^2}(E_y - 2 + E_y^{-1})U(x,y)$$
$$= \frac{1}{h^2}[U(x,\,y+h) - 2U(x,\,y) + U(x,\,y-h)]$$

It follows by addition that

$$\frac{\partial^2 U}{\partial x^2} + \frac{\partial^2 U}{\partial y^2} \sim \frac{1}{h^2}[U(x+h,\,y) + U(x-h,\,y) + U(x,\,y+h) + U(x,\,y-h) - 4U(x,\,y)]$$

Thus the equation

$$\frac{\partial^2 U}{\partial x^2} + \frac{\partial^2 U}{\partial y^2} = 0$$

is replaced by

$$U(x+h,\,y) + U(x-h,\,y) + U(x,\,y+h) + U(x,\,y-h) - 4U(x,\,y) = 0$$

or
$$U(x,\,y) = \tfrac{1}{4}[U(x+h,\,y) + U(x-h,\,y) + U(x,\,y+h) + U(x,\,y-h)] \qquad (1)$$

which is the required difference equation.

The boundary conditions associated with this difference equation are left the same.

6.24. Show how to obtain an approximate solution to the boundary-value problem of Problem 6.23.

We must find $U(x,y)$ at all points inside the square $0 \leqq x \leqq 1$, $0 \leqq y \leqq 1$. Using a suitable value of N and h such that $Nh = 1$, we subdivide the square into smaller squares of sides $h = 1/N$ as in Fig. 6-9. The result is often called a *grid* or *mesh*. Instead of finding $U(x,y)$ at all points inside the square we find the values of $U(x,y)$ at the points of the grid such as $A, B, \ldots$ in Fig. 6-9. To accomplish this we make use of the difference equation (*1*) of Problem 6.23 which is, as we have seen, an approximation to the partial differential equation. By referring to Fig. 6-10, which represents an enlargement of a part of Fig. 6-9, we see that this difference equation, i.e.

$$U(x,\,y) = \tfrac{1}{4}[U(x-h,\,y) + U(x,\,y+h) + U(x+h,\,y) + U(x,\,y-h)] \qquad (1)$$

states that the value of U at P is the arithmetic mean of the values at the neighboring points A, B, C and D. By using this difference equation together with the values of U on the boundary of the square as shown in Fig. 6-9, we can obtain approximate values of U at all points in the grid.

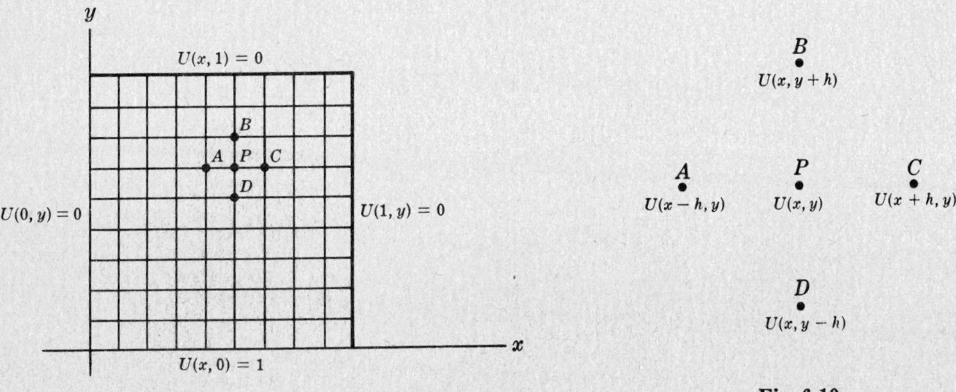

Fig. 6-9

Fig. 6-10

In order to illustrate the procedure we shall use the grid of Fig. 6-11 in which we have chosen for the purposes of simplification $h = 1/3$. In this case there are four points in the grid denoted by $1, 2, 3, 4$ of Fig. 6-11. We wish to find the values of U at these points denoted by U_1, U_2, U_3, U_4 respectively.

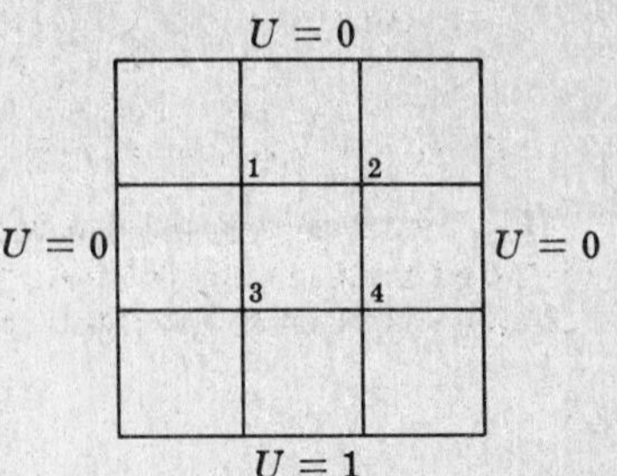

Fig. 6-11

By applying equation (*1*) we see that U_1, U_2, U_3, U_4 must satisfy the equations

$$\begin{aligned} U_1 &= \tfrac{1}{4}(0 + 0 + U_2 + U_3) \\ U_2 &= \tfrac{1}{4}(U_1 + 0 + 0 + U_4) \\ U_3 &= \tfrac{1}{4}(0 + U_1 + U_4 + 1) \\ U_4 &= \tfrac{1}{4}(U_3 + U_2 + 0 + 1) \end{aligned} \qquad (2)$$

The first equation for example is obtained by realizing that U_1, i.e. the value of U at point 1, is the arithmetic mean of the values of U at the neighboring points, i.e. $0, 0, U_2, U_3$ the zeros being the values of U at points on the boundary of the square.

The values U_1, U_2, U_3 and U_4 can be found by solving simultaneously the above system of equations. For a grid with a large number of subdivisions this procedure is often less desirable than a method called the *method of iteration* which can take advantage of machine calculators.

In this method of iteration we first assume a trial set of solutions. The simplest is to take $U_1 = 0,\ U_2 = 0,\ U_3 = 0,\ U_4 = 0$. We refer to this as Trial Number 1 or the first trial as recorded in Fig. 6-12.

Using the values of U_1, U_2, U_3, U_4 from the Trial 1 in equation (*2*) we then calculate the new values

$$U_1 = 0, \quad U_2 = 0, \quad U_3 = 0.2500, \quad U_4 = 0.2500$$

These are entered in Fig. 6-12 for Trial 2.

Trial Number	U_1	U_2	U_3	U_4
1	0	0	0	0
2	0	0	0.2500	0.2500
3	0.0625	0.0625	0.3125	0.3125
4	0.0938	0.0938	0.3438	0.3438
5	0.1094	0.1094	0.3594	0.3594
6	0.1172	0.1172	0.3672	0.3672
7	0.1211	0.1211	0.3711	0.3711
8	0.1231	0.1231	0.3731	0.3731
9	0.1241	0.1241	0.3741	0.3741
10	0.1246	0.1246	0.3746	0.3746
11	0.1248	0.1248	0.3748	0.3748
12	0.1249	0.1249	0.3749	0.3749
13	0.1250	0.1250	0.3750	0.3750
14	0.1250	0.1250	0.3750	0.3750

Fig. 6-12

Using the values of U_1, U_2, U_3, U_4 from the Trial 2 in equations (2) we then calculate the new values

$$U_1 = 0.0625, \quad U_2 = 0.0625, \quad U_3 = 0.3125, \quad U_4 = 0.3125$$

These are entered in Fig. 6-12 for Trial 3.

By continuing in this manner the remaining entries in Fig. 6-12 are obtained. The process is continued until the values for U_1, U_2, U_3, U_4 no longer change as indicated by the last two lines. We see that the values are

$$U_1 = 0.1250, \quad U_2 = 0.1250, \quad U_3 = 0.3750, \quad U_4 = 0.3750$$

As a check the values can be substituted back into equations (2) and will be found to satisfy them.

The equation

$$\frac{\partial^2 U}{\partial x^2} + \frac{\partial^2 U}{\partial y^2} = 0 \tag{3}$$

is called *Laplace's equation in two dimensions.* This equation arises in various problems on heat flow, potential theory and fluid flow. In problems on heat flow for example, $U = U(x, y)$ represents the steady-state temperature in a plane region such as a thin plate whose faces are insulated so that heat cannot enter or escape. In the problem above, the region is a square, one of whose sides is maintained at temperature 1 while the remaining sides are maintained at temperature 0. In such case the values of U represent the steady-state temperatures at various points in the square.

Supplementary Problems

APPLICATIONS TO VIBRATING SYSTEMS

6.25. Show that if we assume that the tension of the string in Problem 6.1 is constant but that the displacements are not necessarily small, then the equation of motion for the kth particle is given by

$$m\frac{d^2y_k}{dt^2} = \frac{\tau}{h}(w_{k+1} - w_k)$$

where

$$w_k = \frac{(y_k - y_{k-1})/h}{\sqrt{1 + (y_k - y_{k-1})^2/h^2}}$$

6.26. Show that if $|(y_k - y_{k-1})/h| < 1$ the equation of Problem 6.25 can be approximated by

$$m\frac{d^2y_k}{dt^2} = \frac{\tau}{h}(y_{k-1} - 2y_k + y_{k+1}) + \frac{\tau}{2h^3}[(y_k - y_{k-1})^3 - (y_{k+1} - y_k)^3]$$

6.27. Show that if the string of Problem 6.1 vibrates in a vertical plane so that gravity must be taken into account, then assuming that vibrations are small the equation of the kth particle is given by

$$\frac{d^2y_k}{dt^2} = \frac{\tau}{mh}(y_{k-1} - 2y_k + y_{k+1}) + g$$

where g is the acceleration due to gravity.

6.28. By letting $y_k = u_k + F(k)$ in the equation of Problem 6.27 and choosing $F(k)$ appropriately, show that the equation can be reduced to

$$\frac{d^2u_k}{dt^2} = \frac{\tau}{mh}(u_{k-1} - 2u_k + u_{k+1})$$

and discuss the significance of the result.

6.29. Suppose that the string of Problem 6.27 fixed at points P and Q [see Fig. 6-1] is left to hang vertically under the influence of gravity. Show that the displacement y_k of the kth particle is given by

$$y_k = -\frac{mg}{2\tau h} x_k(L - x_k) \qquad k = 1, 2, \ldots, N$$

where $x_k = kh$ is the distance of the kth particle from P.

6.30. Suppose that in the string of Problem 6.1 the number of particles N becomes infinite in such a way that the total mass and length are constant. Show that the difference equation in this limiting case becomes the *wave equation*

$$\frac{\partial^2 y}{\partial t^2} = \frac{\tau}{\rho}\frac{\partial^2 y}{\partial x^2}$$

where τ is the tension and ρ is the density [mass per unit length] of the string, both assumed constant.

6.31. What does the equation of Problem 6.30 become if the vibration takes place in a vertical plane and gravity is taken into account?

6.32. Suppose that in Problem 6.1 the tension in the string is not constant but that in the portion of the string immediately to the right of particle k the tension is τ_k. Show that the equation of motion of the kth particle is

$$m\frac{d^2 y_k}{dt^2} = \frac{1}{h}\Delta(\tau_{k-1}\Delta y_{k-1})$$

and write corresponding boundary conditions.

(a) By letting $y_k = A_k \cos \omega t$ in the equation of Problem 6.32 show that we obtain the Sturm-Liouville difference equation

$$\Delta(\tau_{k-1}\Delta A_{k-1}) + m\omega^2 h A_k = 0$$

(b) Discuss the theory of the solutions of this equation from the viewpoint of eigenvalues and eigenfunctions as presented in Chapter 5.

(c) Illustrate the results of this theory by considering the special case where the tension is constant.

6.33. Suppose that at time $t = 0$ the position and velocity of the kth particle of Problem 6.1 are given by $y_k(0)$ and $\dot{y}_k(0)$. Show that the position of the kth particle at any time $t > 0$ is given by

$$y_k = \sum_{n=1}^{N} c_n \sin\frac{nk\pi}{N+1}\cos(\omega_n t + \alpha_n) = \sum_{n=1}^{N}(a_n \cos\omega_n t + b_n \sin\omega_n t)\sin\frac{nk\pi}{N+1}$$

where a_n, b_n or c_n and α_n are determined from

$$a_n = c_n \cos\alpha_n = \frac{2}{N+1}\sum_{k=1}^{N} y_k(0)\sin\frac{nk\pi}{N+1}$$

$$b_n = -c_n \sin\alpha_n = \frac{-2}{N+1}\sum_{k=1}^{N}\dot{y}_k(0)\sin\frac{nk\pi}{N+1}$$

6.34. (a) Show that if the $(N+1)$th mass of the string of Problem 6.1 is free while the first mass is fixed, then the frequency of the lowest mode of vibration is

$$f_1 = \sqrt{\frac{\tau}{mh}}\sin\frac{\pi}{2N-1}$$

(b) What are the frequencies of the higher modes of vibration? [*Hint.* If the force on the $(N+1)$th mass is zero then $y_N = y_{N+1}$.]

6.35. (a) Give a physical interpretation of the boundary conditions $\alpha_0 y_0 + \alpha_1 y_1 = 0$, $\alpha_N y_N + \alpha_{N+1} y_{N+1} = 0$ in Problem 6.1.

(b) Obtain the natural frequencies of the system for this case.

6.36. An axis carries N equally spaced discs which are capable of torsional vibrations relative to each other [see Fig. 6-13]. If we denote by θ_k the angular displacement of the kth disc relative to the first disc show that

$$I\frac{d^2\theta_k}{dt^2} = \sigma(\theta_{k+1} - 2\theta_k + \theta_{k-1})$$

where I is the constant moment of inertia of each disc about the axis and σ is the *torsion constant.*

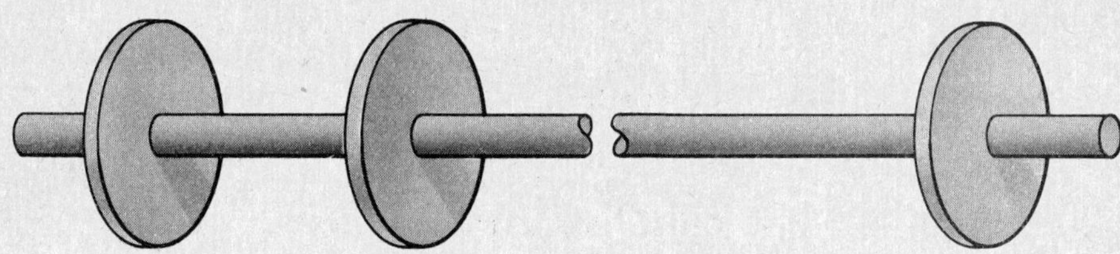

Fig. 6-13

6.37. (*a*) Using appropriate boundary conditions obtain the natural frequencies for the system of Problem 6.36.

(*b*) Explain how you would find the motion for the kth disc by knowing the state of the system at time $t = 0$.

(*c*) Compare with results for the vibrating system of Problem 6.1.

6.38. Work Problem 6.37 if the first disc is fixed while the last is free.

6.39. A frictionless horizontal table AB has on it N identical objects of mass m connected by identical springs having spring constant K as indicated in Fig. 6-14. The system is set into vibration by moving one or more masses along the table and then releasing them.

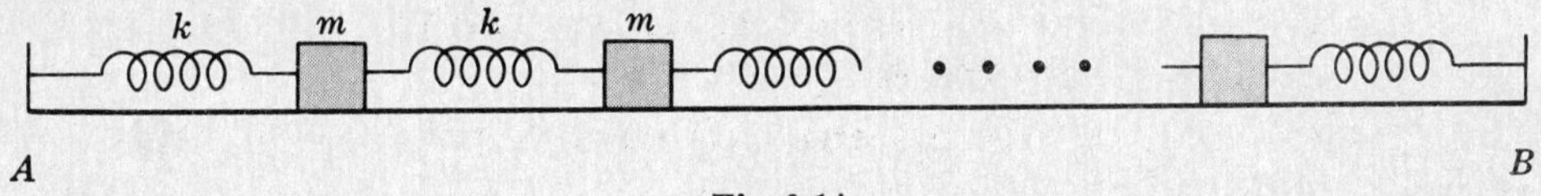

Fig. 6-14

(*a*) Letting x_k denote the displacement of the kth object, set up a difference equation.

(*b*) Find the natural frequencies of the system.

(*c*) Describe the analogies with the vibrating string of Problem 6.1.

6.40. Suppose that the string of Problem 6.1 with the attached particles is rotated about the axis with uniform angular speed Ω.

(*a*) Show that if y_k is the displacement of the kth particle from the axis, then

$$y_{k+1} - 2\left(1 - \frac{m\Omega^2 h}{2\tau}\right)y_k + y_{k-1} = 0$$

(*b*) Show that the *critical speeds* in case $m\Omega^2 h/4\tau < 1$ are given by

$$\Omega_n = \sqrt{\frac{4\tau}{mh}}\sin\frac{n\pi}{2(N+1)} \qquad n = 1, 2, \ldots, N$$

and discuss the physical significance of these.

6.41. Discuss the limiting case of Problem 6.40 if $N \to \infty$ in such a way that the total mass and length remain finite.

6.42. Explain how you would find a solution to the partial differential equation in Problem 6.30 subject to appropriate boundary conditions by applying a limiting procedure to the results obtained in Problem 6.2.

APPLICATIONS TO ELECTRICAL NETWORKS

6.43. An electrical network consists of N loops with capacitors having constant capacitance C_1, C_2 in each loop as indicated in Fig. 6-15. An alternating voltage given by $V_N = K \sin \omega t$ where K and ω are constants is applied across $A_N B_N$. If it is required that the voltage across $A_0 B_0$ be $V_0 = 0$, show that the voltage V_k across $A_k B_k$ is given by

$$V_k = K \frac{\sinh \mu k}{\sinh \mu N} \sin \omega t$$

where

$$\mu = \cosh^{-1}\left(1 + \frac{C_2}{2C_1}\right)$$

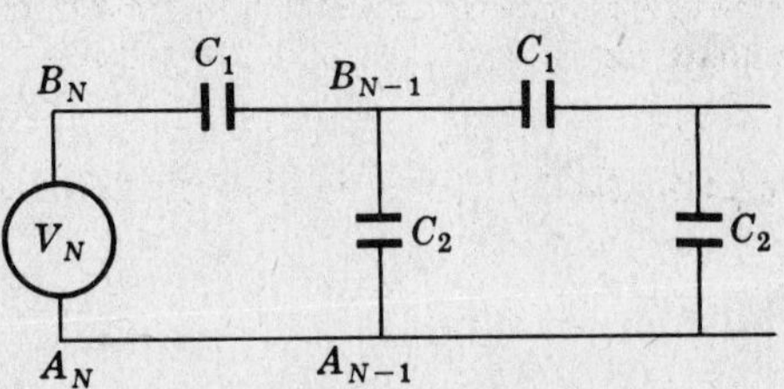

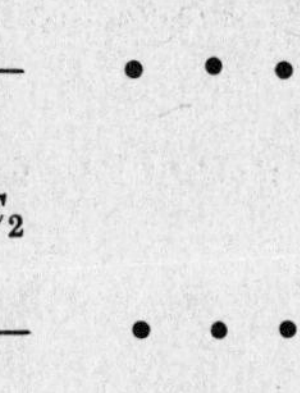
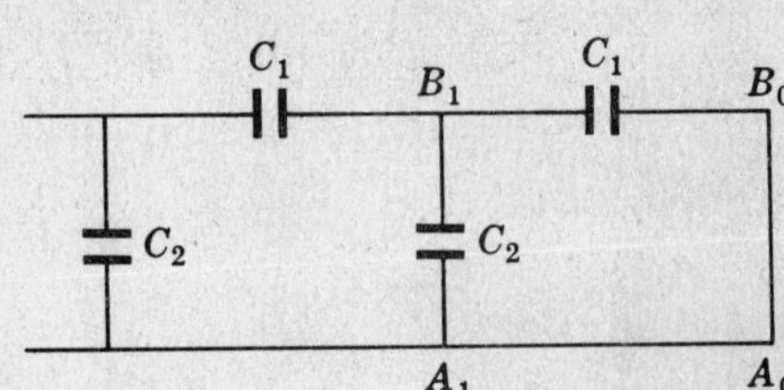

Fig. 6-15

6.44. Work Problem 6.43 if the capacitors C_1 and C_2 are replaced by inductors with constant inductance L_1, L_2.

6.45. Work Problem 6.43 if the capacitors C_1 are replaced by resistors having constant resistance R_1.

6.46. Can Problem 6.43 be worked if it is required that the voltage across A_0B_0 be $V_0 > 0$? Explain.

APPLICATIONS TO BEAMS

6.47. Suppose that in Problem 6.6 the beam is of finite length Nh. Show that the bending moment at the nth support is given by

$$M_n = -Wh\left[\frac{r_1^{n+1-N} - r_2^{n+1-N}}{r_1^{1-N} - r_2^{1-N}}\right]$$

where $r_1 = -2 + \sqrt{3}$, $r_2 = -2 - \sqrt{3}$.

6.48. Show that as $N \to \infty$ the result in Problem 6.47 reduces to that of Problem 6.6.

6.49. In Fig. 6-16 AB represents a continuous uniform beam on which there is a triangular load distribution where the total load on the beam is W. Suppose that the beam is simply supported at the $N+1$ points $P_0, P_1, \ldots, P_N$ at distances h apart. (*a*) Show that the three moment equation can be written as

$$M_{k-1} + 4M_k + M_{k+1} = -\frac{Wh}{N^2}k$$

where the bending moments at the ends A and B are zero, i.e. $M_0 = 0$, $M_N = 0$. (*b*) Show that the bending moment at the kth support is given by

$$M_k = \frac{Wh}{6N}\left[(-1)^{N+k}\left\{\frac{(2+\sqrt{3})^k - (2-\sqrt{3})^k}{(2+\sqrt{3})^N - (2-\sqrt{3})^N}\right\} - \frac{k}{N}\right]$$

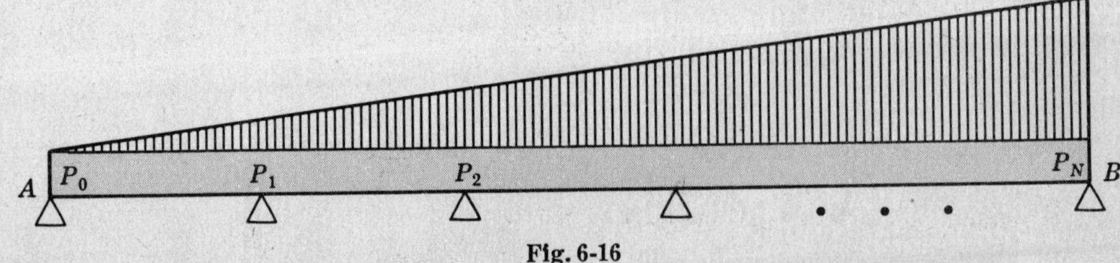

Fig. 6-16

6.50. Does $\lim_{N\to\infty} M_k$ exist in Problem 6.49? Explain.

APPLICATIONS TO COLLISIONS

6.51. Show that the speeds of the balls in Problem 6.7 follow a periodic pattern and determine this period.

6.52. (*a*) Work Problem 6.7 if the two masses are given by m_1 and m_2 where $m_1 \neq m_2$. (*b*) Determine the values of m_1/m_2 for which the speeds will be periodic and determine the periods in such cases. (*c*) Discuss the motion for the case $m_1 = 2m_2$.

6.53. A billiard ball is at rest on a horizontal frictionless billiards table which is assumed to have the shape of a square of side l and whose sides are assumed to be perfectly elastic. The ball is hit so that it travels with constant speed S and so that the direction in which it travels makes the angle θ with the side. Determine the successive points at which the ball hits the sides.

6.54. What additional complexities arise in Problem 6.53 if the billiard table has the shape of a rectangle? Explain.

APPLICATIONS TO PROBABILITY

6.55. Suppose that in Problem 6.9 we let $u_{k,m}$ be the probability of A winning the game when he has k pennies while B has m pennies.

(*a*) Show that the difference equation and boundary conditions for $u_{k,m}$ are given by

$$u_{k,m} = pu_{k+1,\,m-1} + qu_{k-1,\,m+1}$$

$$u_{M,\,N-M} = 1, \quad u_{N-M,\,M} = 0$$

(*b*) Solve the difference equation in (*a*) subject to the conditions and reconcile your results with those given in Problem 6.9. [*Hint.* See Problem 5.59, page 189.]

6.56. Suppose that in a single toss of a coin the probabilities of heads or tails are given by p and $q = 1 - p$ respectively. Let $u_{k,n}$ denote the probability that exactly k heads will appear in n tosses of the coin.

(*a*) Show that

$$u_{k+1,\,n+1} = pu_{k,n} + qu_{k+1,\,n}$$

where

$$u_{k,0} = \begin{cases} 1 & k = 0 \\ 0 & k > 0 \end{cases}$$

(*b*) By solving the problem in (*a*) show that

$$u_{k,n} = \binom{n}{k} p^k q^{n-k}$$

This is called *Bernoulli's probability formula for repeated trials.*

6.57. Suppose that A and B play a game as in Problem 6.9 and that A starts out with 10 pennies while B starts out with 90 pennies. (*a*) Show that if $p = q = 1/2$, i.e. the game is fair, then the probability that A will lose all of his money, i.e. A will be "ruined", is 0.9. (*b*) Show that if $p = 0.4$, $q = 0.6$ then the probability that A will be "ruined" is 0.983. (*c*) Discuss the significance of the result in (*b*) if A represents a "gambler" and B represents the "bank" at a gambling casino. (*d*) What is the significance of the result in (*b*) if A is the "bank" and B is the "gambler"?

6.58. Suppose that in Problem 6.9 $p > q$, i.e. the probability of A's winning is greater than that of B. Show that if B has much more capital than A then the probability of A escaping "ruin" if he starts out with capital a is approximately

$$1 - (q/p)^a$$

What is this probability if $p = 0.6$, $q = 0.4$ and (*a*) $a = 5$, (*b*) $a = 100$?

6.59. Let D_k denote the *expected duration* of the game in Problem 6.9 when the capital of A is k.

(*a*) Show that D_k satisfies the equation

$$D_k = pD_{k+1} + qD_{k-1} + 1, \quad D_0 = 0, \quad D_{a+b} = 0$$

(*b*) Show that if $p \neq q$, $\quad D_k = \dfrac{k}{q-p} - \dfrac{a+b}{q-p}\left[\dfrac{(q/p)^k - 1}{(q/p)^{a+b} - 1}\right].$

(*c*) Show that if $p = q = 1/2$, $D_k = k(a+b-k)$ and thus that if A and B start out with initial capitals a and b respectively the expected duration is ab.

6.60. Work Problems 6.9-6.11 in case a tie can occur at any stage with constant probability $\epsilon > 0$ so that $p + q + \epsilon = 1$.

THE FIBONACCI NUMBERS

6.61. Let F_k denote the kth Fibonacci number and let $R_k = F_k/F_{k+1}$.

(*a*) Show that $R_k = \dfrac{1}{1 + R_{k-1}}, \quad R_0 = 0.$

(*b*) Show that

$$R_k = \cfrac{1}{1 + \cfrac{1}{1 + \cfrac{1}{\ddots + \cfrac{1}{1}}}}$$

i.e. a *continued fraction* with k divisions.

6.62. (*a*) Solve the difference equation of part (*a*) in Problem 6.61 for R_k.

(*b*) Show that $\lim\limits_{k \to \infty} R_k = \dfrac{\sqrt{5} - 1}{2}$ and thus obtain the limiting value of the infinite continued fraction in part (*b*) of Problem 6.61.

6.63. If F_k denotes the kth Fibonacci number prove that

$$F_{k+1}F_{k-1} - F_k^2 = (-1)^k \qquad k = 1, 2, 3, \ldots$$

6.64. Generalize Problem 6.63 by showing that

$$F_{k+m}F_{k-m} - F_k^2 = (-1)^{k+m-1}F_m^2 \qquad k = m, m+1, m+2, \ldots$$

6.65. A planted seed produces a flower with one seed at the end of the first year and a flower with two seeds at the end of two years and each year thereafter. Assuming that each seed is planted as soon as it is produced show that the number of flowers at the end of k years is the Fibonacci number F_k.

6.66. Find the number of flowers at the end of k years in Problem 6.65 if a planted seed produces a flower with four seeds at the end of the first year and a flower with ten seeds at the end of two years and each year thereafter.

6.67. Generalize Problems 6.65 and 6.66 by replacing "four seeds" and "ten seeds" by "a seeds" and "b seeds" respectively. What restriction if any must there be on a and b?

6.68. Find the sum of the series $\displaystyle\sum_{k=0}^{\infty} \frac{F_{k+2}}{F_{k+1}F_{k+3}}$ where F_k is the kth Fibonacci number.

6.69. Show that

$$F_k = \frac{1}{2^{k-1}}\left[\binom{k}{1} + \binom{k}{3}5 + \binom{k}{5}5^2 + \cdots\right]$$

MISCELLANEOUS PROBLEMS

6.70. In *Pascal's triangle* [see Fig. 6-17] each number is obtained by taking the sum of the numbers which lie immediately to the left and right of it in the preceding row. Thus for example in the sixth row of Fig. 6-17 the number 10 is the sum of the numbers 6 and 4 of the fifth row.

```
                  1
                1   1
              1   2   1
            1   3   3   1
          1   4   6   4   1
        1   5   10  10  5   1
      1   6   15  20  15  6   1
```

Fig. 6-17

(a) Show that if $u_{k,m}$ denotes the kth number in the mth row, then

$$u_{k+1,\,m+1} = u_{k+1,\,m} + u_{k,m} \qquad \text{where} \qquad u_{k,0} = \begin{cases} 1 & k > 0 \\ 0 & k = 0 \end{cases}$$

(b) Solve the difference equation of (a) subject to the given conditions and thus show that

$$u_{k,m} = \binom{m}{k}$$

i.e. the numbers represent the binomial coefficients.

6.71. The *Chebyshev polynomials* $T_k(x)$ are defined by the equation

$$T_{k+1}(x) - 2xT_k(x) + T_{k-1}(x) = 0$$

where $\qquad T_0(x) = 1, \quad T_1(x) = x$

(a) Obtain the Chebyshev polynomials $T_2(x)$, $T_3(x)$, $T_4(x)$.

(b) By solving the difference equation subject to the boundary conditions show that

$$T_k(x) = \cos(k \cos^{-1} x)$$

6.72. The *Bessel function* $J_k(x)$ of order k can be defined by

$$J_k(x) = \frac{1}{\pi}\int_0^{\pi} \cos(k\theta - x\sin\theta)\,d\theta$$

Show that $J_k(x)$ satisfies the difference equation

$$xJ_{k+1}(x) - 2kJ_k(x) + J_{k-1}(x) = 0$$

6.73. Show that the Bessel functions of Problem 6.72 have the property that

$$J_k'(x) = \tfrac{1}{2}[J_{k-1}(x) - J_{k+1}(x)]$$

6.74. The *Laplace transform* of a function $F(x)$ is denoted by

$$\mathcal{L}\{F(x)\} = f(s) = \int_0^{\infty} e^{-sx} F(x)\,dx$$

(a) Show that $\mathcal{L}\{J_0(x)\} = \dfrac{1}{\sqrt{s^2+1}}$.

(b) Show that $\mathcal{L}\{J_k(x)\} = \dfrac{[\sqrt{s^2+1} - s]^k}{\sqrt{s^2+1}}$.

6.75. Suppose that a principal P is deposited in a bank and kept there. If the interest rate per period is given by i, (a) set up a difference equation for the total amount A_k in the bank after k periods and (b) find A_k.

6.76. A man deposits a sum S in a bank at equal periods. If the interest rate per period is i, (*a*) set up a difference equation for the amount A_n in the bank after n periods and (*b*) show that

$$A_n = S\left[\frac{(1+i)^n - 1}{i}\right] = Ss_{\overline{n}|i}$$

Such a set of equal periodic payments is called an *annuity.*

6.77. Work Problem 6.76 if the successive deposits are given by $S, S+a, S+2a, S+3a, \ldots$. Discuss the significance of the cases (*a*) $a > 0$, (*b*) $a < 0$.

6.78. In a particular town the population increases at a rate $\alpha\%$ per year [as measured at the end of each year] while there is a loss of β individuals per year. Show that the population after k years is given by

$$P_k = \left(P_0 - \frac{100\beta}{\alpha}\right)\left(1 + \frac{\alpha}{100}\right)^k + \frac{100\beta}{\alpha}$$

where P_0 is the initial population.

6.79. Does the result of Problem 6.78 hold if (*a*) $\alpha < 0$, $\beta < 0$, (*b*) $\alpha > 0$, $\beta < 0$, (*c*) $\alpha < 0$, $\beta > 0$? Interpret each of the cases.

6.80. Let S_k denote the sum of the kth powers of the n roots of the equation

$$r^n + a_1 r^{n-1} + a_2 r^{n-2} + \cdots + a_n = 0$$

(*a*) Show that

$$S_1 + a_1 = 0, \quad S_2 + a_1 S_2 + 2a_2 = 0, \quad \ldots, \quad S_n + a_1 S_{n-1} + \cdots + a_{n-1}S_1 + na_n = 0$$

(*b*) Show that if $k > n$

$$S_k + a_1 S_{k-1} + a_2 S_{k-2} + \cdots + a_n S_{k-n} = 0$$

6.81. Obtain the generating function for the sequence S_k of Problem 6.80.

6.82. A box contains n identical slips of paper marked with the numbers $1, 2, \ldots, n$ respectively. The slips are chosen at random from the box. We say that there is a *coincidence* if the number M is picked at the Mth drawing. Let u_k denote the number of permutations of k numbers in which there are no coincidences. (*a*) Prove that the number of permutations in which there is 1 coincidence is $\binom{k}{1} u_{k-1}$. (*b*) Prove that the number of permutations in which there are m coincidences is $\binom{k}{m} u_{k-m}$ where $m = 0, 1, \ldots, k$.

6.83. (*a*) Prove that u_k in Problem 6.82 satisfies the equation

$$u_k + \binom{k}{1}u_{k-1} + \binom{k}{2}u_{k-2} + \cdots + \binom{k}{k-1}u_1 + \binom{k}{k}u_0 = k!, \qquad u_0 = 1$$

(*b*) By solving the difference equation in (*a*) prove that

$$u_n = n!\left[\frac{1}{2!} - \frac{1}{3!} + \frac{1}{4!} - \cdots + \frac{(-1)^n}{n!}\right]$$

(*c*) Show that the probability of getting no coincidences in the drawing of the n slips from the box of Problem 6.82 is given by

$$\frac{1}{2!} - \frac{1}{3!} + \cdots + \frac{(-1)^n}{n!}$$

(*d*) Show that for large values of n the probability in (*c*) is very nearly

$$1 - \frac{1}{e} = 0.63212\ldots$$

(*e*) Discuss how large n must be taken in (*d*) so that the probability differs from $1 - 1/e$ by less than 0.01.

6.84. An inefficient secretary puts n letters at random into n differently addressed envelopes. Show that the probability of getting at least one letter into its correct envelope is given by

$$1 - \frac{1}{2!} + \frac{1}{3!} - \frac{1}{4!} + \cdots + \frac{(-1)^{n+1}}{n!}$$

and thus show that for large n this probability is about $1/e = 0.36787\ldots$. Can you explain from a qualitative point of view why this probability does not change much with an increase of n as for example from 100 to 200?

6.85. Find the probability that at least two people present at a meeting will have their birth dates the same [although with different ages].

6.86. A group of players plays a game in which every time a player wins he gets m times the amount of his stake but if he loses he gets to play again. Let u_k denote the stake which a player must make in the kth game so that he gets back not only all that he has lost previously but also a profit P.

(a) Show that u_k satisfies the *sum equation*

$$mu_k = P + \sum_{p=1}^{k} u_p$$

(b) Show that the sum equation of (a) can be written as the difference equation

$$u_{k+1} - \frac{m}{m-1}u_k = 0$$

(c) Using appropriate boundary conditions show that

$$u_k = \frac{P}{m}\left(\frac{m}{m-1}\right)^k$$

6.87. State some limitations of the "sure profit" deal in Problem 6.86 illustrating the results by taking $m = 2$.

6.88. A box contains slips of paper on which are written the numbers $1, 2, 3, \ldots, n$ [compare Problem 6.82]. The slips are drawn one at a time from the box at random and are not replaced. Let $u_{k,n}$ denote the number of permutations [out of the total of $n!$ permutations] in which there is no coincidence in k given drawings.

(a) Show that $\quad u_{k,m} = u_{k-1,m} - u_{k-1,m-1}, \qquad u_{k,0} = k!$

(b) Show that the solution to the difference equation of part (a) is

$$u_{k,m} = \binom{k}{0}m! - \binom{k}{1}(m-1)! + \binom{k}{2}(m-2)! - \cdots + (-1)^k\binom{k}{k}0!$$

where $0! = 1$.

(c) Show that the probability that there will be no coincidence in the k drawings is $u_{k,m}/m!$.

(d) Show that for $k = m$ the probability obtained in (c) reduces to that obtained in Problem 6.82.

6.89. A and B play a game with a deck of n cards which are faced downwards. A asks B to name a card. Then A shows B the card and all the cards above it in the deck and all these cards are removed from the deck. A then asks B to name another card and the process is repeated over and over again until there are no cards left. If B does not name any of the top cards then A wins the game, otherwise B wins the game. Find the probability that A wins the game.

6.90. If the deck of Problem 6.89 has 52 cards, determine whether the game is fair if B offers A 2 to 1 odds.

6.91. A sequence has its first two terms given by 0 and 1 respectively. Each of the later terms is obtained by taking the arithmetic mean of the previous two terms. (a) Find the general term of the sequence and (b) find the limit of the sequence.

6.92. An urn has b blue marbles and r red marbles. Marbles are taken from the urn one at a time and each one is returned following the next marble taken. Determine the probability that the kth marble taken from the urn is blue if it is known that the nth marble taken is blue.

6.93. In the preceding problem what is the probability if nothing is known about the nth marble taken?

6.94. (a) Show that the Stirling numbers of the second kind satisfy the equation

$$u_{k+1,m+1} = (k+1)u_{k+1,m} + u_{k,m} \qquad \text{where} \quad u_{k,0} = \begin{cases} 1 & k=0 \\ 0 & k \neq 0 \end{cases}$$

(b) Use (a) to show that

$$u_{k,m} = \frac{(-1)^{k-1}}{k!}\left[\binom{k}{1} - \binom{k}{2}2^m + \binom{k}{3}3^m - \cdots + (-1)^{k-1}\binom{k}{k}k^m\right]$$

and verify this for some special values of k and m.

6.95. There are three jugs which have wine in them. Half of the wine in the first jug is poured into the second jug. Then one-third of the wine in the second jug is poured into the third jug. Finally one-fifth of the wine in the third jug is poured back into the first jug. Show that after many repetitions of this process the first jug contains about one-fourth of the total amount of wine.

6.96. Generalize Problem 6.95.

6.97. Work Problem 6.16 if (a) $|b| > 2|a|$, (b) $|b| = 2|a|$.

6.98. (a) Show that Newton's method [Problem 2.169, page 78] is equivalent to the difference equation

$$x_{n+1} = x_n - \frac{f(x_n)}{f'(x_n)}, \qquad x_0 = a$$

(b) Can you find conditions under which $\lim_{n\to\infty} x_n$ exists? Justify your conclusions.

6.99. Let $x_{n+1} = \frac{1}{2}\left(x_n + \frac{A}{x_n}\right), \quad x_n > 0, \; x_0 = a, \; A > 0$

(a) Prove that $\lim_{n\to\infty} x_n$ exists and is equal to $\sqrt{A}$.

(b) Show how the result of (a) can be used to find $\sqrt{2}$ to six decimal places.

6.100. Can you formulate a method for finding $\sqrt[3]{2}$ analogous to that of Problem 6.99.

6.101. Work Problem 6.20 by using $h = 0.05$ and discuss the accuracy.

6.102. (a) Solve the differential equation

$$\frac{dy}{dx} = x + y, \qquad y(0) = 1$$

for $y(0.5)$ using $h = 0.1$ by using the Euler method and compare with the exact value.

(b) Work part (a) using $h = 0.05$.

6.103. Work Problem 6.22 by using $h = 0.05$ and discuss the accuracy.

6.104. Given the differential equation

$$\frac{dy}{dx} = xy^{1/2}, \qquad y(1) = 1$$

(a) Use the Euler method to find $y(1.5)$ and compare with the exact value.

(b) Find $y(1.5)$ using the modified Euler method.

6.105. (a) Obtain from the equation $\frac{dy}{dx} = f(x,y)$ the approximate "predictor" formula

$$y_{k+1} = y_{k-1} + \frac{h}{3}(y'_{k-1} + 4y'_k + y'_{k+1})$$

(b) Explain how the result in (a) can be used to solve Problem 6.20 and compare the accuracy with the modified Euler method.

6.106. (a) Explain how the method of Taylor series can be used to solve a differential equation. (b) Illustrate the method of (a) by working Problem 6.20. [*Hint.* If $y' = x + y$, then by successive differentiations $y'' = 1 + y'$, $y''' = y''$, ... Now use the Taylor series $y(x) = y(a) + (x-a)y'(a) + \frac{(x-a)^2 y''(a)}{2!} + \cdots$.]

6.107. Work (a) Problem 6.102, (b) Problem 6.104 by the method of Taylor series.

6.108. Explain the advantages and disadvantages of (a) the Euler method, (b) the modified Euler method, (c) the Taylor series method for solving differential equations.

6.109. (a) Explain how the Euler method can be used to solve the second order differential equation

$$\frac{d^2y}{dx^2} = f\left(x, y, \frac{dy}{dx}\right)$$

(b) Use the method of (a) to solve

$$\frac{d^2y}{dx^2} = x + y, \qquad y(0) = 1, \quad y'(0) = 0$$

for $y(0.5)$ and compare with the exact value.

6.110. Show how the differential equation of Problem 6.109(a) can be solved using the modified Euler method and illustrate by solving the differential equation of Problem 6.109(b).

6.111. Work Problem 6.109(b) by using the Taylor series method.

6.112. Explain how you would solve numerically a system of differential equations. Illustrate your procedure by finding the approximate value of $x(0.5)$ and $y(0.5)$ given the equations

$$\frac{dx}{dt} + y = e^t, \qquad x - \frac{dy}{dt} = t, \qquad x(0) = 0, \quad y(0) = 0$$

and compare with the exact value.

6.113. Work Problem 6.24 by choosing a grid which is subdivided into 16 squares.

6.114. Explain how you could use the results of Problem 6.24 or 6.113 to obtain the steady-state temperature in a square plate whose faces are insulated if three of its sides are maintained at 0°C while the fourth side is maintained at 100°C.

6.115. Explain how you could use the results of Problem 6.24 or 6.113 to obtain the steady-state temperature in a square plate if the respective sides are kept at 20°C, 40°C, 60°C and 80°C.

6.116. Solve the equation $\frac{\partial^2 U}{\partial x^2} + \frac{\partial^2 U}{\partial y^2} = 0$ for the plane region shown shaded in Fig. 6-18 with the indicated boundary conditions for U.

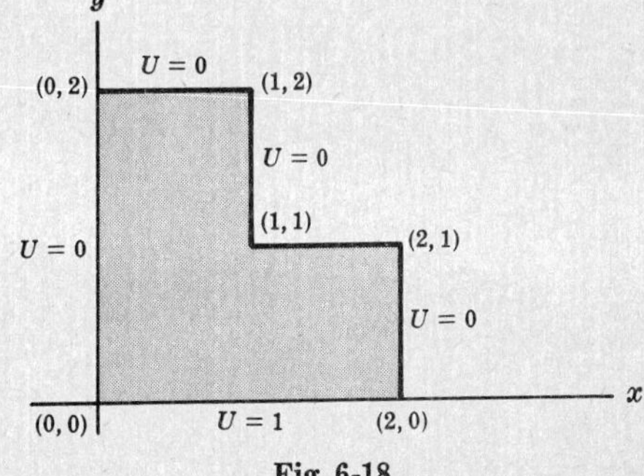

Fig. 6-18

6.117. The equation

$$\frac{\partial^2 U}{\partial x^2} + \frac{\partial^2 U}{\partial y^2} = \rho(x, y)$$

where $\rho(x, y)$ is some given function of x and y is called *Poisson's equation.* Explain how you would solve Poisson's equation approximately for the case where $\rho(x, y) = 100/(1 + x^2 + y^2)$ if the region and boundary conditions are the same as those of Problem 6.23.

6.118. Explain how you would solve approximately *Laplace's equation in three dimensions* given by

$$\frac{\partial^2 U}{\partial x^2} + \frac{\partial^2 U}{\partial y^2} + \frac{\partial^2 U}{\partial z^2} = 0$$

inside a cube whose edge is of unit length if $U = U(x, y, z)$ is equal to 1 on one face and 0 on the remaining faces.

6.119. Explain how you would find an approximate solution to the boundary-value problem

$$\frac{\partial U}{\partial t} = \frac{\partial^2 U}{\partial x^2}, \qquad U(0, t) = 0, \quad U(1, t) = 0, \quad U(x, 0) = 1$$

Appendix A

Stirling Numbers of the First Kind s_k^n

n \ k	1	2	3	4	5	6	7	8	9	10	11	12
1	1											
2	−1	1										
3	2	−3	1									
4	−6	11	−6	1								
5	24	−50	35	−10	1							
6	−120	274	−225	85	−15	1						
7	720	−1,764	1,624	−735	175	−21	1					
8	−5,040	13,068	−13,132	6,769	−1,960	322	−28	1				
9	40,320	−109,584	118,104	−67,284	22,449	−4,536	546	−36	1			
10	−362,880	1,026,576	−1,172,700	723,680	−269,325	63,273	−9,450	870	−45	1		
11	3,628,800	−10,628,640	12,753,576	−8,409,500	3,416,930	−902,055	157,773	−18,150	1,320	−55	1	
12	−39,916,800	120,543,840	−150,917,976	105,258,076	−45,995,730	13,339,535	−2,637,558	357,423	−32,670	1,925	−66	1

Appendix B Stirling Numbers of the Second Kind S_k^n

$n \backslash k$	1	2	3	4	5	6	7	8	9	10	11	12
1	1											
2	1	1										
3	1	3	1									
4	1	7	6	1								
5	1	15	25	10	1							
6	1	31	90	65	15	1						
7	1	63	301	350	140	21	1					
8	1	127	966	1,701	1,050	266	28	1				
9	1	255	3,025	7,770	6,951	2,646	462	36	1			
10	1	511	9,330	34,105	42,525	22,827	5,880	750	45	1		
11	1	1,023	28,501	145,750	246,730	179,487	63,987	11,880	1,155	55	1	
12	1	2,047	86,526	611,501	1,379,400	1,323,652	627,396	159,027	22,275	1,705	66	1

Appendix C Bernoulli Numbers

All values $B_3, B_5, B_7, \ldots$ are equal to zero.

$$B_0 = 1$$

$$B_1 = -\frac{1}{2} = -0.500000000\ldots$$

$$B_2 = \frac{1}{6} = 0.166666666\ldots$$

$$B_4 = -\frac{1}{30} = -0.0333333333\ldots$$

$$B_6 = \frac{1}{42} = 0.02389523895\ldots$$

$$B_8 = -\frac{1}{30} = -0.0333333333\ldots$$

$$B_{10} = \frac{5}{66} = 0.075757575\ldots$$

$$B_{12} = -\frac{691}{2{,}730} = -0.2531135531135\ldots$$

$$B_{14} = \frac{7}{6} = 1.1666666666\ldots$$

$$B_{16} = -\frac{3{,}617}{510} = -7.09215686274509803921 5\ldots$$

$$B_{18} = \frac{43{,}867}{798} = 54.971177944862155388\ldots$$

$$B_{20} = -\frac{174{,}611}{330} = -529.1242424242\ldots$$

$$B_{22} = \frac{854{,}513}{138} = 6{,}192.1231884057898550\ldots$$

$$B_{24} = -\frac{236{,}364{,}091}{2{,}730} = -86{,}580.2531135531135\ldots$$

$$B_{26} = \frac{8{,}553{,}103}{6} = 1{,}425{,}517.16666\ldots$$

$$B_{28} = -\frac{23{,}749{,}461{,}029}{870} = -27{,}298{,}231.067816091954\ldots$$

$$B_{30} = \frac{8{,}615{,}841{,}276{,}005}{14{,}322} = 601{,}580{,}873.90064236838\ldots$$

Appendix D Bernoulli Polynomials

$$B_0(x) = 1$$

$$B_1(x) = x - \frac{1}{2}$$

$$B_2(x) = x^2 - x + \frac{1}{6}$$

$$B_3(x) = x^3 - \frac{3}{2}x^2 + \frac{1}{2}x$$

$$B_4(x) = x^4 - 2x^3 + x^2 - \frac{1}{30}$$

$$B_5(x) = x^5 - \frac{5}{2}x^4 + \frac{5}{3}x^3 - \frac{1}{6}x$$

$$B_6(x) = x^6 - 3x^5 + \frac{5}{2}x^4 - \frac{1}{2}x^2 + \frac{1}{42}$$

$$B_7(x) = x^7 - \frac{7}{2}x^6 + \frac{7}{2}x^5 - \frac{7}{6}x^3 + \frac{1}{6}x$$

$$B_8(x) = x^8 - 4x^7 + \frac{14}{3}x^6 - \frac{7}{3}x^4 + \frac{2}{3}x^2 - \frac{1}{30}$$

$$B_9(x) = x^9 - \frac{9}{2}x^8 + 6x^7 - \frac{21}{5}x^6 + 2x^3 - \frac{3}{10}x$$

$$B_{10}(x) = x^{10} - 5x^9 + \frac{15}{2}x^8 - 7x^6 + 5x^4 - \frac{3}{2}x^2 + \frac{5}{66}$$

Appendix E Euler Numbers

All values $E_1, E_3, \ldots$ are equal to zero.

$$E_0 = 1$$

$$E_2 = -1$$

$$E_4 = 5$$

$$E_6 = -61$$

$$E_8 = 1{,}385$$

$$E_{10} = -50{,}521$$

$$E_{12} = 2{,}702{,}765$$

$$E_{14} = -199{,}360{,}981$$

$$E_{16} = 19{,}391{,}512{,}145$$

$$E_{18} = -2{,}404{,}879{,}675{,}441$$

$$E_{20} = 370{,}371{,}188{,}237{,}525$$

$$E_{22} = -69{,}348{,}874{,}393{,}137{,}901$$

$$E_{24} = 15{,}514{,}534{,}163{,}557{,}086{,}905$$

Appendix F Euler Polynomials

$$E_0(x) = 1$$

$$E_1(x) = x - \frac{1}{2}$$

$$E_2(x) = \frac{1}{2}x^2 - \frac{1}{2}x$$

$$E_3(x) = \frac{1}{6}x^3 - \frac{1}{4}x^2 + \frac{1}{24}$$

$$E_4(x) = \frac{1}{24}x^4 - \frac{1}{12}x^3 + \frac{1}{24}x$$

$$E_5(x) = \frac{1}{120}x^5 - \frac{1}{48}x^4 + \frac{1}{48}x^2 - \frac{1}{240}$$

Appendix G Fibonacci Numbers

$F_0 = 0$
$F_1 = 1$
$F_2 = 1$
$F_3 = 2$
$F_4 = 3$
$F_5 = 5$
$F_6 = 8$
$F_7 = 13$
$F_8 = 21$
$F_9 = 34$
$F_{10} = 55$
$F_{11} = 89$
$F_{12} = 144$
$F_{13} = 233$
$F_{14} = 377$
$F_{15} = 610$
$F_{16} = 987$
$F_{17} = 1{,}597$
$F_{18} = 2{,}584$
$F_{19} = 4{,}181$
$F_{20} = 6{,}765$
$F_{21} = 10{,}946$
$F_{22} = 17{,}711$
$F_{23} = 28{,}657$
$F_{24} = 46{,}368$
$F_{25} = 75{,}025$
$F_{26} = 121{,}393$
$F_{27} = 196{,}418$
$F_{28} = 317{,}811$
$F_{29} = 514{,}229$
$F_{30} = 832{,}040$
$F_{31} = 1{,}346{,}269$
$F_{32} = 2{,}178{,}309$
$F_{33} = 3{,}524{,}578$
$F_{34} = 5{,}702{,}887$
$F_{35} = 9{,}227{,}465$
$F_{36} = 14{,}930{,}352$
$F_{37} = 24{,}157{,}817$
$F_{38} = 39{,}088{,}169$
$F_{39} = 63{,}245{,}986$
$F_{40} = 102{,}334{,}155$
$F_{41} = 165{,}580{,}141$
$F_{42} = 267{,}914{,}296$
$F_{43} = 433{,}494{,}437$
$F_{44} = 701{,}408{,}733$
$F_{45} = 1{,}134{,}903{,}170$
$F_{46} = 1{,}836{,}311{,}903$
$F_{47} = 2{,}971{,}215{,}073$
$F_{48} = 4{,}807{,}526{,}976$
$F_{49} = 7{,}778{,}742{,}049$
$F_{50} = 12{,}586{,}269{,}025$

Answers to Supplementary Problems

CHAPTER 1

1.46. (*a*) $1 + 2\sqrt{x} + x$
(*b*) $2x^4 - 4x^3 + 2x^2 + 6x - 3$
(*c*) 9
(*d*) $6(3x+2)$
(*e*) $16x^4 - 32x^3 + 32x^2 - 22x + 6$
(*f*) $2x^4 + 4x^3 - 6x^2 - 6x$
(*g*) $4x^4 + 8x^3 - 2x^2 - 6x$
(*h*) $2x(4x+1)$
(*i*) 0
(*j*) $16x^7(1-x)$

1.52. (*a*) $8xh + 4h^2 - 2h$
(*b*) $\sqrt[3]{5x + 5h - 4}$
(*c*) $4h^2$
(*d*) $3x^2 + 12xh + 12h^2 + 3$
(*e*) $(x + 2h + 1)^2$
(*f*) $3x^3 + 8x^2h + 6xh^2 + x^2$
(*g*) 0
(*h*) $2x^2 + 14xh + 19h^2$
(*i*) $2x^2 + 14xh + 19h^2$
(*j*) $4x^2h^2 + 9xh^3$

1.53. (*a*) Yes (*b*) No

1.55. (*a*) Yes (*b*) Yes (*c*) Yes

1.59. (*a*) Yes (*b*) No

1.61. (*a*) $(3x^2 - 6x + 2)\,dx$ (*b*) $6(dx)^2$

1.68. (*a*) $h(15x^{(4)} + 20x^{(3)} - 14x^{(1)} + 3)$ (*b*) $-3x^{(-4)} + 6x^{(-3)}$ (*c*) $h(x^{(1)} - x^{(-3)})$

1.69. (*a*) $h^2(24x^{(-5)} - 18x^{(-4)} + 8)$ (*b*) $24x^{(1)} - 120x^{(-7)}$

1.70. (*a*) $3x^{(2)} - 2x^{(1)} + 2$ and $3x^{(2)} + (3h-5)x^{(1)} + 2$
(*b*) $2x^{(4)} + 12x^{(3)} + 19x^{(2)} + 3x^{(1)} + 7$ and $2x^{(4)} + 12hx^{(3)} + (14h^2+5)x^{(2)} + (2h^3 + 5h - 4)x^{(1)} + 7$

1.71. (*a*) $4x^3 + 6x^2h - 12xh^2 - 4x + 25h^3 - 2h + 5$
(*b*) $12x^2 + 24xh + 14h^2 - 4$

1.72. (*a*) $(2x-1)(2x-5)(2x-9)(2x-13)$
(*b*) $(3x+5)(3x+2)(3x-1)$
(*c*) $\dfrac{1}{(4x-1)(4x+3)}$
(*d*) $\dfrac{1}{(5x+12)(5x+22)(5x+32)(5x+42)}$

1.73. (*a*) $(3x-2)^{(3)}$, $h = 7/3$
(*b*) $(2x+2)^{(4)}$, $h = 3/2$
(*c*) $(x-2)^{(-3)}$, $h = 2$
(*d*) $(2x-5)^{(-4)}$, $h = 2$

1.76. $s_1^1 = 1,\ s_1^2 = -1,\ s_1^3 = 2,\ s_2^1 = 0,\ s_2^2 = 1,\ s_2^3 = -3,\ s_3^1 = 0,\ s_3^2 = 0,\ s_3^3 = 1$

1.77. $S_1^1 = 1,\ S_1^2 = 1,\ S_1^3 = 1,\ S_2^1 = 0,\ S_2^2 = 1,\ S_2^3 = 3,\ S_3^1 = 0,\ S_3^2 = 0,\ S_3^3 = 1$

1.78. $S_1^6 = 1,\ S_2^6 = 31,\ S_3^6 = 90,\ S_4^6 = 65,\ S_5^6 = 15,\ S_6^6 = 1$

1.81. (*a*) $3x^{(2)} - 2x^{(1)} + 2$ and $3x^{(2)} + (3h-5)x^{(1)} + 2$

(*b*) $2x^{(4)} + 12x^{(3)} + 19x^{(2)} + 3x^{(1)} + 7$ and $2x^{(4)} + 12hx^{(3)} + (14h^2+5)x^{(2)} + (2h^3+5h-4)x^{(1)} + 7$

1.87. $2^x(x^2 + 12x + 30)$

1.88. $(a^h-1)^n xa^x + nh(a^h-1)^{n-1}a^{x+h}$

1.89. $(a^h-1)^n x^2 a^x + 2nh(a^h-1)^{n-1}xa^{x+h} + nh^2(na^h-1)(a^h-1)^{n-2}a^{x+h}$

1.91. (*a*) $4xh - 2h^2 + 3h$ (*b*) $4xh + 3h$ (*c*) $4h^2$ (*d*) $4h^2$

1.92. 0

1.94. Yes

1.97. Yes

1.116. (*a*) $18h^2x - 4h^2 + 18h^3$ (*b*) $24h^3x + 12h^3 + 36h^4$

CHAPTER 2

2.52. (*a*) $y_{k+n} = y_k + n\Delta y_k + \frac{n(n-1)}{2!}\Delta^2 y_k + \cdots + \Delta^n y_k$

(*b*) $\Delta^n y_k = y_{k+n} - ny_{k+n-1} + \frac{n(n-1)}{2!}y_{k+n-2} - \cdots + (-1)^n y_k$

2.54. (*a*) $y_k = k^{(3)} - k^{(2)} - k^{(1)} - 3$

(*b*) $\Delta y_k = 3k^{(2)} - 2k^{(1)} - 1 = 3k^2 - 5k - 1$

(*c*) $\Delta^2 y_k = 6k^{(1)} - 2 = 6k - 2$

(*d*) $\Delta^3 y_k = 6$

(*e*) $\Delta^4 y_k = 0$

2.56. $60k^2 + 108k + 54$

2.58. 1489, 2053, 2707, 3451, 4285, 5209

2.59. (*b*) 2366, 3994, 6230, 9170, 12910

2.61. 5

2.63. $A = 4,\ B = 0,\ C = -2,\ D = 4,\ E = 4,\ F = -1,\ G = -5,\ H = 2,\ J = 3,\ K = -1,\ L = 0$

2.64. (*a*) 6 (*b*) 7 (*c*) 5

2.68. (*a*) $u_k = 3k^2 - 2k + 4$ (*b*) $u_k = \frac{1}{6}k(k+1)(k+2)$

2.69. (*a*) $u_k = k^3 - \frac{9}{2}k^2 + \frac{21}{2}k - 4$ (*b*) 651

2.71. $y = x^3 - 4x^2 + 5x + 1$

2.72. $y = \frac{1}{8}(x^3 - 5x^2 + 7x + 21)$

2.74. Exact value = 0.70711, Computed value = 0.71202

2.76. (*a*) Exact value = 0.71911 (*b*) Exact value = 0.73983

2.77. (*a*) Exact value = 0.87798 (*b*) Exact value = 0.89259

2.79. (*a*) Exact value = 0.93042 (*b*) Exact value = 0.71080
Computed value = 0.90399 Computed value = 0.70341

2.80. (*a*) Exact value = 0.49715 (*b*) 1.75×10^{-4}
Computed value = 0.49706

2.81. (*a*) Computed value = 0.49715

2.82. Exact values (*a*) 1.27507 (*b*) 1.50833

2.83. Exact value = 0.92753

2.84. Exact values (*a*) 0.30788 (*b*) 0.62705 (*c*) 0.84270

2.85. (*b*) Computed value = 0.71911

2.86. Computed values (*a*) 0.87798, 0.89725 (*b*) 0.49715

2.87. Computed values (*a*) 0.71940, 0.73983 (*b*) 0.87800, 0.89726 (*c*) 0.49715

2.88. Computed values (*a*) 0.71915, 0.73983 (*b*) 0.87798, 0.89726 (*c*) 0.49715

2.92. Computed value = 0.92752

2.93. Computed values (*a*) 0.71912, 0.73983 (*b*) 0.87798, 0.89720 (*c*) 0.49715 (*d*) 0.92756

2.98. (*a*) $y = x^3 - 4x^2 + 5x + 1$ (*b*) 0.49715

2.99. (*a*) $y = 2x^3 - 3x^2 + 5x + 2$ (*b*) 44 (*c*) 202 (*d*) 874

2.100. (*a*) $4n^3 - 27n^2 + 59n - 35$ (*b*) 7, 85

2.101. (*a*) $\sin x = 2.80656(x/\pi)^4 - 6.33780(x/\pi)^3 + 0.21172(x/\pi)^2 + 3.12777(x/\pi)$

(*b*) Computed values and exact values are given respectively by 0.25859, 0.25882; 0.96612, 0.96593; 0.64725, 0.64723

2.103. 15

2.104. 106, 244

2.105. Exact values are 0.25882, 0.96593

2.106. (*a*) 44, 202 (*b*) 874

2.107. 6.2, 7.6, 11, 17

2.108. 8, 14, 22, 32

2.111. Exact values are 0.97531, 0.95123, 0.92774

2.113. (*b*) $f(x) = 3x^2 - 11x + 15$ (*c*) 19, 159

2.114. (*a*) $f(x) = x^3 - 20x + 40$ (*b*) 65, 840

2.121. (*b*) 1.32472 (*c*) $-0.66236 \pm 0.56228i$

2.122. 1.36881, $-1.68440 \pm 3.43133i$

2.123. 2.90416, $0.04792 \pm 1.31125i$

2.124. 0.5671

2.125. 0, 4.9651

2.126. Exact value = 3.67589

2.127. 2.2854%

2.128. 3.45

2.129. Exact value = 0.2181

2.131. Exact value = 0.08352

2.132. Exact values are (*a*) 0.4540 (*b*) -0.8910

2.133. Exact values are (*a*) 1.2214 (*b*) 1.2214

2.134. Exact values are (*a*) -0.31683 (*b*) -0.29821

2.135. Exact values are (*a*) 0.91230 (*b*) 0.72984

2.138. $A = 6$ $B = 8$ $C = 6$ $D = 3$ $E = 2$ $F = -2$ $G = -4$ $H = -4$ $J = -2$ $K = -3$

2.140. Exact value = 3.80295

2.141. Exact values are (*a*) 1.7024 (*b*) 2.5751

2.142. 11.72 years

2.143. (*a*) 0.9385 (*b*) 0.6796

2.144. Exact value = 22.5630

2.145. 2.285%

2.146. (*a*) 38.8 (*b*) 29.4

2.148. Exact values are 0.74556, 0.32788, 0.18432, 0.13165, 0.08828, 0.05261

2.150. 0.97834, 0.66190, 0.47590, 0.34600, 0.24857, 0.17295, 0.11329, 0.06604, 0.02883

2.151. $f(x) = 3^{2x^2+x+1}$

2.157. $\frac{1}{6}k(k+1)(2k+1)$

2.158. (*a*) $\frac{1}{2}k(k+1)$ (*b*) $\frac{1}{4}k^2(k+1)^2$

2.159. $\frac{1}{2}k(6k^2+3k-1)$

2.160. (*a*) $2k^2 - 5k + 3 + 2^k$ (*b*) 1177

2.165. $k = 10$

2.166. All values of k

2.168. At $x = 6$, $f(x) = 0.68748$ should be 0.68784

CHAPTER 3

3.55. (*a*) $\dfrac{3x^5}{5} - \dfrac{x^4}{2} + \dfrac{x^2}{2} - x + c$

(*b*) $\frac{4}{3}x^{3/2} - \frac{9}{4}x^{4/3} + c$

(*c*) $\frac{2}{3}(2\sin 3x + \cos 3x) + c$

(*d*) $2e^{2x} - \frac{3}{4}e^{-4x} + c$

(*e*) $4\ln(x-2) + c$

(*f*) $\dfrac{2^x}{\ln 2} + c$

3.57. (*a*) $-\frac{1}{3}e^{-3x}(x + \frac{1}{3}) + c$

(*b*) $2\sin x - x\cos x + c$

(*c*) $\frac{1}{2}x^2(\ln x - \frac{1}{2}) + c$

(*d*) $\frac{1}{3}x^3\ln^3 x - \frac{2}{9}x^3\ln x + \frac{2}{27}x^3 + c$

3.58. (a) $\frac{1}{2}x \sin 2x - \frac{1}{2}x^2 \cos 2x + \frac{1}{4}\cos 2x + c$

(b) $\frac{1}{8}e^{2x}(4x^3 - 6x^2 + 6x - 3) + c$

3.59. (a) $\ln(e^x + \sin x) + c$

(b) $2e^{\sqrt{x}}(\sqrt{x} - 1) + c$

(c) $6\cos\sqrt[3]{x} + 6\sqrt[3]{x}\sin\sqrt[3]{x} - 3\sqrt[3]{x^2}\cos\sqrt[3]{x} + c$

(d) $\frac{1}{2}\tan^{-1}\left(\dfrac{x-2}{2}\right) + c$

(e) $\dfrac{-1}{2(2x-3)} + c$

3.65. (a) $\dfrac{x \sin r(x - \frac{1}{2}h)}{2 \sin \frac{1}{2}rh} + \dfrac{h \cos rx}{4 \sin^2 \frac{1}{2}rh}$

(b) $\dfrac{-x \cos r(x - \frac{1}{2}h)}{2 \sin \frac{1}{2}rh} + \dfrac{h \sin rx}{4 \sin^2 \frac{1}{2}rh}$

3.66. (a) 768 (b) 0 (c) 1 (d) $\dfrac{7}{660}$

3.67. $\dfrac{1}{12} - \dfrac{1}{2(n+2)(n+3)}$

3.68. $(n+1)2^{n+1} - 2^{n+2} + 2$

3.72. (a) $\dfrac{x \sin r(x - \frac{1}{2}h)}{2 \sin \frac{1}{2}rh} + \dfrac{h \cos rx}{4 \sin^2 \frac{1}{2}rh}$

(b) $\dfrac{-x \cos r(x - \frac{1}{2}h)}{2 \sin \frac{1}{2}rh} + \dfrac{h \sin rx}{4 \sin^2 \frac{1}{2}rh}$

3.74. (a) $\frac{1}{2}n(n+1)$ (b) $\frac{1}{2}n(3n-1)$

3.75. (a) $\frac{1}{3}n(n+1)(n+2)$ (c) $\frac{1}{3}n(4n^2 + 6n - 1)$

(b) $\frac{1}{4}n(n+1)(n+2)(n+3)$ (d) $n(3n^2 + 6n + 1)$

3.76. (a) $\dfrac{n}{n+1}$ (b) $\dfrac{n}{2n+1}$ (c) $\dfrac{1}{24} - \dfrac{1}{6(3n+1)(3n+4)}$ (d) $\dfrac{1}{4} - \dfrac{1}{2(n+1)(n+2)}$

3.78. $(n-1)2^{n+1} + 2$

3.81. (a) 1 (b) $\frac{1}{2}$ (c) $\frac{1}{24}$ (d) $\frac{1}{4}$

3.82. $\frac{1}{12}n(n+1)(n+2)(3n+5)$

3.83. $2^{n+1}(n^2 - 2n + 3) - 6$

3.84. $\frac{1}{3}n(2n-1)(2n+1)$

3.85. (a) $\dfrac{n^2}{(n+1)(n+2)}$ (b) 1

3.86. (a) $\dfrac{5}{12} - \dfrac{2n+5}{2(n+2)(n+3)}, \; \dfrac{5}{12}$ (b) $\dfrac{23}{480} - \dfrac{2n^2+14n+23}{4(n+2)(n+3)(n+4)(n+5)}, \; \dfrac{23}{480}$

3.89. $\frac{1}{12}n(n+1)(n+2)(3n+1)$

3.90. (a) $\dfrac{5n^2+n}{12(n+2)(n+3)}$ (b) $\dfrac{5}{12}$

3.91. (a) $\dfrac{21}{2} + \dfrac{1}{10}(2n+7)(2n+5)(2n+3)(2n+1)(2n-1)$ (b) $\dfrac{1}{90} - \dfrac{1}{6(2n+1)(2n+3)(2n+5)}$

3.93. (a) $\dfrac{11n^2+19n}{6(2n+1)(2n+3)}$ (b) $\dfrac{11}{24}$

3.94. $2 - \dfrac{n+2}{2^n}$

3.95. $\frac{1}{6}n(n+1)(2n+1)$

3.96. $\dfrac{a}{(1-a)^3}[1 + a - (n+1)^2a^n + (2n^2+2n-1)a^{n+1} - n^2a^{n+2}]$

3.102. (a) 6 (b) $\dfrac{3\sqrt{\pi}}{4}$ (c) $\dfrac{-8\sqrt{\pi}}{15}$

3.103. (a) $\frac{3}{4}$ (b) $\dfrac{\sqrt{\pi}}{4}$

3.104. $\frac{1}{3}\Gamma(\frac{2}{3})$

3.105. $\frac{2}{9}$

3.108. $\left\{\dfrac{\Gamma'(x+n+1)}{\Gamma(x+n+1)}\right\}^2 - \dfrac{\Gamma''(x+n+1)}{\Gamma(x+n+1)} - \left\{\dfrac{\Gamma'(x+1)}{\Gamma(x+1)}\right\}^2 + \dfrac{\Gamma''(x+1)}{\Gamma(x+1)}$

In general, $\displaystyle\sum_{k=1}^{n} \frac{1}{(x+k)^r} = \frac{(-1)^{r-1}}{(r-1)!}\frac{d^{r-1}}{dx^{r-1}}\left[\frac{\Gamma'(x+n+1)}{\Gamma(x+n+1)} - \frac{\Gamma'(x+1)}{\Gamma(x+1)}\right]$

3.110. (a) $\beta_4(x) = \dfrac{x^4}{24} - \dfrac{x^3}{12} + \dfrac{x^2}{24} - \dfrac{1}{720}, \quad \beta_5(x) = \dfrac{x^5}{120} - \dfrac{x^4}{48} + \dfrac{x^3}{72} - \dfrac{x}{720}$ (b) $-1/30, \; 0$

3.111. $-1/30, 1/42, -1/30$

3.114. (a) $\dfrac{1}{4}n^2(n+1)^2$ (b) $\dfrac{1}{30}n(n+1)(2n+1)(3n^2+3n-1)$

3.115. $3B_0(x) - 4B_1(x) + B_2(x) + 2B_3(x) + 5B_4(x)$

3.119. (a) $e_4 = 0, \; e_5 = -1/240$

(b) $E_4(x) = \dfrac{1}{24}x^4 - \dfrac{1}{12}x^3 + \dfrac{1}{24}x, \; E_5(x) = \dfrac{1}{120}x^5 - \dfrac{1}{48}x^4 + \dfrac{1}{48}x^2 - \dfrac{1}{240}$

(c) $E_4 = 5, \; E_5 = 0$

3.125. $2^x(x^3 - 6x^2 + 18x - 26)$

3.126. $(n+1)^3 \cos 2(n+1) - (3n^2+9n+7) \cos 2(n+2)$

$$+ 6(n+2) \cos 2(n+3) - 6 \cos 2(n+4) - \cos 2$$

$$+ 7 \cos 4 - 12 \cos 6 + 6 \cos 8$$

3.132. $\frac{1}{12}n^2(2n^4+6n^3+5n^2-1)$

3.146. (a) $\pi/4$ (b) $\pi/32$ (c) $\pi/\sqrt{2}$

3.162. $B_1 = -1/2,\ B_2 = 1/6,\ B_3 = 0,\ B_4 = -1/30,\ B_5 = 0,\ B_6 = 1/42$

CHAPTER 4

4.40. (a) n^2 (b) $\frac{1}{3}n(2n-1)(2n+1)$ (c) $\frac{1}{2}n(6n^2+3n-1)$ (d) $\frac{4}{3}n(n+1)(n+2)$ (e) $\frac{1}{4}n^2(n+1)^2$ (f) $\frac{1}{8}(2n-1)(2n+1)(2n+3)(2n+5) + \frac{15}{8}$ (g) $\frac{1}{12}n(n+1)(n+2)(3n+17)$ (h) $\frac{1}{6}n(3n^3+46n^2+255n+596)$

4.41. $\dfrac{x+1}{(1-x)^3}$

4.42. $[1 + x - (n+1)^2x^n + (2n^2+2n-1)x^{n+1} - n^2x^{n+2}]/(1-x)^3$

4.43. 9/32

4.44. (a) $\dfrac{x}{(1-x)^2}$ (b) $\dfrac{x(x+1)}{(1-x)^3}$ (c) $\dfrac{3+6x-x^2}{(1-x)^3}$

4.45. 359/500

4.46. $5e - 1$

4.48. $(x^4 - 7x^2) \sin x + (x - 6x^3) \cos x$

4.49. $\frac{3}{4}x \cosh \sqrt{x} + \frac{1}{4}\sqrt{x}\,(1+x) \sinh \sqrt{x}$

4.50. $\dfrac{x}{3^n(3-x)^3}[3^{n+2} + 3^{n+1}x - 9(n+1)^2x^n + 3(2n^2+2n-1)x^{n+1} - n^2x^{n+2}]$

4.59. Exact value = 1.09861

4.61. Exact value = 1

4.62. 0.63212

4.64. (a) $\frac{1}{6}n(n+1)(2n+1)$ (c) $\frac{1}{3}n(n+1)(n+2)$

(b) $\frac{1}{4}n^2(n+1)^2$ (d) $\frac{1}{30}n(n+1)(2n+1)(3n^2+3n-1)$

4.67. $\gamma = 0.577215$

4.70. 1.20206

4.78. Exact value $=$ 3,628,800

4.79. $2^{100}/5\sqrt{2\pi} \approx 4.036 \times 10^{28}$

4.83. $xe^{-x}(2-x)$

4.84. (a) $\dfrac{11}{180} - \dfrac{6n+11}{12(2n+1)(2n+3)(2n+5)}$ (b) $\dfrac{11}{180}$

4.86. 1.092

4.88. (a) 1.0888

4.94. $\frac{1}{2}(-1)^{n-1}n(n+1)$

4.95. $2^n(n^2-2n+3)-3$

4.99. (a) 1.09861 (b) 0.785398 (c) 0.63212

CHAPTER 5

5.60. (b) $y = 5e^x - x^2 - 2x - 2$

5.61. $y(x) = 5(1+h)^{x/h} - x^2 + (h-2)x - 2$

5.63. $y = 2\cos x + 5\sin x + 3x^2 - 5x - 2$

5.66. (a) $y_{k+1} - (h+1)y_k = h^3k^{(2)}$

(b) $y_{k+2} - 2y_{k+1} + (1+h^2)y_k = 3h^4k^2 - (3h^3+5h^2)hk + 4h^2$

5.67. (a) $y_{k+3} - 3y_{k+2} + 3y_{k+1} - y_k = h^3k^{(3)} - 2h^2k^{(2)}$

(b) $y_{k+4} + y_k = \cos hk$

5.68. (a) $2f(x+2h) - 3f(x+h) + f(x) = 0$

(b) $f(x+3h) + 4f(x+2h) - 5f(x+h) + 2f(x) = \dfrac{x^2}{h^2} - 4\dfrac{x}{h} + 1$

5.69. (a), (c), (d), (e), (f), (g) linearly independent; (b), (h) linearly dependent

5.75. (*a*) $y_k = c_1(-3)^k + c_2(-2)^k$

(*b*) $y_k = 2^k$

(*c*) $y_k = c_1(-2/3)^k + c_2(-2)^k$

(*d*) $y_k = \frac{5}{6}(3)^k + \frac{7}{6}(-3)^k$

(*e*) $y_k = 4(-3/2)^k$

(*f*) $y_k = 2^{k/2}\cos(3\pi k/4)$

(*g*) $y_k = 2^{k-1}\sin(k\pi/2)$

(*h*) $y_k = (5/2)^k[c_1 \sin(k\pi/2) + c_2\cos(k\pi/2)]$

(*i*) $y_k = 3^k(c_1 + c_2 k)$

(*j*) $y_k = (-1/2)^k(c_1 + c_2 k)$

5.76. $y_k = 5\cdot 2^k - 2\cdot 3^k - 3$

5.77. $$y_k = c_1\cos\frac{k\pi}{4} + c_2\sin\frac{k\pi}{4} + c_3\cos\frac{3k\pi}{4} + c_4\sin\frac{3k\pi}{4}$$

5.78. $$y_k = c_1(5/2)^{k/2} + c_2(-5/2)^{k/2} + (5/2)^{k/2}\left(c_3\cos\frac{k\pi}{2} + c_4\sin\frac{k\pi}{2}\right)$$

5.79. $$y_k = 2^k\left(c_1 + c_2\cos\frac{2\pi k}{3} + c_3\sin\frac{2\pi k}{3}\right)$$

5.80. $$y_k = c_1 2^k + c_2(-2)^k + 4^k\left(c_3\cos\frac{k\pi}{2} + c_4\sin\frac{k\pi}{2}\right)$$

5.81. $$y_k = c_1 + 2^{k/2}\left(c_2\cos\frac{3\pi k}{4} + c_3\sin\frac{3\pi k}{4}\right)$$

5.82. $$y_k = (c_1 + c_2 k)2^k + (c_3 + c_4 k + c_5 k^2 + c_6 k^3)(-2)^k + c_7 4^k + 5^k(c_8\cos k\theta + c_9\sin k\theta)$$

where $\theta = \tan^{-1}(-4/3)$

5.83. $$y_{k+9} + 6y_{k+8} + 5y_{k+7} - 136y_{k+6} - 484y_{k+5} + 464y_{k+4} + 3376y_{k+3} + 1792y_{k+2} - 6336y_{k+1} - 6400y_k = 0$$

or $$(E-2)^2(E+2)^4(E-4)(E^2+6E+25)y_k = 0$$

5.85. (*a*) $y_k = c_1 + c_2 2^k + \frac{1}{6}\cdot 4^k$

(*b*) $y_k = 2^k(c_1 + c_2 k) + k^2 + 4k + 8$

(*c*) $y_k = 2^{-k}\left(c_1\cos\frac{k\pi}{2} + c_2\sin\frac{k\pi}{2}\right) + \frac{2k^2}{5} - \frac{57k}{25} + \frac{371}{125}$

(*d*) $y_k = \frac{1}{2}k^2 - \frac{1}{2}k + 1$

(*e*) $y_k = (-1)^k(c_1 + c_2 k) + \frac{1}{4}k - \frac{1}{4} + \frac{1}{9}\cdot 2^k$

(*f*) $y_k = \frac{3}{2}k(k-1)\cdot 4^{k-2}$

(*g*) $y_k = c_1 + c_2 2^{-k} + \frac{k^3}{3} - \frac{9k^2}{2} + \frac{127k}{6}$

(*h*) $y_k = 2^k\left(c_1\cos\frac{k\pi}{2} + c_2\sin\frac{k\pi}{2}\right) + \frac{5}{13}(-3)^k + 2k - \frac{4}{5}$

(*i*) $y_k = c_1 + c_2(-1)^k - \frac{9}{8}\cdot 3^{-k}$

5.86. $$y_k = \begin{cases} c\alpha^k + \beta/(1-\alpha) & \alpha \neq 1 \\ c + \beta k & \alpha = 1 \end{cases}$$

5.87. $y_k = 2[\cos 2k + \cos(2k-4)]/(1 + \cos 4)$

5.88. $y_k = c_1 + c_2 2^k + c_3 3^k + 2k + k^2 - \frac{3}{2}k \cdot 2^k - \frac{1}{24}5^k$

5.89. $y_k = c_1 2^k + c_2(-2)^k + 2^k\left(c_3 \cos\frac{k\pi}{2} + c_4 \sin\frac{k\pi}{2}\right) - \frac{k^2}{15} + \frac{67k}{225} - \frac{422}{3375} - \frac{4 \cdot 3^k}{65}$

5.90. $y_k = c_1(-1)^k + c_2 \cos\frac{k\pi}{3} + c_3 \sin\frac{k\pi}{3} + \frac{2^k}{65 + 16\cos 9}[8\cos(3k-9) + \cos 3k]$

5.91. $y_k = c_1 \cdot 3^{-k/2} + c_2(-3)^{-k/2} + c_3(-1)^k + c_4 + 3^k\left(\frac{k}{208} - \frac{52}{11{,}041}\right)$

5.92. (*a*) $y_k = c_1 + c_2 2^k + 2 \cdot 3^k - \frac{1}{6} \cdot 5^k$

(*b*) $y_k = (-2)^k(c_1 + c_2 k) + \frac{k^2}{7} - \frac{13k}{27} + \frac{7}{9}$

(*c*) $y_k = \left(c_1 + \frac{k}{30}\right)3^k + c_2(-3)^k + \frac{k}{4} - \frac{3}{16}$

(*d*) $y_k = 4^k(c_1 + c_2 k) + \frac{3}{32}k^2 \cdot 4^k - \frac{5}{36}(-2)^k$

(*e*) $y_k = (\frac{1}{2})^k\left(c_1 \cos\frac{k\pi}{4} + c_2 \sin\frac{k\pi}{4}\right) + k^3 - 6k^2 + 2k + 18 + \frac{5^k}{41}$

(*f*) $y_k = 2^k(c_1 + c_2 k) + \frac{2^k}{48}(k^4 - 4k^3 + 5k^2 - 2k)$

(*g*) $y_k = c_1 + c_2 k + c_3 k^2 + \frac{1}{6}k^3 + \frac{1}{12}k^4 - \frac{1}{8} \cdot 3^{k-1}$

(*h*) $y_k = c_1 2^k + c_2(-2)^k + 2^k\left(c_3 \cos\frac{k\pi}{4} + c_4 \sin\frac{k\pi}{4}\right) - \frac{k}{5} - \frac{4}{75} + \frac{k \cdot 2^k}{64} + \frac{3^k}{65}$

5.94. $y_k = c_1(-6)^k + c_2 2^k + \frac{1}{258}\left(25\sqrt{3}\sin\frac{\pi k}{3} - 105\cos\frac{\pi k}{3}\right)$

5.95. (*c*) $y_k = c_1 2^k + c_2 4^k - 3^k + \frac{k^2}{3} + \frac{8k}{9} + \frac{38}{27}$

5.96. $y_k = c_1 + c_2 k + c_3(-2)^k + \left(c_4 + \frac{k}{60}\right)3^k - \frac{k^4}{72} + \frac{7k^3}{54} - \frac{77k^2}{108}$

5.97. $y_k = (c_1 + c_2 k)\cos\frac{k\pi}{2} + (c_3 + c_4 k)\sin\frac{k\pi}{2} + \frac{1}{2}k - \frac{7}{4}$

5.100. (*a*) $y_k = c_1 2^k + c_2 3^k + \frac{1}{2}k^2 + \frac{3}{2}k - \frac{5}{2}$

(*b*) $y_k = c_1 + c_2 k + k^2 + \frac{1}{6}k^3 + \frac{1}{9} \cdot 4^k$

(*c*) $y_k = (c_1 + c_2 k + \frac{3}{2}k^2)2^{-k}$

(*d*) $y_k = c_1 \cos\frac{k\pi}{2} + c_2 \sin\frac{k\pi}{2} + \sum_{r=0}^{\infty} \frac{(-1)^r}{k + 2r}$

(*e*) $y_k = c_1 + c_2(\frac{2}{5})^k - \frac{51k}{98} + \frac{3k^2}{14} + \frac{1}{24}(-2)^k$

(*f*) $y_k = c_1 \cos\frac{k\pi}{3} + c_2 \sin\frac{k\pi}{3} + \frac{2}{\sqrt{3}} \sum_{r=0}^{k}\left[\frac{1}{(k-r-2)!}\sin\frac{\pi(r+1)}{3}\right]$

5.103. $y_k = c_1 + \frac{29k}{6} + \frac{k^2}{2} + \frac{k^3}{6} + (c_2 - \frac{5}{2}k)2^k + c_3 3^k$

5.106. (*a*) $y_k = 2^k(2^k - 1)$

(*b*) $y_k = 1 - 1/2^k$

(*c*) $y_k = 2^k(1-k)$

(*d*) $y_k = c_1 3^{-k} + c_2 2^{-k}$

(*e*) $y_k = 2(-1)^k(1-k)$

(*f*) $y_k = c_1 + c_2 k + c_3 k^2$

(*g*) $y_k = (-1)^k(c_1 + c_2 k) + \frac{1}{6}(2k-1)$

(*h*) $y_k = 2^{-k-1}(k^2 + 3k)$

5.107. $y_k = \frac{1}{5}\sum_{r=0}^{k}\left[\frac{(-1)^r 4^{r-1}}{(k-r)!} + \frac{1}{r!}\right] + \frac{1}{5} - \frac{1}{5}(-4)^k - \frac{1}{4k!}$

5.108. (a) $\sum_{k=0}^{\infty} kt^k = \frac{t}{(1-t)^2}$ (b) $\sum_{k=0}^{\infty} k^2t^k = \frac{t(t+1)}{(1-t)^3}$ (c) $\sum_{k=0}^{\infty} k^3t^k = \frac{t(t^2+4t+1)}{(1-t)^4}$

5.111. (a) $y_k = e(k-1)! - \sum_{r=0}^{\infty}(k-1)^{(-r)}$

(b) $y_k = \frac{c(-2)^k}{(k-1)!} + \sum_{r=0}^{\infty}\frac{(-1)^r k^{(r)}}{2^{r+1}}$

(c) $y_k = c(-1)^k(k-1)! + k - 1 + \sum_{r=0}^{\infty}(-1)^{r+1}(k-1)^{(-r)}$

(d) $y_k = \frac{1}{5}(k-1)!\,[1 + (-4)^{k-1}]$

(e) $y_k = \frac{1}{(k-1)!}\sum_{r=0}^{k}[2^{r+1}r! - 2^{r+k}(k+r-1)!]$

5.112. $y_k = (1+e)\sum_{r=1}^{k-1}(r-1)! - \sum_{s=1}^{k-1}\sum_{r=0}^{\infty}(s-1)^{(-r)}$

5.113. $y_k = A(k-1)!\sum_{r=1}^{k-1}\frac{1}{r} + B(k-1)!$

5.114. $y_k = (k-1)!\sum_{s=1}^{k-1}\left[\frac{1}{s}\sum_{r=1}^{s-1}\frac{2^r}{r!} + \frac{A}{s}\right] + B(k-1)!$

5.115. $y_k = c_1\left(1 + \frac{k^{(2)}}{1\cdot 3} + \frac{k^{(4)}}{1\cdot 3^2\cdot 5} + \frac{k^{(6)}}{1\cdot 3^2\cdot 5^2\cdot 7} + \cdots\right)$

$+ c_2\left(k^{(1)} + \frac{k^{(3)}}{2\cdot 4} + \frac{k^{(5)}}{2\cdot 4^2\cdot 6} + \frac{k^{(7)}}{2\cdot 4^2\cdot 6^2\cdot 8} + \cdots\right)$

5.116. $y_k = c_1(3/2)^k(k-3) + c_2(2)^{-k}(k+1)$

5.119. (a) $y_k = c(k+1) + (k+1)\sum_{r=0}^{k-1}\frac{r(-2)^r}{(r+2)!}$

(b) $y_k = c(k+1) - 1 + (k+1)\sum_{r=0}^{k-1}\frac{r(-2)^r}{(r+2)!}$

5.122. $y_k = \frac{c_1\alpha^k}{b_k}\left\{1 + \sum_{r=1}^{\infty}\left[\left(\frac{a_{k+r}}{b_{k+r}}\right)\left(\frac{a_{k+r-1}}{b_{k+r-1}}\right)\cdots\left(\frac{a_{k+1}}{b_{k+1}}\right)\right]\right\} + c_2\left\{\left(\frac{b_{k-1}}{a_k}\right)\left(\frac{b_{k-2}}{a_{k-1}}\right)\cdots\left(\frac{b_1}{a_2}\right)\right\}$

5.133. $y_k = 2/[1 + c(-1)^k]$

5.134. $y_k = 1 - \sqrt{2} + 2\sqrt{2}/[1 + c(2\sqrt{2}-3)^{-k}]$

5.135. $y_k = \tan\left[c_1\cos\frac{2\pi k}{3} + c_2\sin\frac{2\pi k}{3}\right]$

5.136. $y_k = K^{1/2}\tan\left[c_1\cos\frac{2\pi k}{3} + c_2\sin\frac{2\pi k}{3}\right]$

5.137. $y_k = e^{3^{-k}}$

5.138. (a) $y_k = e^{c_1\alpha^k + c_2\beta^k}$ where $\alpha = \frac{1}{2}(1+\sqrt{5})$, $\beta = \frac{1}{2}(1-\sqrt{5})$

(b) $y_k = \cos(c \cdot 2^k)$

5.139. $y_k = e^{(-1)^{k-1}}(c_1 + c_2 k)$

5.140. $y_k = c_1 3^k$ or $y_k = c_2 2^k$

5.141. (a) $y_k = 2\cos(\pi/2^{k+1})$ (b) 2

5.144. $u_k = c_1(-2)^k + c_2, \quad v_k = -c_1(-2)^k + \frac{1}{2}c_2$

5.145. $y_k = \frac{1}{4}(k-1)!\,[c_1 2^{k+1} + c_2(-2)^{k+1}], \quad z_k = (k-1)!\,[c_1 2^k + c_2(-2)^k]$

5.146. $u_k = \{q(u_0+v_0) + [(1-p)u_0 - qv_0](p-q)^k\}/(1-p+q)$

$v_k = \{(1-p)(u_0+v_0) + [qv_0 - (1-p)u_0](p-q)^k\}/(1-p+q)$

5.147. $u_k = u_0 + (u_0+v_0)q^k, \quad v_k = v_0 - (u_0+v_0)q^k$

5.148. $u_k = (c_1 + c_2 k)(\sqrt{8})^k + (c_3 + c_4 k)(-\sqrt{8})^k$

$v_k = (2c_1 + 6c_2 + 2c_2 k)(\sqrt{8})^{k-1} + (2c_3 + 6c_4 + 2c_4 k)(-\sqrt{8})^{k-1}$

5.149. $y_k(t) = 1 + t + \dfrac{t^2}{2!} + \dfrac{t^3}{3!} + \cdots + \dfrac{t^k}{k!}$

5.150. (a) e^t

5.153. $u(x,y) = (-2)^x F(x+y)$

5.154. $u(x,y) = 2^x(3x+3y+2)$

5.155. $u(x,y) = (-2)^{-x}F(x+y) + xy - \frac{5}{3}x - \frac{2}{3}y + 2$

5.156. $u(x,y) = 3^{x/h}F\left(\dfrac{x}{h} + \dfrac{y}{k}\right) - \frac{1}{2}x - \frac{1}{2}y - \frac{1}{4}h + \frac{3}{4}k$

5.157. $u(x,y) = 2^{x/h}F\left(\dfrac{x}{h} - \dfrac{y}{k}\right) + G\left(\dfrac{x}{h} + \dfrac{y}{k}\right)$

5.158. $u_{k,m} = F(2k-m) - 2^{k-m+1}$

5.159. $u_{k,m} = \dbinom{m}{k}$

5.160. $u_{k,m} = 2^k F(k-m) + G(k-m) + 2^m H(k-m)$

5.161. $u_{k,m} = F(k-m)$

5.162. $u_{k,m} = F(k-m)\cos\dfrac{\pi k}{3} + G(k-m)\sin\dfrac{\pi k}{3} + H(k-m)\cos\dfrac{\pi m}{3} + J(k-m)\sin\dfrac{\pi m}{3}$

5.163. $u_k(y) = 3^{k+1}e^{-2y/3}$

5.167. (c) $y = c_0\left(1 + x + \frac{x^4}{1\cdot 3} + \frac{x^6}{1\cdot 3\cdot 5} + \cdots\right) + c_1\left(x + \frac{x^3}{2} + \frac{x^5}{2\cdot 4} + \frac{x^7}{2\cdot 4\cdot 6} + \cdots\right)$

5.169. (a) $y = c_1(1+h)^{x/h} + c_2(1+3h)^{x/h} + \frac{x}{3} + \frac{10}{9}$

(b) $y = c_1 e^x + c_2 e^{3x} + \frac{x}{3} + \frac{10}{9}$

5.170. $y = e^{-x}\left(c_1 + c_2 x + \frac{x^2}{2}\right)$

5.177. $y_k = \begin{cases} (-1)^{\frac{1}{2}(k+1)}2^{\frac{1}{2}k(k-1)} & k \text{ odd} \\ (-1)^{\frac{1}{2}k}2^{\frac{1}{2}k(k-1)} & k \text{ even} \end{cases}$

5.178. $y_k = \tan\left[c + \sum \tan^{-1}(1/2^k)\right]$

5.181. $y_k = \frac{c_1\alpha^{k+1} + c_2\beta^{k+1}}{c_1\alpha^k + c_2\beta^k} - b$ where $\alpha, \beta = \frac{1}{2}[b \pm \sqrt{b^2 + 4a}\,]$

5.182. $y_k = \frac{2\cdot 4\cdot 6\cdots(k-1)ac}{1\cdot 3\cdot 5\cdots k}$ k odd

$y_k = \frac{1\cdot 3\cdot 5\cdots(k-1)}{2\cdot 4\cdot 6\cdots kc}$ k even

5.183. $u_{k,m} = F(k+m) + kG(k+m) + k^2H(k+m)$

5.184. $u_{k,m} = s_m^k =$ Stirling numbers of the first kind

5.185. $u_{k,m} = S_m^k =$ Stirling numbers of the second kind

5.187. $y_k = \cos\left(4\pi \sum_{r=1}^{k-1} \frac{1}{r}\right)$

5.188. $Y(x) = (x-1)/e$

5.189. $Y(x) = \frac{x^2 - 4x + 1 + 2x\tan 1}{\cos 1}$

5.190. $u_k = \frac{1}{1!} - \frac{1}{2!} + \frac{1}{3!} - \cdots + \frac{(-1)^{k-1}}{k!}$

5.191. (a) $u(x,y) = \frac{1}{4}(3x+2y)^2 + e^{-(3x+2y)/2}$

(b) $u(x,y) = x^2 + x - \frac{5y^2}{18} - \frac{y}{3} - \frac{xy}{3} + 1$

5.192. (a) $u(x,y) = x^2 + y + \frac{2}{3}xy$

(b) $u(x,y) = y\left[1 + \left(\frac{x}{2} - A\right)x - \frac{xy}{8}\right]$

CHAPTER 6

6.62. (a) $\frac{\sqrt{5}-1}{2}\left(\frac{1-a^k}{1-a^{k+1}}\right)$ where $a = \frac{1-\sqrt{5}}{1+\sqrt{5}}$

6.66. $\dfrac{1}{2\sqrt{19}}\{(4+\sqrt{19})^k - (4-\sqrt{19})^k\}$

6.68. 2

6.71. (a) $T_2(x) = 2x^2-1,\ T_3(x) = 6x^3 - 3x,\ T_4(x) = 8x^4 - 8x^2 + 1$

6.75. (a) $A_{k+1} = A_k + iA_k$ (b) $A_k = P(1+i)^k$

6.76. $A_{n+1} - (1+i)A_n = S$

6.77. $A_{n+1} - (1+i)A_n = S + na, \qquad A_n = \dfrac{a+Si}{i^2}[(1+i)^n - 1] - \dfrac{an}{i}$

6.89. $\dfrac{1}{2!} - \dfrac{1}{3!} + \dfrac{1}{4!} - \cdots + \dfrac{(-1)^n}{n!}$

6.91. (a) $\dfrac{2}{3} + \dfrac{(-1)^{n-1}}{3 \cdot 2^{n-1}}$ (b) $\dfrac{2}{3}$

6.92. $\dfrac{b}{b+r} + \dfrac{(-1)^{k-n}r}{(b+r)(b+r-1)^{k-n}}$

6.93. $\dfrac{b}{b+r}$

6.102. (a) Exact value $= 1.7974$

6.104. (a) Exact value $= \dfrac{441}{256} = 1.72265625$

6.109. (b) Exact value $= 1.1487$

6.112. Exact values: $x(0.5) = 0.645845$
$y(0.5) = 0.023425$

6.113.

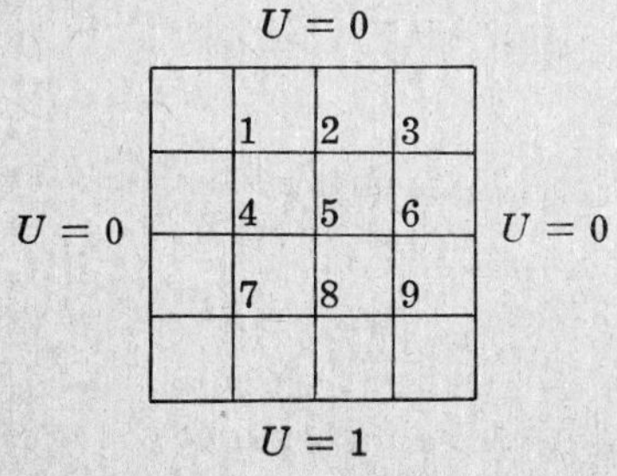

$U_1 = 0.071428571,\ U_2 = 0.098214285,$
$U_3 = U_1,\ U_4 = 0.1875,\ U_5 = 0.2500,$
$U_6 = U_4,\ U_7 = 0.428571428,$
$U_8 = 0.526785714,\ U_9 = U_7$

INDEX

Catalog

If you are interested in a list of SCHAUM'S OUTLINE SERIES in Science, Mathematics, Engineering and other subjects, send your name and address, requesting your free catalog, to:

SCHAUM'S OUTLINE SERIES, Dept. C
McGRAW-HILL BOOK COMPANY
1221 Avenue of Americas
New York, N.Y. 10020